Ferrite

Nanostructures with Tunable Properties and Diverse Applications

Edited by

Gaurav Sharma[1,2], Amit Kumar[1,2], Pooja Dhiman[1]

[1] International Research Centre of Nanotechnology for Himalayan Sustainability (IRCNHS), Shoolini University, India

[2] College of Materials Science and Engineering, Shenzhen Key Laboratory of Polymer Science and Technology, Guangdong Research Center for Interfacial Engineering of Functional Materials, Nanshan District Key Lab. for Biopolymers and Safety Evaluation, Shenzhen University, Shenzhen 518060, PR China

Published by **Materials Research Forum LLC**
Millersville, PA 17551, USA

Published as part of the book series
Materials Research Foundations
Volume 112 (2021)
ISSN 2471-8890 (Print)
ISSN 2471-8904 (Online)

Print ISBN 978-1-64490-158-8
eBook ISBN 978-1-64490-159-5

Distributed worldwide by

Materials Research Forum LLC
105 Springdale Lane
Millersville, PA 17551
USA
https://www.mrforum.com

Manufactured in the United States of America
10 9 8 7 6 5 4 3 2 1

Table of Contents

Preface

Day-by-day up gradation of technology is demanding promising materials which can serve for diverse applications. Among various nano-hybrid structures, there exist some wonder host materials for bio-medical, technological and at the same time environmental remedial applications also.

Ferrite materials belong to such category and their properties can be tuned for wide range of applications. Furthermore, ferrites exhibit unexpected properties at nano-scale making them suitable for high frequency devices, storage applications, sensing applications, bio-medical applications, and photo-catalytic degradation of organic pollutants. Currently, magnetic and semiconducting ferrites at nano-scale have become fascinating materials with intriguing results. Today, Ferrites are truly utilized in electronic gadgets, but still ferrites have much more hidden possibilities. Industrial people are continuously starving for wonder materials as future technology requires compact devices with multi-applications. Ferrites are most promising materials in harnessing the technology for large scale applications. The quest for changing technology is also associated with some challenges like stability, toxicity, compatibility, and reusability of materials for device formations. It is the need of the time to accumulate the enormous applications potential of ferrites at a one spot knowledgeable resource which was the aim of this book.

This book consists of 10 chapters which cover almost all segments of ferrites including basics of ferrites, their properties and diverse applications. Chapter 1 describes the introduction of the ferrites, their classification, their properties as well as their applications. Chapter 2 includes the latest advancements in the processing of ferrites and also it is focused on bio-medical applications of ferrites. Chapter 3 focuses on another important application of ferrites "photo-catalytic activity" for the degradation of organic pollutants. Chapter 4 highlights the suitability of ferrites for biomedical applications of ferrites. Chapter 5 covers the review on most important Ferrite Ni-Zn ferrite suitable for high frequency applications and much more. Chapter 6 and 7 involves the photo-catalytic and adsorption applications of spinel ferrites for wastewater remediation with the focus on removal of organic dyes and drugs from aquatic systems. Chapter 8 focuses on most suitable ferrites for solar cell applications. Chapter 9 presents the comparison of Aurivillius based ceramics and ferrites for multi-ferroic properties. Chapter 10 describes the hexagonal ferrites and their synthesis routes along with the latest advancement in their properties and applications.

Overall, this book covers a wide are of knowledge starting from basics to applications and recent trends in the field of ferrites.

This book traces the important advances, challenges, and processing of ferrites for countless applications, where the scientists can share their knowledge, experiences, and ideas. Industry people can find ideas of tailored and engineered materials for new era devices with sustainable growth. Finally, this book aims to cover bountiful knowledge of physics, chemistry and biology associated with ferrites along with author's experimental experiences with ferrites.

Materials Research Foundations **112** (2021) 1-61 https://doi.org/10.21741/9781644901595-1

Chapter 1

"Ferrites": Synthesis, Structure, Properties and Applications

Shanta Mullick [1*], Garima Rana [2], Amit Kumar [3], Gaurav Sharma[4], Mu. Naushad[4]

[1] Department of Physics, IEC University, Baddi, Himachal Pradesh, India

[2] School of Physics & Materials Science, Shoolini University of Biotechnology and Management Sciences, Bajhol, Solan (H.P.) 173229

[3] International Research Centre of Nanotechnology for Himalayan Sustainability (IRCNHS), Shoolini University, India

[4] Department of Chemistry, College of Science, Building#5, King Saud University, Riyadh, 11451, Saudi Arabia

*shantakapoor@gmail.com

Abstract

Ferrites synthesis method and characterization techniques are attracting huge attentions of researchers because of their wide scope of uses in numerous areas. The ferrites include high resistivity, saturation magnetization, permeability, coercivity and low power losses. The above-mentioned useful ferrites characteristics make them appropriate for use in different applications. These ferrites are used in biomedical field for cancer cure and MRI. Electronic applications are transformers, transducers, and inductors which are also made using ferrites and also used in making magnetic fluids, sensors, and biosensors. Ferrite is a profoundly helpful material for many electrical and electronic applications. It has applications in pretty much every domestic device like LED bulb, mobile charger, TV, microwave, fridge, PC, printer, etc. This review mainly focus on the synthesis method, characterization techniques, and implementation of FNPs. This Chapter presents various methods used for ferrites preparation with distinctive examples, their advantages as well as limitations in detail. Ferrites properties like structural, optical, electrical and magnetic with their characterization techniques and various applications in the areas of biomedical, electronics, and environment are also discussed.

Keyword

Ferrites, Magnetic-Nanoparticles, Synthesis, Characterization, Biomedical Applications

Materials Research Foundations **112** (2021) 1-61 https://doi.org/10.21741/9781644901595-1

Contents

1. Introduction

The name ferrite comes from the Latin word 'Ferrum' meaning iron [1]. Magnetite (Fe_3O_4) also called Lodestone, a natural mineral is a genuine ferrite. Ferrites are derived from magnetite and hematite (Fe_2O_3) along with other metal oxides. A ferrite is a ceramic-like magnetic material with ferromagnetic ordering (in macroscopic phase) mainly composed of a small proportion of one or two metallic elements with a large iron oxide (α-Fe_2O_3) proportion. In modern days, ferrites have gained immense attention as magnetic nanoparticles because of their distinctive structural properties, surface reactivity, electrical properties and magnetic properties and the nanostructured phase greatly influenced all these properties [2-10] which are not the same as the bulk material. Nanoparticles are ranged between 1 and 100 nm in size.

At the nanoscale nanoparticles display exceptional physical, chemical, and biological properties contrasted with their individual particles at high scales. This occurrence is because of a relatively larger surface area to the volume, increased reactivity or stability in a chemical process, enhanced mechanical strength, etc. Macro-structured have the same properties as the bulk material. **On the surface, the number of particles increases as the particle size decreases.** Surface atoms can move easily because they have less coodination number as compared to inner atoms. quantum size effects appears when the size of nanoparticles approaches the de Broglie wavelength limit and diameter reaches smaller than the quasiparticle interaction [11]. Controlling size of nanoparticles during synthesis by utilizing various methods, controlling of nanomaterials properties is feasible. The electronic property dependes upon electronic band structure of a metal. The band structure depends upon the size of the particle. The delocalized bands are seen in the molecular states. The band structure of nanocrystal is in between the discrete states of atoms (and molecules) and continuous bands of crystals. The adjacent lines energy separation depends on the particle size. As the size decreases, the separation of energy levels increases. The metallic nature decreases and gradually changes to the semiconductor nature. There are various kinds of magnetic materials (dia, para, ferro, antiferro, ferromagnetic materials) [12]. Magnetic material are distinguished as the soft and hard materials on the bases of their coercivity. Low coercivity belongs to soft magenetic material with a small hysteresis area and these are magnetized by a low magnetic field and vice-versa. The coercivity will change with the particle size. Generally, on decreasing grain size and increasing the surface area the coercivity and also the saturation magnetization increases. So, the nanoparticles have an additional surface area with less grain size with high magnetic strength. If further, size of the particle decreased then the ferromagnetic particles changed to paramagnetic due to instability. The behaviour of these paramagnetic materials is different than the bulk magnetic material due to which they called superparamagnetic. The material interaction with light is called optical property. This property arises because of surface plasmon and electrons quantum confinement. The surface plasmon resonance energy will evaluated by dielectric medium and free electron density of the nanomaterial. This property also depends upon the size, shape, doping, surface characteristics, and also on the surrounding environment interaction. All properties generally rely upon the size and state of nanomaterials. So the controlling and control of nanomaterials properties is conceivable by controlling their size during their preparation by utilizing various techniques [13, 14]. **Figure 1** depicts the immense research attention in the research community from various researchers for ferrites.

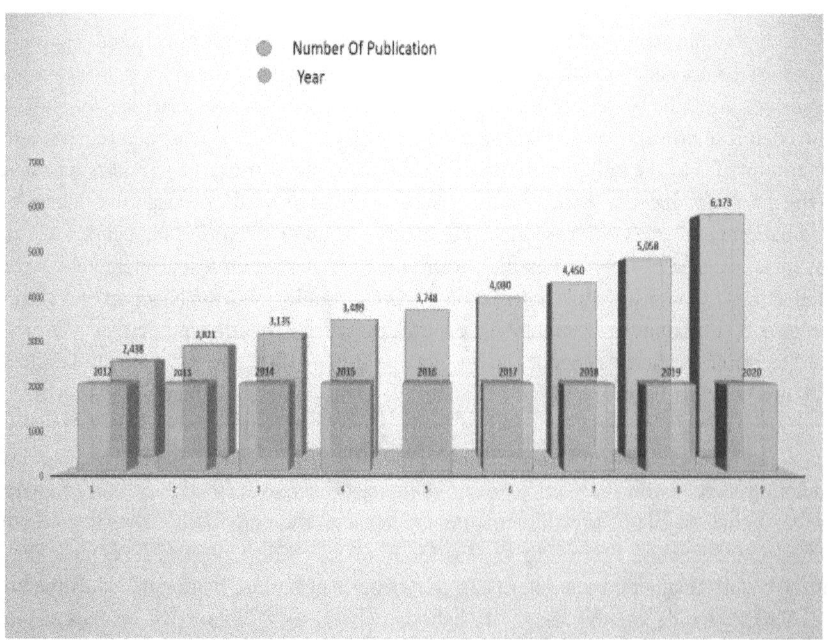

Figure1: As per Scopus data, the number of publications for the keyword "ferrites" from 2012 to 2020.

Materials Research Foundations **112** (2021) 1-61 https://doi.org/10.21741/9781644901595-1

2. Magnetic material and types of magnetism

Classification of materials can be done by how they respond in an applied external magnetic field [4, 12]. When the magnetic field is applied externally, the atomic current loops, which arise due to the electrons orbital motion counter to oppose it. Depictions of the magnetic moments orientations help to classify different type of magnetism. Diamagnetism, paramagnetism, ferromagnetism, antiferromagnetism and ferrimagnetism are the five fundamental kinds of magnetism. This classification is associated to the material net magnetization and magnetic dipole, the behaviour of which is schematically represented in **Figure 2**.

a) Diamagnetism: The materials show a kind of weak repulsion in an externally applied magnetic field well-known for diamagnetism. Nonetheless, diamagnetism is feeble, and accordingly some other type of magnetic behaviour that material may have usually overcome the effects of the atomic current loops. In terms of materials electronic configuration, materials which have filled electronic sub-shells and so have paired magnetic moments and on the whole cancel each other. Diamagnetic materials possess a negative susceptibility $\chi < 0$ so examples of diamagnetic materials are quartz SiO_2 and calcite $CaCO_3$. The effects of atomic current loops can be overpowered if the material exhibits magnetic moments with long-range order or a net magnetic moment.

b) Paramagnetism: All other kinds of magnetic behaviour shown by materials have unpaired electrons, frequently in the 3d and 4f atomic shells of each atom. Materials whose atomic magnetic moments are unpaired show paramagnetism. Thus, paramagnetic materials moments have a little positive magnetic susceptibility $\chi \approx 0$ and have no long-range ordering e.g., montmorillonite and pyrite.

c) Ferromagnetism: Materials that own ferromagnetism have aligned equal atomic magnetic moments and there is direct pairing interactions among the moments in their crystalline structure, which can strongly boost the flux density. Some examples are Fe, Ni, and Co. Furthermore, even in the absence of externally applied magnetic field, in ferromagnetic materials the aligned atomic moments can give a spontaneous magnetization. The Materials that preserve permanent magnetization even if the applied field is absent are hard magnets.

d) Antiferromagnetism: Materials in which equal atomic magnetic moments in magnitude arranged antiparallelly show antiferromagnetism for instance troilite FeS, ilmenite $FeTiO_2$ etc. Due to antiparallel exchange interaction coupling of moments net magnetization is zero. Above the Neel temperature (TN), the material exhibits paramagnetic behaviour because there is enough thermal energy which randomly fluctuate the antiparallely aligned atomic moments which leads to the vanishing of long-range order of materials.

e) **Ferrimagnetism:** Ferrimagnetism is shown by materials in which atoms or ions have non-parallel ordered arrangement in absence of applied magnetic field below the Neel temperature. In the typical case, a significant net magnetization is due to the antiparallel arrangement of ions within a magnetic domain. These have higher saturation magnetization as compared to ferromagnetic materials. Ferrites (e.g., Fe_3O_4) are ferrimagnetic materials where cations (metal) and anions (oxygen) arranged with this fashion in the crystal lattice.

f) **Superparamagnetism:** Superparamagnetism arises when size is reduced in magnetic materials leading the creation of single-domain of particles. Briefly, superparamagnetism arises when an applied field or thermal fluctuations can easily displace the magnetic moments to the chosen crystallographic axes where the magnetic moment have to point along.

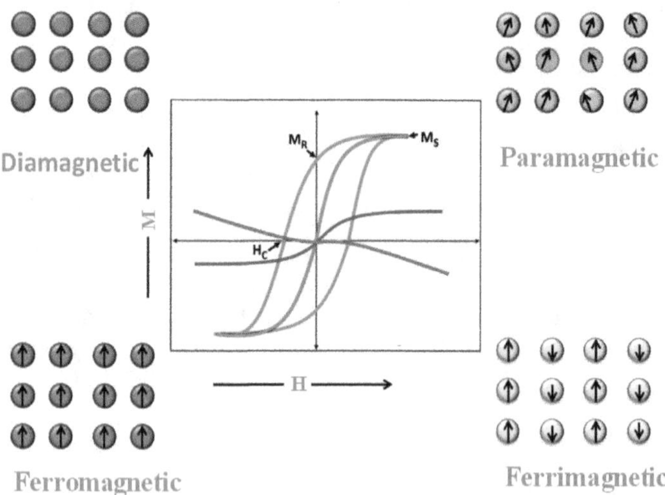

Figure 2: Schematic illustration of the arrangement of the magnetic dipoles of various types of magnetic materials (diamagnetic, paramagnetic, ferromagnetic, and ferrimagnetic) and representative hysteresis loops that illustrate the magnetic behavior of materials when an external field is applied.

Materials Research Foundations **112** (2021) 1-61 https://doi.org/10.21741/9781644901595-1

2.1 Classification of ferrites based on response in a magnetic field

Based on the ability of magnetization and demagnetization, there are two types of ferrites; soft and hard [9, 15]. Coercive force (H_c) which is the main property of these magnetic substances.

Soft ferrites: Materials with low H_c are termed as magnetically soft, easy to magnetize, and demagnetize. With a narrow hysteresis loop these materials exhibit soft magnetic properties for example high permeability, low coercivity, and high value of saturation magnetization and are free of magnetic field amplitude. They utilized the field of electronics to make effective magnetic cores for transformers, high-frequency inductors, antennas and in different microwave parts. Ferrites used in ectromagnetic cores containing Ni, Zn, or Mn are soft ferrites. They are widely used to make the cores of RF transformers, Switched-Mode Power Supply (SMPW) and inductors due to their relatively low losses at high frequencies.

Hard ferrites: Ferrites with higher H_c are termed magnetically hard materials. Once magnetized hard ferrites retain their magnetization. With a wide hysteresis loop hard ferrites have high coercive strength [7]. These ferrites are used as permanent magnets for applications such as refrigerator magnets, loudspeakers, and small electric motors. In contrast to soft ferrites, hard ferrites (permanent ferrite magnets) have a high value of remanence after magnetizatized and are prepared using iron oxide with barium/strontium oxides. These ferrites conduct magnetic flux pretty well in a state of magnetically saturation, and have a high value of magnetic permeability. This makes these ceramic magnets ideal for the storage of strong magnetic fields as compared to pure iron [15].

2.2 Classification of ferrites based on crystal structures:

Ferrites generally have classification on the bases of spatial arrangement of oxygen atom around the metal ion i.e cubic close-packed and hexagonal arrangement of oxygen atoms. In the structure the cation must satisfy both size and charge neutrality consideration. Cubic ferrites have two kinds as spinel ferrites and garnet ferrites. **Figure 3** shows different crystal structures of ferrites. A brief explanation of ferrites crystal structures is given below:

Spinel ferrite:

Various anions including sulphur (thiospinels), chlorine (halospinels), and oxygen can form spinel structures. Spinel ferrites with the general formula $MeFe_2O_4$ or $A[B_2O_4]$ where Me refers to the metal, can be depicted as a cubic close-packed arrangement of oxygen atoms at two different crystallographic sites with Me^{2+} and Fe^{3+}. These sites have tetrahedral (A-sites) and octahedral (B-sites), oxygen coordination and hence, both sites have different resultant local symmetries. In the spinel structure metal cation occupies cation sites. One is 8 A-sites where the metal cations with oxygen are coordinated tetrahedrally and the second is 16 B-sites that have octahedral coordination. Depending upon the arrangement of metal cation Me^{2+}, ferrospinels further classified as:

Normal spinel structure $A[B_2O_4]$: In this all Me^{2+} cations occupy A(tetrahedral) sites and Fe^{3+} catiions in B (octahedral) sites; the structural formula for these ferrites is $Me^{2+}[Fe_2^{3+}]\ O_4^{2-}$. This kind of arrangement occurs in $CdFe_2O_4$ and $ZnFe_2O_4$.

Inverse spinels, $B[ABO_4]$: Here the Me^{2+} ions occupy the B (octahedral) sites and Fe^{3+} ions are arranged over both B octahedral and A tetrahedral sites. Fe^{3+}cations occupy the A-sites completely and Me^{2+} and Fe^{3+}cations occupy the B-sites randomly; the structural formula of inverse ferrites is $Fe^{3+}[Me^{2+}Fe^{3+}]\ O_4^{2-}$. Magnetite Fe_3O_4, ferrites $MgFe_2O_4$, $NiFe_2O_4$ and $CoFe_2O_4$ have inversed spinel structure.

Mixed spinels, $A_{1-x}B_x[A_xB_{2-x}O_4]$ or $(Me_{1-x}Fe_x)[Fe_{2-x}Me_x]O_4$ with Me^{2+} and Fe^{3+} ions distributed over both octahedral and tetrahedral sites e.g. $MnFe_2O_4$. Where the degree of inversion is x. In the majority of spinels, the cation arrangement have an intermediate value of degree of inversion and where both sites have a fraction of both the cations (Me^{2+} and Fe^{3+})

Garnet Ferrites:

Garnet ferrites own the silicate mineral garnet strucure, $X_3Y_2(SiO_4)_3$, where divalent cations [Ca^{2+}, Mg^{2+}, Fe^{2+}] generally occupy the X site and trivalent cations [Al^{3+}, Fe^{3+}, Cr^{3+}] occupy the Y site in the octahedral/tetrahedral framework where $[SiO4]^{4-}$ provides the tetrahedra. The general formula ($A_3B_5O_{12}$), or $M_3Fe_5O_{12}$ is for garnet ferrites where M is a rare-earth (yettrium) cation. Crystal structure of garnet ferrites possess three crystallographic lattices sites (a, b, and c). With the octahedral and tetrahedral sites, which are seen in spinels, garnets possess dodecahedral (12- coordinated) sites. 24Fe^{3+} ions take place in tetrahedral sites among three lattice sites, 16Fe^{3+} ions take place octahedral sites and 24R^{3+} ions go on the dodecahedral sites, while oxygen ions are arranged in the interstitial sites [16]. The net ferrimagnetism is therefore a complex result of spin alignment among these three types of sites. Garnets are magnetically hard. While spinels have superior magnetization properties, garnets have superior dielectric properties, as these are more stress-sensitive than spinels.

Hexagonal ferrites:

The magnetic mineral magneto plumbite $PbFe_{12}O_{19}$ was first described in 1925 have a hexagonal structure, so they are called as hexagonal ferrites and possess the general composition, $MeFe_{12}O_{19}$ where (Me= Ba, Pb, Sr). These ferrimagnetic oxides contain a principal component, Fe_2O_3 in combination with divalent oxides (BaO, PbO, or SrO). Hexagonal ferrites have high uniaxial anisotropy, high magnetization values, and good chemical stability, consequently fitting very well in the necessities of recording technology. Films of hexaferrites of $BaFe_{12}O_{19}$ are capable candidates for microwave wave devices and high-density recording media [17, 18].

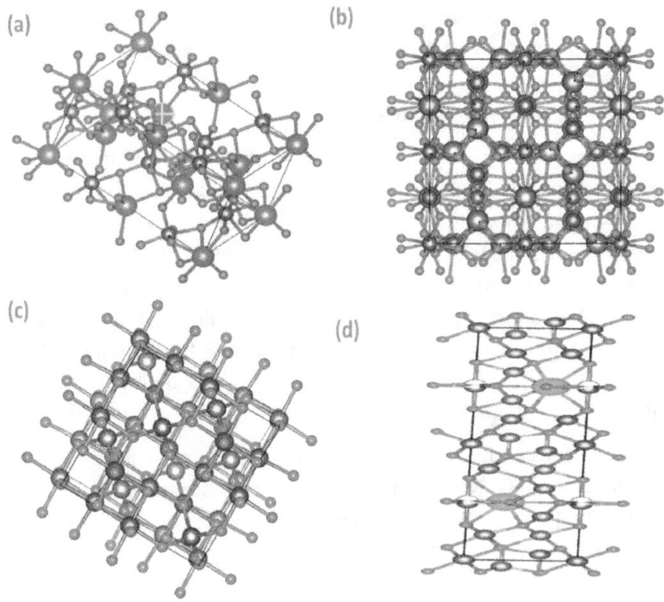

Figure 3: Schematic representation of (a) perovskite ferrite (b) garnet ferrite (c) spinel ferrite and (d) hexagonal ferrite structure.

Perovskites ferrites:

Cubic perovskite or ortho ferrites structure has the general formula ABX_3, where cations A and B have different oxidation numbers and with anion X. The cation A resides at the center, while the arrangement of cation B and the anions X are in the corners and at the edges of the unit cell, respectively [19]. Rare earth ferrites $RFeO_3$ compounds such as orthoferrites of lanthanum and samarium are two most considered materials [20, 21]. $BiFeO_3$ [22]. Perovskite ferrites nanoparticles had shown strong absorption at nearly 590 nm arising from the bandgap absorption, which has applications of environmental purification and/or water splitting.

The commonly known soft ferrites Mn–Zn ferrite, Ni– Zn ferrite, etc. whereas $BaFe_{12}O_{19}$ and $SrFe_{12}O_{19}$ are examples of hard ferrites. It is well established that the ferrites magnetic properties depend upon chemical compositions, microstructure, crystal structure, particle size, and also the surface area of the material. The methods of material preparation greatly influence these parameters.

3. Formation of ferrites nanoparticles

Ferrites nanoparticles (FNPs) formation can be done by two kind of approaches top-down and bottom up as shown in **Figure 4**. To get nanoparticles the bulk material is broken down in the **top-down**. Top down has many limitations for example the constraint of high temperature, inhomogeneity, presence of impurities, crystal defects, broad size distribution, surface imperfection. On the other side, the **bottom-up** puts together small atomic building blocks to form nanoparticles and is the mainly approving approach as the nanoparticles formed in these methods are homogeneous, highly pure and narrow size distribution. **Table 5** shows the electrical, magnetic, and optical properties of ferrites with different methods.

Fabrication of NPs further includes a range of methods such as **physical-chemical, biological and hybrid** methods to fabricate the various kinds of ferrites, the NPs formed by each method exhibit particular properties. Usually, there are four basic steps in ferrite preparation. First step is to prepare the materials to generate an close blend of the metal ions in a ratio which we will require in the final yield. The second step is calculations of the blend to yield a ferrite. In the third step, there is yielding a materials powder and shape it into the requisite form and the fourth one involves the sintering of the processed material to get the final produce. These four steps cannot be isolated; however, only the first step has a different operation. The property and application aspects of the synthesized ferrite depend strongly upon the particle size, composition, and morphology, which are very sensitive to preparation methodology used in their synthesis as given in **Table 1.** Hence ferrite preparation of desired size, composition and morphology has attracted more attention by researchers in the past few decades. Every method possess some advantages and limitations as described in **Table 2** and some important methods which are used for ferrite preparation are discussed below.

Materials Research Foundations **112** (2021) 1-61 https://doi.org/10.21741/9781644901595-1

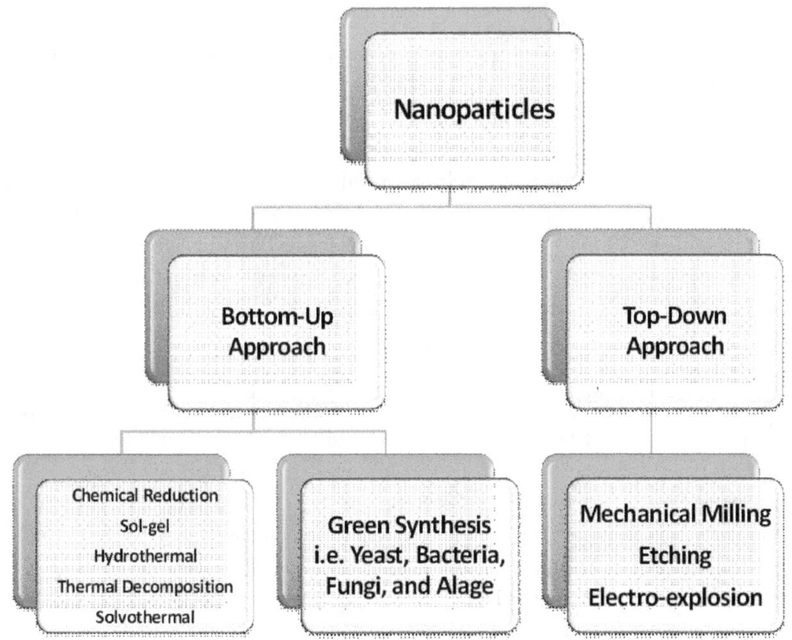

Figure 4:Bottom-up and the top-down approaches in ferrites nanomaterials.

3.1 Mechanical milling

Among the various top-down methods, mechanical milling is extensively used to yield various nanoparticles. Mechanical milling utilized for milling and post-annealing of NPs through synthesis by milling different elements in an inert atmosphere. The influencing factors in mechanical milling are plastic deformation that leads to particle shape, crack leads to reduction in particle size, and cold-welding is used to enhance the particle size [23]. Zhigang Zhang et al. [24] have prepared pure $NiFe_2O_4$ nanoparticles using a planetary ball milling assisted solid-state reaction process. By planetary ball milling one can be lower the temperature to 750 °C to obtain impurity-free NPs of $NiFe_2O_4$ having wide range of size (35–85 nm) distribution and double morphology (sphere for smaller grains and polygon for larger ones). The v saturation magnetization is improved from 33.4 emu/g at 750 °C to 42.7 emu/g at 800 °C, and coercivity is reduced from 143.2 Oe to 116.6 Oe. But [25] saturation magnetization and coercivity NPs of the $NiFe_2O_4$, got from calcining the precursor ground by hands at 800 °C for 1 h, are 46.5 emu/g and 107.6 Oe respectively. E. Manova et al. [26] synthesized $CuFe_2O_4$ by two-step procedure, including first co-precipitation and second mechanical milling which together is called mechanochemical synthesis of the co-precipitation precursors. Co-precipitated hydroxide carbonates, used as precursors were prepared from an aqueous solution of $Fe(NO_3)_3$ $9H_2O$ and $Cu(NO_3)_2$ $3H_2O$ by dropwise addition of 1 M sodium carbonate up to pH 9 at continuous stirring. The obtained precursor powders were milled by using a Fritsch Planetary miller in a hardened steel vial together with grinding balls with the different diameters (from 3 to 10 mm). Precursors were mechanically milled for 1, 2, 3, and 5h. The phase composition of the as obtained materials depends upon the milling duration. The ferrites catalytic properties are strongly affected by their phase transformations under the reaction medium.

3.2 Thermal decomposition

To control the synthesis of ferrite, solid-state thermal decomposition method is a simple, general and effective method. Thermal decomposition of organometallic precursors for example metallic acetylacetonates and carbonyls in organic solvents and surfactants (oleic acid and hexadecyl amine) for the production of ferrites nanoparticles shown in **Figure 8(b).** On the bases of the kind of precursor materials, the temperature used can be high or low. Generally, the method can be used for larger production (at the 40 g level per each synthesis) of high-quality crystalline FNPs with size- and shape-controlled output. The method can be used to prepare the FNPs required for industrial as well as medical applications. The strontium ferrite [27] NPs was prepared by a solid-state thermal decomposition without using any catalyst, surfactant, and solvent at 800 °C for 2 h in a horizontal tube furnace having an inner diameter of 10 cm and length of 120 cm. In the typical procedure, a mixture including 1 mmol of the obtained SrC_2O_4 and 2 mmol of Fe $(NO_3)_3$ $9H_2O$, was put into a small quartz boat. The quartz boat entered into tube furnace in a gas blend of 85% Ar and 15% H_2. The quartz boat kept in the heating zone at 800 °C for 2 h and afterthat was slowly dragged out of the furnace. The sample was cooled down natuarally in ambient conditions. Obtained ferrites particles have nano size of about 40 nanometer and the strontium FNPs showed a ferromagnetic behavior. $MnFe_2O_4$ [28] nano-ferrites prepared using thermal decomposition method. The stoichiometric amounts of iron (III) acetylacetonate Fe (acac)3 (6 mmol) and manganese (II) acetylacetonate Mn (acac)2 (3 mmol) were dissolved in solvent benzyl ether (40 ml) under mechanical stirring. Equimolar amounts of oleylamine(surfactant), oleic acid, and 1, 2-hexadecandiol were mixed with the metal salt solution. The molar ratio of oleic acid, oleyl amine to 1, 2-hexadecandiol was 3:3:5. Solution was transferred to an autoclave reactor bomb and was placed into a preheated oven at 270 °C for 90 min. After cooling, the nanoparticles were obtained. Produced particles are spherical and the mean particle size ranged ~18.9 nm. The influence of aging of the oleate precursor is studied to obtain cobalt ferrite [29] nanoparticles with narrow size distribution and which respond to changes in the magnetic field exclusively by the Brownian relaxation mechanism. It was observed that aging for up to 30 days resulted in changes in the coordination mode between the oleate molecules and the iron/cobalt ions, which in turn influenced the thermal behavior of the iron-cobalt oleate precursor and resulted in the formation of the correct crystal ferrite structure and cobalt ferrite nanoparticles synthesized using the iron-cobalt oleate aged for only 2 days showed poor agreement with the expected ferrite structure, possibly due to the presence of a contaminant phase which may be a cobalt oxide.

3.3 Co-precipitation method

Because of its many advantages, such as being less time-consuming, cost-effective, easy to accomplish, and dependable, the co-precipitation approach is widely utilised for the synthesis of ferrite nanoparticles. It generates a homogenous powder with the highest yield while requiring no organic fuels such as citric acid. Aqueous solutions containing a mixture of divalent and trivalent transition metals (Soluble salts containing Fe(III) are usually utilised as a source of trivalent metal ions) are combined together at a mole ratio of 1:2 in this procedure. On most cases, the experiment is carried out in an alkaline media. To protect these nanoparticles from oxidation and agglomeration in the ambient atmosphere oxygen, they are usually coated with a surfactant such as sodium dodecyl sulphate (NaDS) or oleic acid and then dispersed in a carrier liquid such as ethanol, methanol, ammonia, or other liquids based on the nature of the materials (nanoparticles) to be dispersed. To make high-quality FNPs, the synthesis technique necessitates precise pH adjustment and control. The solution pH is usually adjusted using ammonium solution or sodium hydroxide solution. The solution will then be vigorously stirred in the absence or presence of heat under inert conditions.

The advantage of this technique over the others is that the production rate of ferrite nanoparticles and the disadvantage is the low crystallinity of FNPs [32, 33]. This method is used by many researchers to prepare FNPs such as $BiFeO_3$ [34], $CoFe_2O_4$ [35], $ZnFe_2O_4$ [36]. **Figure 7(b)** describes the fabrication of AFe_2O_4 NPs via the co-precipitation method.

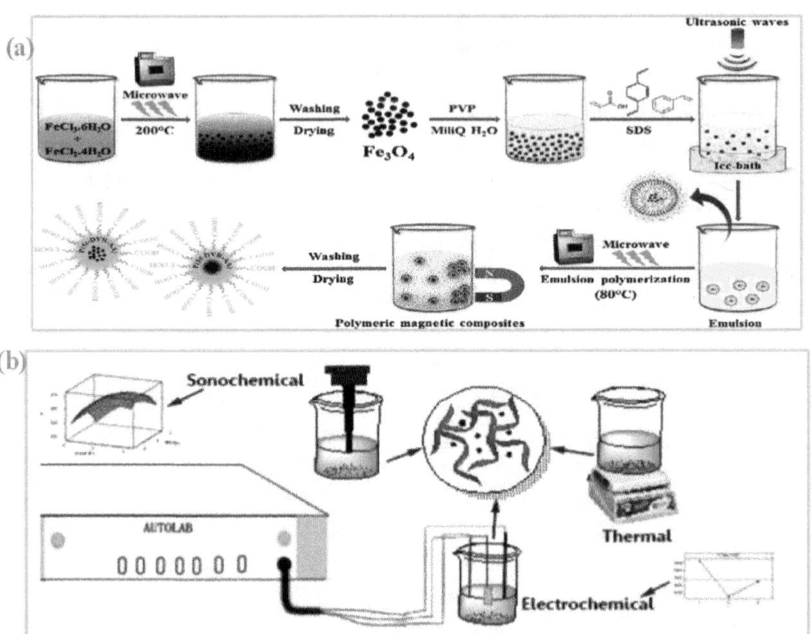

Figure 5: Synthesis of nanoparticles via (a) microwave-assisted [30] (Copyright with permission from Copyright Clearance Centre, License Number, 5024660680757) (b) sonochemical and thermal treatment method adopted by Ashassi-Sorkhabiet al. for the synthesis of ferrites [31]

3.4 Combustion method

The solution combustion technique is of great importance because of its potential advantages like simple, fast production rate, low temperature/cost and produces pure and homogeneous nanoparticles. The combustion process is an exothermic redox reaction between an oxidizer and a fuel. Oxidizer (O) and fuel (F) are used to carry out combustion. The fuel chosen has a very important effect on ferrite prepared. The fuel in this process is either urea or glycine [37, 38] or their mixture [39]. By balancing the oxidising (O) and reducing valency (F) of the reactants, the Stoichiometry $\phi e = (O/F)$ is maintained at unity. The purity, crystallite size, shape, porosity, and carbon content of as-prepared samples are all influenced by the F/O ratio [40]. Both raw materials are dissolved in deionized water with constant stirring and heating at some moderate temperature. Heating will lead to complete oxidation in the open air which results in fine particles having homogenous particles symmetry **(Figure 6(b))**. Some examples of ferrites prepared using this method are α-$LiFe_5O_8$ and β-$LiFe_5O_8$ [41], Co^{2+} doped Mg–Mn ferrites [42] etc.

3.5 Hydrothermal method

The hydrothermal synthesis method [43, 44], established at the end of the nineteenth century, is one of the most promising approaches because of its many benefits over others, including its simplicity and the ability to create high crystallinity products at low temperatures. The structure and performance of the hydrothermal products during the process are affected by the parameters, such as the ratio of raw materials, pH value of the solution, hydrothermal temperature, and time [45]. Improvements to hydrothermal treatment, with an emphasis on improving reaction kinetics or optimising experimental conditions, are still a popular area of research. More recently, the inclusion of external energy to hydrothermal treatment, such as microwave energy, sonar, mechanochemical, magnetic, electrical and so on, has created a whole new zone in materials processing known as multienergy processing of naomaterials. Microwave hydrothermal method [46], organic solvent hydrothermal method, ball milling assisted hydrothermal technique, and magnetic field-assisted hydrothermal technique are examples of multi-energy systems. Examples of ferrites prepared by this method are Hexagonal M type $SrFe_{12}O_{19}$ (SrM) ferrites [47] and $NiFe_2O_4$ nanoparticle [48]. **Figure 6(a)** shows the preparation of ferrites via hydrothermal method for the preparation of ferrites.

3.6 Sol-gel method

The sol-gel method has many advantages, including better starting material, excellent chemical homogeneity, great stoichiometric control, and the ability to produce ultrafine particles with a narrow size distribution in a comparatively short reaction time at lower temperatures [49, 50]. Sol-gel is one of the wet chemical methods to produce soft magnetic ferrites **(Figure 6(a))**. The reaction starts with a chemical solution that acts as a precursor for either network polymers or discrete particles to form a gel. Metal chlorides and metal alkoxides are commonly utilised as precursors in various forms of polycondensation and hydrolysis. Metal oxides are made by connecting metal centers together with hydroxo (M–OH–M) or oxo (M–O–M) bridges in solution, resulting in metal-hydroxo or metal-oxo polymers. As a result, the sol evolves into a gel-like granulating system that includes both solid and liquid phases with morphologies ranging from discrete particles to continuous polymer networks.

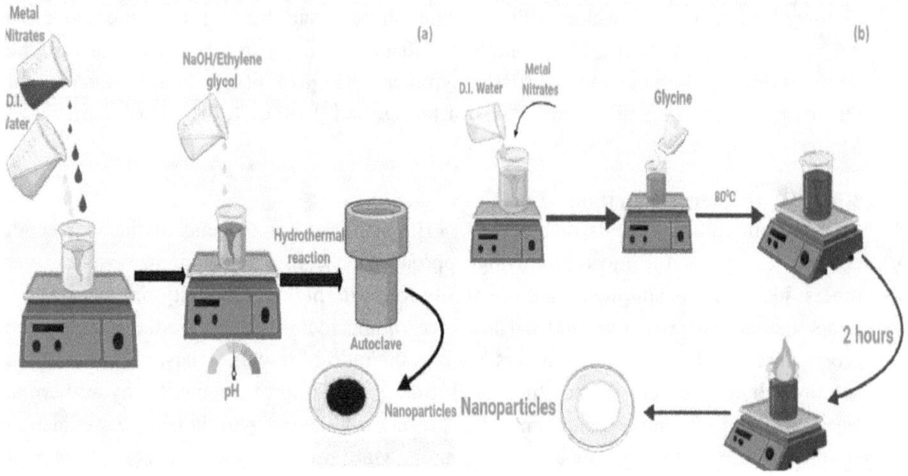

Figure 6: Fabrication of nanomaterial's via (a) Hydrothermal (b) combustion method.

The major advantages of this method were a low synthesis temperature, high purity, energy-saving, small and uniformed particle sizes as well as homogeneous distribution. From the other side, it has drawbacks, such as the fact that this approach is environmentally unfriendly due to the toxicity of some of the alcohol used. Another disadvantage of this process is the expensive cost of raw chemical components, as well as the possibility of segregation in the utilised dopants during gel formation. In addition, high temperatures are required to remove remaining hydroxides from the products [51, 52]. Akhtar et al. [53] used sol-gel technique to prepare Al-doped spinel and garnet nanostructured ferrites for microwave frequency applications. NiCuZn [54] nano ferrite was effectively manufactured employing the sol-gel method with polyvinyl alcohol as a chelating agent. The production of spinel phase at low annealing temperature (500 °C) with tiny spherical particles was observed. The study reveals that material qualities are sensitive to microstructure and can be improved by increasing the sintering temperature by 900, 950, or 1,030 degrees. Ms increases with sintering temperature up to 950 °C, with a highest value of saturation magnetization Ms of 76.76 emu/g, and thereafter diminishes to 72.73 emu/g. The enhanced grain size was due to the increased magnetism with sintering temperature (up to 950 °C). The decrease of M_s after the sintering temperature of 950 °C is due to Zn evaporation at higher sintering temperature resulting in the formation of excess antiferromagnetic hematite (α-Fe_2O_3) which is partly converted to magnetite (Fe_3O_4).

3.7 Microemulsion method

Over the last 30 years, microemulsion methods have been widely employed to produce high-quality metal and semiconductor nanoparticles with a narrow size distribution. Size control is also simple to achieve by making slight changes to the experimental parameters. Microemulsions are isotropic, macroscopically homogeneous, and thermodynamically stable solutions made up of at least three components: a polar phase (often water), a nonpolar phase (often oil), and a surfactant [55]. Normal micelle methods, also known as oil-in-water (o/w) techniques, and reverse micelle techniques, also known as water-in-oil (w/o) techniques, are two types of microemulsion procedures. Surfactants are utilised in both cases, and their concentrations are higher than the critical micelle concentration (CMC) **Figure 8(a)** [4]. $ZnFe_2O_4$ [56] nanoparticles were fabricated via using co-precipitation in water-in-oil microemulsions with a narrow size distribution.

3.8 Green synthesis

As of now, for the preparation of nanoparticles utilizing naturally occurring reagents, for example, organic products, microorganisms as reductants and capping agents, biodegradable polymers (chitosan, and so on), marine green growth, sugars, plant parts (roots, seed, leaf, stem, and latex) could be assessed as best materials for nanotechnology. Moreover, the green blend of nanoparticles offers headway over other ordinary strategies as they are one stage, straightforward, savvy, environment friendly, and often lead to preparing more stable materials and comparatively reproducible [57, 58].

3.9 Other methods

There are many other methods such as sol–gel auto-combustion method [59], standard ceramic process [60], conventional solid-state reaction method [61], citrate-gel route [62], electro spinning method [63] etc. and hybrid methods.

The average size of cobalt ferrite ($CoFe_2O_4$) nanoparticles produced by combustion, coprecipitation, and precipitation techniques. The average particle size of 69.5 nm for combustion, 49.5 nm for coprecipitation, and 34.7 nm for precipitation samples, which resembled SEM pictures [35]. In all three samples, XRD analysis showed a reverse cubic spinel structure with the space group *Fd-3m*. VSM data revealed a magnetic field saturation value of less than 15 kOe.

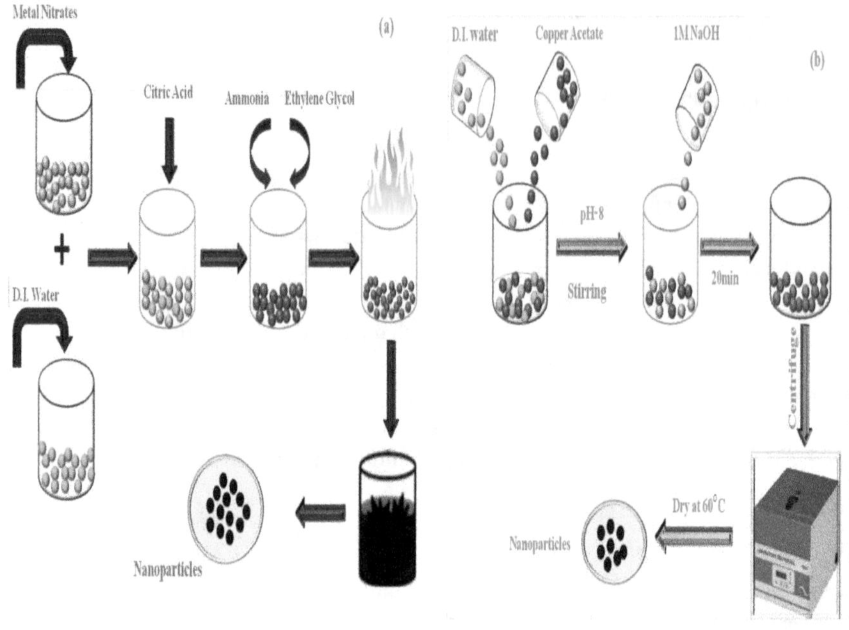

Figure 7: Preparation of ferrite using (a) sol-gel (b) co-precipitation method.

The value of magnetization saturation (M_s) was 56.7 emu/g for combustion synthesized samples, 55.8 emu/g for coprecipitation samples, and 47.2 emu/g for precipitation samples. Coercivity (H_c) was 2002 Oe for combustion synthesized samples, 850 Oe for coprecipitation samples, and 233 Oe for precipitation samples. These findings reveal that different NPs production processes can result in diverse particle sizes and magnetic properties. The combustion process has the highest Hc and Ms, whereas the precipitation approach has the lowest. One analysis revealed that the sintering process parameters and chemical compositions have a significant impact on the magnetostrictive capabilities of pure and substituted cobalt ferrites. Controlling the sintering temperature and substitution levels is essential when manufacturing magnetostrictive cobalt ferrites for specific sensors and actuators applications.

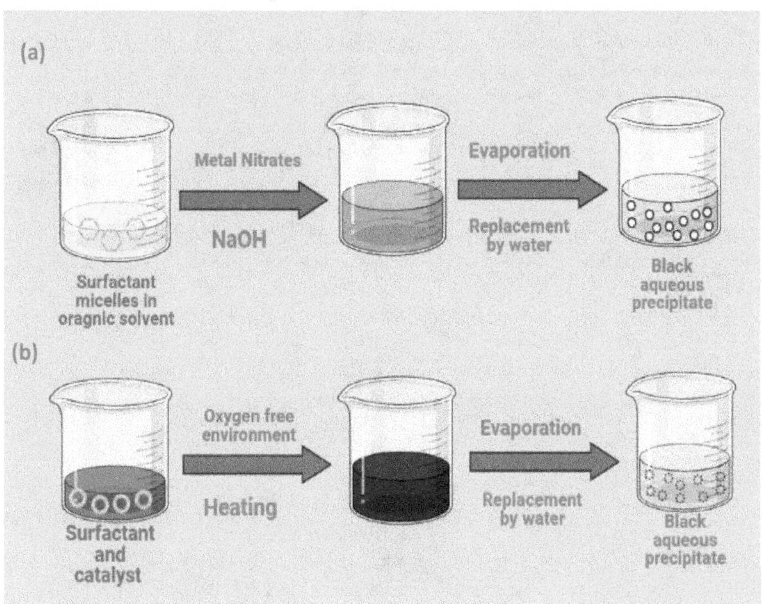

Figure 8: Synthesis of nanomaterials via (a) micro-emulsion (b) thermal decomposition method

Table1: Various synthesis techniques and applications of Ferrites.

Ferrites	Synthesis method	Solvent/acid/ base used	Temp (°C)	Size	Ms	H_C	Applications	Ref.
$MnFe_2O_4$	Water in-toluene reverse micelles.	sodium dodecyl benzene sulfonate (NaDBS)	heat treatment at 350 °C for 12 h	7.7 ± 0.7 nm			contrast enhanceme nt agents of MRI and magnetic carriers for site-specific drug delivery	[64]
$MnFe_2O_4$	combustion method (glycine)	tetra ethylene glycol [TEG] assisted	80°C for the dehydration reaction 300 °C for igniting combustion	30-35	46 emu/ g	64G	gas sensor	[65]

			reaction.					
$MnZnFe_2O_4$	Microwave–Hydrothermal (M–H) method	sodium hydroxide (NaOH)	annealed at 600 1C/20 min using the microwave sintering method	40 nm.	340-335 mT		construction of transformer with high-frequency application.	[66]
$MnFe_2O_4$ and Mn/Fe_3O_4 nanowires	Thermal decomposition of Fe–Mn–NA organometallic polymer precursors	deionized water and isopropyl alcohol, nitrilotriaceticacid (NA) as a chelating agent	heated at 180 C in an electric oven for 6 h, dried at 60 C for 12 h in an oven, heated at 450 C for 4 h to producing porous nanostructures	(5–10 nm)	45.9 and 48.7 emu g1		water treatment	[67]
$NiFe_2O_4$	co-precipitation NaOH solution,	oleic acid	a reaction temperature of 80 ^{0}C and stirred for 40 min	8–28 nm	9 and 40.5 emu/g			[68]
NiFe2O4	Mechanical Milling Method		1100^{0}C for 3 h	10–31 nm	24	0		[69]
NiFe2O4	Hydrothermal process.	Utrica plant extract	200◦C for 2 h dried at 40◦C for 24 h	~56 ±1 nm	16.10 emu g −1		Therapeutic properties Cancer biomedical	[70]
$NiFe_2O_4$	Sol gel auto combustion method		heating in a water bath at 75 C thermally treated at 500^0 C, 700^{0}C, 900^0 C	65 nm			humidity sensors.	[71]
$NiFe_2O_4$	thermal treatment of the precursors		>800 ◦C for 120 min in the studied	8.5 nm	32 emu g−1		catalyst for CO oxidation.	[72]

			pH range, from 7 (neutral) to 12 (highly alkaline).					
$CoFe_2O_4$	co-precipitation	NH_4OH	80°C	6.0±1.7	35.3 -41.0 emu/g	2-5 Oe	biomedical applications	[73]
$CoFe_2O_4$	polymer pyrolysis method.	Polyacrylate salt precursor,	calcined at temperatures of 800, 900, and 1000 °C for 3 h in air	84 to 320 nm	3.42 μB	1100 Oe	spintronic and transparent nanodevices	[74]
fcc-$CaFe_2O_4$	sol-gel assisted combustion technique	Fuel citric acid	~125°C for approximately two hours	76 (±5) Å	1.16 μB	150 Oe	-	[75]
$CaFe_2O_4$	sol-gel method		dried in the oven at 70-80°C overnight, calcined at 550°C for 2 hours	13.59 nm	88.3 emu/g	85.331 G	targeted drug delivery	[76]
$CuFe_2O_4$	mechanochemical treatment		3-5 h milling	10– 12 nm			catalyst	[26]
$CuFe_2O_4$	Sonochemical method	water and ethanol,	irradiated with high-intensity ultrasound radiation under argon at room temperature (25 °C) for 3.5 or 6.5 h dried in vacuum for 10 h at 110 °C	60 nm				[77]
Zn ferrites	Co-precipitation Method	sodium hydroxide (NaOH) and acetone	(20–80 °C) At 40 °C	5– 10 nm	7.3 emu/g			[36]

$BaFe_{12}O_{19}$	solid-state method hydrothermal synthesis	-	-	1–2.5 μm	40.0 emu/g	1039 Oe		[78]
	microwave hydrothermal synthesis	-	-	s 2.0–4.0 μm	53.6 emu/g	623.8 Oe	LTCC technology and the fabrication of chip microwave gyromagnetic devices.	[78]
$SrFe_2O_4$ and $SrFe_{0.75}M_{0.25}O_{3d}$ (M =Cr, Mo, W, Nb, Ti and Zr)	solid state reaction	distilled water, while the oxides WO_3 and MoO_3 were dissolved in diluted ammonia	powders were fired at 800°C to remove the carbonaceous species and to achieve crystallization.	-	-	-	symmetric electrode materials for Solid Oxide Fuel Cells.	[79]

Table 2: List of methods with advantages and disadvantages.

Methods	**Advantages**	**Disadvantages**
Mechanical Milling Method:	Simple, Low cost, short time, large scale production	The requirement of high energy, Contamination by milling tool,
Thermal Decomposition Method	Small size distribution, Size control and Scalable, high crystallinity	The requirement of high temperature,long time, organic solvent, and ligand exchange
Co-precipitation Method	Aqueous media, cost-effective, simple to do and reliable method, a homogeneous powder with maximum yield, does not require any organic fuels, easily functionalized, scalable size and morphology control	Very long reaction time Required, Broad size distribution, low crystallinity
Hydrothermal Method	High yield Aqueous media, Size control, Scalable	High pressure (P2000 PSI) special reactor required High-temperature Longer reaction time
Sol-gel method	Controlled size and shape Low cost	Takes longer time Yield is

		medium
Combustion method	Less time and energy required Simple and effective method Versatile and fast Nanoparticles produced are pure and homogeneous	Very high temperature is required,
Thermal treatment method	Controlled size and shape, scalable	costly and complicated methodology, the high reaction temperatures and the long reaction times, the toxic reagents
Sonochemical method	Less time, Small size distribution, Size control, and Scalable	Poor shape control and medium yield
Microwave-Assisted method	uniform heating, fast reaction under a controlled environment, the product of high purity, high yield, small size distribution	Organic solvent
Microemulsion method	small size distribution, sizecontrol, highcrystallinity, low temperature	Low yield, long time, excess organic solvent
Green synthesis	one stage, straightforward, savvy, environment friendly	

Table3: Summary of characterization techniques.

Properties	**Parameters**	**Ferrite**	**Characterization Techniques**	**Uses**	**Ref.**
Structural Properties	Particle size, shape	$MnFe_2O_4$	XRD	The high intensity and well defined peaks infers the good crystallinity of the $MnFe_2O_4$ nanoparticles diffraction peaks at 30.10°, 35.42°, 43.10°, 53.72°, 56.94° and 62.46° corresponds to the lattice planes of (220), (311), (400), (511), (440) and (444) the calculated lattice constant of 8.420 Å reveals the cubic structure of $MnFe_2O_4$	[65]

		($Zn_xCo_{1-x}Fe_2O_4$	SEM	The scanning electron micrographs (SEM) and uniform distribution of $CoFe_2O_4$ and Zn-doped $CoFe_2O_4$ NPs (with x = 0.0 and x = 0.4) shown in **Fig. 3(a & b).** The images indicate the formation of agglomerated particles with spherical shape and the grain size is affected by the dopant at nano-regime. The internal heat energy produced during combustion causes the interfacial surface friction leads to the agglomeration of particles	[80]
		$ZnxCo_{1-x}Fe_2O_4$	TEM	TEM images of $Zn_{0.4}Co_{0.6}Fe_2O_4$ NPs are represented in **Fig. 3(c–d).** The images confirm the uniform distribution of particles and are in the nanometer range, it clearly shows that the spherical and agglomerated particles	[80]
	Surface Area	$MnFe_2O_4$	BET	The total specific surface area (SSA) is estimated using the Brunauer-Emmett-Teller (BET) and is found to be 33.479 $m^2\ g^{-1}$ with total pore volume of 0.1609 $cm^3\ g^{-1}$. The average pore diameter is 19.23 nm	[65]
	cationic distribution	$ZnFe_2O_4$	Mossbauer spectroscopy	Fe^{3+} species were two which demonstrated 66% occupancy of Fe^{3+} on the octahedral site	[81]

				and 29.8% occupancy at the tetrahedral site	
	Bonds	Li-Zn ferrite	FTIR	The FTIR spectrum for each of the sample consists of two broad absorption bands: one at ~ 510 - 800 cm^{-1} corresponding to the vibrations of tetrahedral complexes (ν1) and the other, at ~ 250 - 510 cm^{-1} corresponding to the vibrations of octahedral-complexes (ν2) in the spinel structure of the Li-Zn ferrites	[82]
	Binding energy	$Mg_xFe_2O_4$ (x = 0.5, 0.95 and 1.0)	XPS	for x = 0.5, the Fe 2p spectra could be deconvoluted into six peaks; two $Fe^{3+}2p_{3/2}$ peaks: at ~711.74 eV, corresponding to Fe^{3+} ions in octahedral sites and at ~714.32 eV, corresponding to Fe^{3+} ions in tetrahedral sites , with a $2p_{3/2}$ satellite peak at ~720.8 eV, two $Fe^{3+}2p_{1/2}$ peaks at 725.16 eV and 728.4 eV with a satellite peak at 733.5 eV for Fe^{3+} $2p_{1/2}$.	[83]
	Elemental composition	$CoFe_2O_4$	EDS	confirmed the presence of elements, such as cobalt, iron, and oxygen, without any additional impurities, with a ratio of 1:2 for the Co:Fe ions in $CoFe_2O_4$ nanoparticle annealed at 300 °C	[84]
		$CaFe_2O_4$	Raman spectroscopy	Raman spectra reveal crystallization of $CaFe_2O_4$ nanoparticles in the fcc type	[75]

				mixed spinel ferrite (space group: Fd-3) structure	
		$CoFe_2O_4$	thermogravimetric analysis (TGA)	TGA data of the as-synthesized $CoFe_2O_4$. Showed Upon heating to 240⁰C there occurs an initial mass loss of about 13% attributed to the desorption of the organic matter adsorbed on the surface of the nanoparticles. Another 3% of the weight was lost above 240⁰C for a total of 16% during the healing process,	
	Intrinsic point defects vacancies	$Zn_{0.04}Fe_2O_4$, $Co_{0.01}Zn_{0.03}Fe_2O_4$, $Co_{0.02}Zn_{0.02}Fe_2O_4$	PL spectra	The observed three peaks in the visible range for all J1, J2, and J3 samples are generally assigned to the charge transfer between Fe^{3+} ion at the octahedral site, Zn^{2+} ion at tetrahedral sites, and the surrounded O^{2-} ions, respectively. The peak centered at 419 nm is attributed to the zinc vacancies in the lattice which corresponds to violet emission.	[85]
Electrical Properties	Permittivity and electrical resistivity measurements		LCR Meter-type impedance analyzer.	At 20 Hz, the relative permittivity values of P1 and P2 ferrites are 31 and 22, respectively for low values of relative humidity, P1 and P2 ferrites, high electrical resistivity values, i.e 4.1 107 Um and 5.3 107 Um, respectively,	[71]

Materials Research Foundations **112** (2021) 1-61 https://doi.org/10.21741/9781644901595-1

Optical Properties	Bandgap	$BiFeO_3$	UV–visible spectra	strong band-gap absorption at ~590 nm (2.1 eV) for BFO nanoparticles suggesting their potential application as visible-light response photocatalyst	[22]
Magnetic Properties	saturation magnetization and coercivity	$CoFe_2O_4$	VSM	ferromagneticbehavior (FM) of the $CoFe_2O_4$ obtained from VSM measurements with a magnetic field of ±10 kOe at room temperature (RT). The saturation magnetization increased from 3.18 to 3.42 mB and the coercivity decreased from 1100 to 650 Oe with increasing calcination temperature 800 to 1000^0 C	[74]
		$Sn_xMn_{1-x}Fe_2O_4$ (x=0.00, 0.25, 0.50, 0.75, 1.00)	SQUID	At applied magnetic field (0 T - 7 T),thecoercivity of as-prepared $Sn_xMn_{1-x}Fe_2O_4$ ferrites are about 35–110 Oe. The saturated magnetization is in the range of 22–68 emu/g	[86]

4. Characterization of ferrites

There are various methods to characterize FNPs. Some techniques are selective for the investigation of a definite property, while in some cases they are combined. The microscopy-based techniques which gives information about NPs morphology, size, and crystal structure are SEM, TEM, HRTEM, AFM, etc. Numerous different techniques which uses different phenomenas like scattering, X-rays, or spectroscopy, for instance e.g., XRD, EDX, RAMAN SPECTROSCOPY, etc. give additional information of the structure, nanoscale elemental composition, the optical properties, and other common and more specific physical properties of the sample of NPs. Some methods are particular for certain magnetic materials, e.g. SQUID, VSM, FMR, XMCD, XPS etc. [87] as summarized in **Table 3.**

4.1 Structural properties

K. Maaz et al. [88] investigated the structural properties of magnetic nanoparticles (nickel doped cobalt ferrite) produced using the co-precipitation method. The full width half maximum (FWHM) of the highest X-ray diffraction (XRD) peak and transmission electron microscopy (TEM) methods are used to obtain particle size in the range of 18–2874 nm. The presence of Co, Ni, Fe, and oxygen, as well as the necessary phases, were confirmed in the synthesized NPs using energy dispersive X-ray (EDX) analysis. The crystalline nature of the synthesized NPs was confirmed by selective area electron diffraction (SAED) analysis. The factors used to determine topography and morphology can be manually determined, as shown below.

a) Lattice Constant: The first characteristic parameter is lattice constant or lattice parameter. As Spinel ferrites possess a cubic structure, all three lattice constants possess the same value ($a = b = c$). Different methods are used by which the lattice parameter can be calculated. These days, most researchers prefer to use software approaches such as PowderX, UnitcellWin for the lattice parameter calculations. But other manual techniques are also used for the lattice constant calculation of ferrites:

i) Average lattice constant: (a) In this method, to calculate the inter-planar spacing (d), for every diffraction peak in XRD analysis Bragg's law is used,
by using the formula:

$$d_{hkl} = \frac{\lambda}{2 \sin \theta_{hkl}} \quad [1]$$

And also $'\ a\ '$ is calculated for each peak using the formula:

$$a = d_{hkl}\sqrt{h^2 + k^2 + l^2} \quad [2]$$

where h, k, and l are Miller indices of crystal planes.

ii) True lattice parameter:(a_0) Using the Nelson-Riley method more accurate values of lattice parameter can be calculated. In this approach, the extrapolation function$F(\theta)$, named Nelson-Riley function, is calculated for each diffraction peak using the formula[89]:

Materials Research Foundations **112** (2021) 1-61 https://doi.org/10.21741/9781644901595-1

$$F(\theta) = \frac{1}{2}\left[\frac{\cos^2\theta}{\sin\theta} + \frac{\cos^2\theta}{\theta}\right] \quad [3]$$

Then the value of a (average lattice constant) calculated in the first method for each diffraction peak is plotted against $F(\theta)$. The straight line passing through the values gives the value of the true lattice constant.

iii) Theoretical lattice constant (a_{th}): K.A. Mohammed et al. [90] study the infrared and structural investigation of $MgZnFe_2O_4$ ferrites nanoparticles. They used the expected/ ideal cationic distribution to determine the mean ionic radii of A sites (r_A) and B sites (r_B) using the formula

$$r_A = \sum \text{ and } r_B = \frac{1}{2}\sum_i \alpha_i r_i \quad [4]$$

where 'α_i' is the concentration of the element 'i' of ionic radius 'r_i' on the respective side. Lattice parameter is calculated using the formula:

$$a_{th} = \frac{8}{3\sqrt{3}}\left[(r_A + R_0) + \sqrt{3}(r_B + R_0)\right] \quad [5]$$

where R_0 is the radius of oxygen ion (1.32 Å)[89, 91].

b) Crystallite Size: There are two methods to calculate the crystallite size of a ferrite material:

i) Scherrer's formula: Scherrer's or X-ray diffraction line half-width method is an excellent method for calculating grain size when the grain size is smaller than 100 nm. Scherrer's equation is used to calculate the average crystallite size (D)[90]:

$$D = \frac{K\lambda}{\beta\cos\theta} \quad [6]$$

where, θ = Bragg's diffraction angle λ = wavelength of X-ray, β = line broadening measured by the full width at half maximum in radians, and K is the shape factor or Scherrer constant, depending on several factors including Miller Index of reflecting plane and shape of the crystal

ii) Williamson-Hall plot: When the grain size is more than 100 nm, this approach is utilised to determine crystallite size using XRD data. The crystallite size (D)[90] is determined using the William-Hall equation in this method:[90]:

$$Bcos\theta = \frac{k\lambda}{D} + 4\varepsilon\sin\theta \qquad [7]$$

where B is full-width at half maximum (in radian) of the peaks, θ is peak position, k is Scherrer constant (0.89), λ is the wavelength of the X-ray, and is ε the lattice strain.

c) Lattice strain: Lattice strain is generally calculated using William-Hall equation[90, 92]:

$$Bcos\theta = \frac{k\lambda}{D} + 4\varepsilon\sin\theta, \qquad [8]$$

where B is full-width at half maximum (in radian) of the peaks, θ is the peak position, k is Scherrer constant (0.89), λ is the wavelength of the X-ray, D is crystallite diameter and ε is the lattice strain of the structure. Simpler formula to calculate lattice strains using Williamson-Hall effect:

$$\varepsilon = \frac{B}{4tan\,\theta} \qquad [9]$$

modified Williamson-Hall equation:

$$\eta = \frac{2d|k-1|}{D} \qquad [10]$$

where d is the lattice spacing, D is the average crystallite size and k is the shape factor. Lattice strain, which arises in the crystal as a result of substitution, can be attributed to the distortion of the tetrahedron with doping.

d) Densities and porosity: Two types of densities are generally measured while performing structural analysis: x-ray density and actual density [90]. X-ray density is estimated using the relation:

$$D_X = \frac{\Sigma A}{NV} = \frac{8M}{Na^3} \qquad [11]$$

where, $\sum A$ = sum of atomic weights of all atoms in the unit cell = 8M; M is the molecular weight of the particular ferrite, V= volume of the unit cell, and N Avogadro's Number. 8 is written because each primitive cell of spinel consists of eight molecules. Since spinel's have cubic structure, hence the volume a^3 , where, is lattice constant [(a = b = c)].The bulk D_{exp} densities of the specimens were evaluated using the measured mass, m, and volume, V, of the samples using the following relation:

$$D_{exp} = \frac{m}{V} \quad [12]$$

In sintering and powder processing ceramics, porosity is an intrinsic phase. Sintering at a high rate with a small crystallite size decreases porosity significantly[90]. Porosity (P) can be calculated using the relation:

$$P = \frac{D_x - D_{exp}}{D_x} \times 100 \quad [13]$$

Where $D_{exp}$is the actual density and D_x is the X-ray density.

Grain size or particle size: Scanning electron microscopy is a technique with which the structure and size of the grains are estimated. The grain size determined using SEM analysis is always greater than the crystallite size determined using Debye Scherrer's formula in XRD analysis. This difference arises because the grain size provided by SEM analysis is the size of agglomerated particles, which are composed of several crystallites by the soft reunion. Synthesis method influences the size and shape of the grains[93].

Cationic-site distribution parameter: The ionic radii, type of bonding, and preparation procedure all influence the cation distribution between A-site and B-site. The cation distribution can be influenced by changing variables such as temperature, pressure, magneto crystalline anisotropy, and metal ion composition. The types of cation and their distribution amongst these two interstitial sites of the spinel lattice affect the magnetic properties of ferrites.

4.2 Electrical properties

Ferrites are also excellent dielectric materials, with a wide range of applications in technology from microwave to radiofrequency. The dielectric and electric properties of ferrites are essential to understand because they determine how these materials are used in the manufacture of electronic devices. The material's high resistivity and low dielectric loss made it suited for high-frequency power applications[62]. Dielectric and electric properties are usually measured using an impedance analyzer. The dielectric constant was (ε') measured using the relation:

$$\varepsilon' = \frac{\text{VRtSsin}\,\varphi}{\varepsilon 0 \omega \text{RAs}} \quad [14]$$

whereε_0 is the permittivity of free space, and t_S, As are the thickness and surface area of the sample respectively. This approach measures the voltage drop V_R across a standard resistance R and, phase angle φ between $V_R$and current I [94].

4.3 Dielectric properties

The dielectric parameters of ferrites are influenced by a variety of elements, including the technique of manufacture, sintering temperature, chemical composition, crystal structure, grain structure or size, additives, applied field frequency, and humidity. The dielectric constant is defined as "the ability to store electric charge". High-dielectric-constant materials are of particular relevance for use in miniaturized memory devices based on capacitive components and energy storage principles. Ferrites, having very high dielectric constant, are useful in designing good microwave devices such as isolators, switches, etc. The polarisation process, or how quickly the polarizable units in a material orient themselves to keep up with the oscillations of an alternating field, determines the dielectric constant of ferrites. There are four types of polarizations: dipolar, electronic, ionic, and interfacial polarization [62]. At lower frequencies, dipolar and interfacial polarization play a significant role, while at higher frequencies, electronic and ionic polarization are the main contributors. The conduction mechanism is connected to the dielectric constant. The fundamental mechanism that triggers the conduction process is the hopping activity between the electrons of Fe^{2+} and Fe^{3+}. Because of the polarization at grain boundaries, this hopping of electrons causes local charges to be displaced [95].

Dielectric constant: The dielectric constant was determined from the formula[96].

$$\varepsilon = \frac{Cd}{\varepsilon o A} \qquad [15]$$

where C is the capacitance of the pellet in farad, d the thickness of the pellet in meter, A the cross-sectional area of the flat surface of the pellet, and ε_o the constant of permittivity of free space.

Dielectric loss: The term "dielectric loss" refers to the amount of energy absorbed by a dielectric material. Impurities and defects in the crystal lattice induce dielectric loss, which occurs when polarization lags behind the applied alternating field. The reduction in dielectric losses occurs as the frequency rises. According to the Maxwell-Wagner model [61] of inhomogeneous double structure, the highly conducting grains are separated by poor conducting grain boundaries, which leads to the accumulation of charge carriers at the boundaries. Grain boundaries, with a large value of resistance, are active at lower frequencies. Therefore, more energy is required for electron exchange between ferric and ferrous ions, hence polarization within the grain boundaries is increased at low ($Fe^{2+} \leftrightarrow Fe^{3+}$)frequencies, which leads to high energy losses. On the other hand, less resistive grains are active at higher frequencies. Hence, relatively low energy is needed for the hopping of electrons, which results in low energy losses at higher frequencies. The Tangent of dielectric loss angle can be calculated using the relation:

$$\tan\delta = \frac{1}{2\pi f R_s C_s} \qquad [16]$$

where δ is the loss angle, f is the frequency, R_s is the equivalent series resistance and C_s is the equivalent series capacitance. The dielectric loss factor (ε') is also measured in terms of tangent loss factor (tanδ) [96] defined by the relation ε' = ε' tan δ,

AC conductivity: Due to their high electrical conductivity compared to other magnetic materials, spinel ferrites are employed in magnetic devices. As a result, one of the major aspects of spinel ferrites is electrical conductivity, which provides essential information on the conduction mechanism. Measuring the frequency dependence of electrical conductivity is an informative method to understand the type and mechanism of the transport properties. The dielectric and magnetic characteristics of ferrites are substantially influenced by the order of magnitude of conductivity. The AC conductivity of ferrites samples is determined using the relationship from the dielectric constant and dielectric loss factor measurements [96]:

$$\sigma_{AC} = \omega \varepsilon' \varepsilon_o \tan \eth \qquad [17]$$

whereσ_{AC} is the AC conductivity, ω is the angular frequency, ε is the permittivity of free space, ε_0 is the dielectric constant and tan ð is the dielectric loss factor of the samples.

DC resistivity: The activation energy and kind of charge carriers responsible for electronic conduction are determined by DC resistivity. The ferrites' semiconductor nature is indicated by a reduction in resistance as temperature rises. DC resistivity [97], as a function of temperature, can be represented by the Arrhenius type (ρ) equation:

$$\rho = \rho_0\, exp\frac{\mathrm{Ea}}{2\mathrm{KB\,T}} = \rho_0 exp\frac{\Delta \mathrm{E}}{\mathrm{KB\,T}} \qquad [18]$$

Where ρ is the resistivity of the sample at room temperature, E_a is the activation energy, k_B is Boltzmann constant and T is the absolute temperature. E_a in the ferromagnetic region can be calculated from the plot using the equation. The value of activation $Ea = slope \times 4.606 \times 8.62 \times 10^{-5} eV$ energy is observed to be higher for the sample with high electrical resistivity in comparison to that in the case of the sample with lower electrical resistivity.

4.4 Optical and absorption properties

To study the effect structure features such as defects, size, composition on the optical UV–Visible diffuse reflectance measurement and PL spectroscopy are used to investigate the effect of structure factors like as defects, size, and composition on the optical properties of ferrite materials. The band edge absorption spectra of ferrites nanoparticles may also be calculated using a diffuse UV–visible spectroscopy method. In general, the direct band-gap energy value, E_g (**table 4**) for each sample can be evaluated using the optical spectral absorption model [91].

Kubelka–Munk function: The bandgap energy can be approximately calculated from the optical reflectance data by the Kubelka–Munk function,

$$F(R) = \frac{(1-R)^2}{2R} \quad [19]$$

where R is the diffuse reflectance A graph (Tauc plot) is plotted between $[F(R)h\upsilon]^2$ versus hʋ (eV) and the bandgap energy is calculated by extrapolating the linear region of plots until they cross with the x-axis [98].

Table 4: Bandgap energy of various ferrites.

Ferrites	Bandgap(eV)	References
$NiFe_2O_4$	1.27 to 1.47	[99]
$BiFeO_3$	2.10	[100]
$NiCr_xFe_{2-x}O_4$	1.39–2.00 eV.	[101]
$Zn_{1-x}Co_xFe_2O_4$ (x = 0.0, 0.1, 0.2, 0.3, 0.4, 0.5)	1.89 to 1.31 eV	[102]
$Mn_xFe_{3-x}O_4$ ($0.0 \leq x \leq 1.0$)	2.05 eV to 1.17 eV	[103]
$CaFe_2O_4$	1.82 eV	[104]
$MgFe_2O_4$	2.18 eV	[105]
$ZnFe_2O_4$	1.92	[106]

4.5 Magnetic properties

The existence of irregular 3d electrons scattered in uneven amounts at the tetrahedral and octahedral sites causes magnetism in ferrites. Neel's two sublattice model describes the ferrimagnetic type ordering of spinel ferrites [107]. Ferrimagnetism is a type of anti-ferromagnetism in which magnetic moments are aligned opposite to each other, but their magnitudes are not equal. Hence, these materials have non-zero net magnetization[108]. Yafet and Kittel developed on Neel's theory, which accounted for the magnetic properties of ferrites very well [109]. Yafet and Kittel showed that in certain cases, tetrahedral and octahedral cationic sites get divided into sub-lattices in such a way that the vector resultant of the magnetic moments of sub-lattices is aligned in the opposite directions to each other, leading to ferrimagnetism.

Saturation magnetization (M_S) The saturation magnetization is the highest induced magnetic moment that can be achieved in a magnetic field (H); beyond this field, no additional magnetization can be generated. The cationic distribution of metal ions at tetrahedral (A) and octahedral (B) sites determines saturation magnetization. The magnetization of spinel ferrites can be explained using "Neel's two sub-lattice models of ferromagnetism". "Magnetic moments at A and B cationic sites are aligned anti-parallel to each other, and their spins have a collinear structure," according to this model. Thus, total magnetization:

$$M = |M_B - M_A| \quad [20]$$

There are three types of exchange interactions between A and B cationic sites: A-A (inter A-site), A-B (between A and B sites), and B-B (inter B-site). All these interactions are negative i.e., anti-ferromagnetic. Among these, A-B interactions take place between Fe^{3+} ions in different sites over oxygen (i.e.,$^{3+} site - A - O^{2+} - Fe^{3+} site - B$), are the strongest. ($|AB| \gg |BB| > |AA|$). These interactions render the pure spinels as ferrimagnetic with A-site moments aligned anti-parallel to B-site moments. The theoretical value of saturation magnetization for ferrites with a single-phase structure can be determined (MS) using the expression:

$$M_S = \frac{8n\mu B}{103a\ 3\rho} \quad [21]$$

where 'n' is the theoretical number of magnetic moments, μ_B is the magnitude of Bohr magnetron, 'a' is the lattice parameter, ρ is the density.

Remanent magnetization: The value of magnetization that remains in the elimination of an induced magnetic field is known as remanent magnetization. According to Neel, if a group of such particles is subjected to a large magnetic field in order to overcome the dipolar field between them, their moments will line up parallel to the applied field. If the field is switched off, the magnetization should obey the following law [110]:

$$M_r = M_s \exp(\frac{-t}{\tau}) \quad [22]$$

where M_s is the saturation magnetization, t is the time after removal of the magnetic field, and τ is the relaxation time which is thermally activated and given by

$$\frac{1}{\tau} = f_0 \exp(\frac{-KV}{kBT}) \quad [23]$$

where f_0 is a frequency factor usually taken to be 10^9 sec^{-1}, T is the temperature, and kB is the Boltzmann constant. Consequently, if V is greater than a critical value VP or T is smaller than the so-called blocking temperature TB, τ is large compared to the time necessary to measure the magnetization. As the external field is decreased, the magnetic moments do not return to random directions, a remnant magnetization subsists in zero fields and the magnetization exhibits a hysteresis. In the opposite sense, if V is smaller than VP or T is higher than TB, τ becomes small enough for the magnetization to fluctuate during the measurement. In a zero applied field, the observed average value is zero, and an ensemble of such particles behaves similarly to a group of paramagnetic atoms in a nonzero field. The phenomenon, called "superparamagnetism", can be described by Langevin's theory of paramagnetism, and the magnetization is given by:

$$M(H,T) = \frac{MsL[MsVH]}{kBT} \quad [24]$$

where M_s is the saturation magnetization of our sample, L is the Langevin function, V is the volume of the particle, k_B is the Boltzmann constant, T is the temperature, and H is the magnetic field. According to this Langevin function the smaller the particles, the higher is the magnetic field necessary to attain the saturation of the system.

Coercivity: (H_C) Coercive force is inversely proportional to grain size. As the size of grains decreases (grains become finer), the volume concentration of grain boundaries increases. This leads to the enhancement of the pinning possibility of magnetic domain walls, resulting in an increase in, which is the measure of reversible changes in magnetization. Sintering temperature affects coercivity (***H**c*). Because lowering the sintering temperature reduces particle size, coercivity, which is proportional to grain size, will rise. Coercivity depends on magnetocrystalline anisotropy, micro-strains, magnetic particle morphology, particle size, and saturation magnetization and can be expressed as:

$$H_C = \frac{2K1}{\mu 0MS} \quad [25]$$

Magneto crystalline anisotropy of samples is a sum of bulk and surface anisotropies which are related by the equation:

$$K_{eff} = \frac{KBulk + 6\ KS}{d} \quad [26]$$

where, d = particle size. Bulk anisotropy (K_{Bulk})is proportional to ion concentration, Fe^{3+} which decreases on replacement of iron. Assuming surface anisotropies of all the samples to be the same (as are prepared in the same condition),decreases. Hence, the coercive field decreases. Surface anisotropy may sometimes change with particle morphology and size.

Curie temperature: Curie temperature is a significant parameter in the study of magnetic materials because it shows the overall strength of inter-sublattice (A-B) interactions in ferrites. It depends on the number of $Fe^{3+}A — O — Fe^{3+}B$ associations per ion per formula unit. Above this temperature, the initial permeability falls sharply and the material Fe^{3+} is no longer magnetic. At T_C, the material gets transformed from ferrimagnetic to paramagnetic. The highest value of initial permeability just beneath Curie temperature is named Hopkinson peak. The temperature where fast T_{pk} fall is observed is referred to as Curie temperature [111].

Yafet-Kittel Angle: Yafet and Kittel Model [109] also explains the magnetic properties of the ferrites: According to this model, a particular composition, the sub-lattice (A or B) starts splitting into two or four different parts and each part is at an angle, called Yafet-Kittel angle with each other. Yafet-Kittel (Y-K) angle is calculated from the magnetic moment per formula unit using the relation:

$$n_B(x) = M_B cosq_{YK} - M_A \quad [27]$$

where x is the substitution content. where $n_B(x)$ is the experimental magnetic moment expressed in the units of Bohr magneton and M_A and M_B are the Bohr magneton on the A site and B sites, respectively. Y–K angles as a function of x [112].

Table 5: Magnetic, optical and electrical properties of ferrites with different methods.

S. No.	Year	Composition	Method	Properties: Magnetic	Optical	Electrical/dielectric	Refs.
1	2015	$NiFe_2O_4$	Sol-gel	M_s =45.902emu/g M_r = 1.506 emu/g H_c = 27.374 Oe	_	$\sigma_{dc} = 7.7\times10^{-9}\Omega\ cm^{-1}$	[113]
2	2015	$CdFe_2O_4$	thermal treatment	M_s =23.34 emu/g M_r =153 emu/g H_c =3235 Oe	Band gap = 2.0614eV	_	[114]
3	2016	$ZnFe_2O_4$	microwave combustion	M_s = 2.598 emu/g M_r = 0.014 emu/g H_c =7.472 Oe	Band gap = 1.88eV	_	[115]
4	2016	$Mn_{0.8}Zn_{0.2}Fe_2O_4$	microwave combustion	M_s =48.09 emu/g M_r =8.473 emu/g H_c =53.93 Oe	Band gap = 1.95 to 2.35eV	_	[115]
5	2017	$CoFe_2O_4$ $NiFe_2O_4$ $ZnFe_2O_4$	Co-precipitation method	_	Band gap = 2.44eV 3.54eV 3.25eV	Resistivity (Ω)= 0.2344×10^6 0.1153×10^6 0.2063×10^6	[116]
6	2017	$ZnFe_2O_4$	Sol-gel	M_s =12.81 emu/g M_r=0.483 emu/g H_c=16.17 Oe	Band gap = 1.90eV	$\sigma_{ac}=1.32\times10^{-9}$ Scm-^{1}	[117]

7	2018	$Co_{1-x}Ca_xFe_2O_4$ (x= 0.05)	Sol-Gel Auto combustion	M_s=18.86 emu/g M_r=54.91 emu/g H_c=858.14 Oe	_	_	[118]
8	2019	$CoFe_2O_4$	co-precipitation	M_s=45.16 emu/g M_r=10.75 emu/g H_c=0.1 Oe	Band gap =1.5 eV	_	[119]
9	2019	$Ni_{0.4}Co_{0.6}Fe_2O_4$	co-precipitation	M_s=31.45 emu/g M_r =13.48 emu/g H_c=0.8 Oe	Band gap =1.36eV	_	[119]
10	2020	$Ni_{0.4}Co_{0.6}ErFe_2O_4$	co-precipitation	M_s= 35.99 emu/g M_r= 12.21 emu/g H_c= 0.16 Oe	_	Resistivity(Ω)= 6.20 ×10^7	[120]
11	2020	$Ni_{0.4}Co_{0.6}Er_{0.015}Fe_{1.98}O_4$	co-precipitation	M_s= 41.1 emu/g M_r= 10.93 emu/g H_c= 0.4Oe	–	Resistivity (Ω) =1.93×10^7	[120]

5. Applications

Because of the efficient magnetic, electrical, and optical properties of FNPs, researchers are paying considerable attention in the synthesis of FNPs and making their use in a a number of applications that include the medical field [14], electronic devices [6], and environment remediation [121].

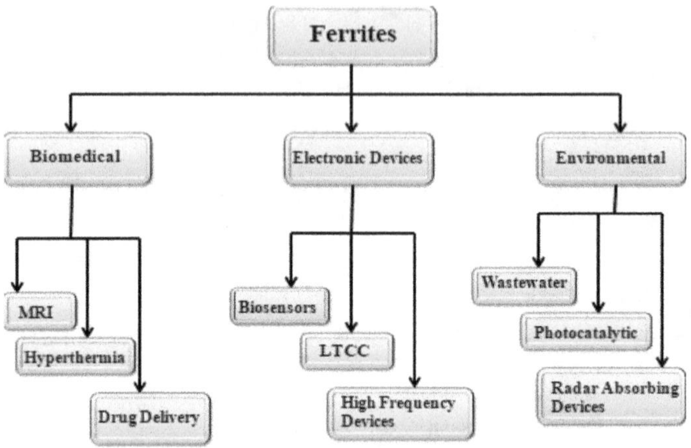

Figure 9: Applications of FNPs.

5.1 Biomedical application

Low toxicity, superparamagnetic, and high magnetization are some essential conditions that demanded the use of magnetic NPs in the field biomedical. In addition, concerns about the nanoparticle's size have to be considered. Therefore, FNPs must have these requirements to be useful [70].

Hyperthermia: Magnetic fluid hyperthermia is a non-invasive treatment that destroys cancer cells by heating up a ferrofluid-impregnated malignant tissue with an AC magnetic field at the same time causing minimal harm to the neighboring healthy tissue [122]. This physical therapy was originated from the observation of some patients with high fever leading to reduction or destruction of the cancer cells. hyperthermia would damage, kill or make the cancerous cells very sensitive to the radiation effects and some certain anticancer drugs. Some techniques like ultrasound, Solid State Phenomena microwave, radiofrequency, and others are currently available to induce the heat. However, the need of homogeneous heat distribution and deepness of the therapeutic temperatures (~45°C) is the key problem for these thermal methodologies [123, 124]. In this sense, the thermal process based on MNPs is a great candidate to solve as mentioned problems. The quantity of induced heat by MNPs is computed in conditions of specific loss power which also depends upon several physical, magnetic properties of the NPs including particle size distribution, size , magneto crystalline anisotropy (K) in addition of several extrinsic factors including the frequency (f) and amplitude magnetic field externally applied. Ferrites nanoparticles (MFe_2O_4 where M= Mn,Co, Zn, Ni) have required magnetic properties because of,their ease of synthesis, low toxicity, physical as well as chemical stabilities [125] they are appropriate as hyperthermia agents. At low magnetic field range hyperthermia applications, maghemite and copper-ferrite NPs are promising materials, whereas in the high field amplitude cobalt-ferrite is the best magnetic hyperthermia agent.

Magnetic resonance imaging MRI: MRI is very important powerful medical imaging methods to get a diagnosis for clinical purposes. This is a non-invasive technique essentially developed on the bases of the interaction of protons and biological tissues in an applied magnetic field. However, the detection of diseased tissue is not easy. Sometimes, the contrast distinction between the target and normal tissue is negligible. Thus, it make complex to detect the disease, which causes the decrease in the cure chances for the patient. To solve this problem, contrast agents were developed to enhance the signal distinction improving the contrast between pathological lesions and normal tissues. However, the requirement of specificity is a problem for MRI contrast agents. Their poor sensitivity becomes unfeasible in accurate detection of small targets for instance tumors in the early stage of cancer, for example. In this consideration, magnetic ferrite NPs appear as an excellent candidate. Their magnetic properties and the easy surface modification and control of parameters like size, composition and shape make them an attractive platform as contrast agents [126, 127] for MRI.

Drug delivery: The growth and development of nanostructured systems for the transport of the active agent (anti-cancer drug, for example) to the target has been the goal of many researchers worldwide. The MNPs applications in drug delivery has gained much attention in the last two decades. The advantages provided by magnetism allow designing a nanosystem easily controlled by an applied magnetic field. This strongly affects selectivity. Several researches have established the specific accumulation in/on the target by enjoying the magnetic and surface properties of NPs. Additionally, a possibility to track the MNPs using the MRI as stated before, which also makes them a potential candidate for carrier drug [128, 129].

5.2 Electronics device application

Ferrite nanoparticles are used in making many electronic devices because of their enhanced electrical properties for instances the high resistivity, low ac conductivity, low power losses, etc. [6].

Gas sensors: Due to sensing properties [130, 131] ferrites are can be used as gas sensors, the response of Mg ferrite ($MgFe_2O_4$) for methane (CH4), hydrogen sulfide (H_2S), liquefied petroleum gas (LPG), and ethanol gas (C_2H_5OH) [132]. Cu- ferrite ($CuFe_2O_4$) [133] and Zn ferrite ($ZnFe_2O_4$) [134,135] showed a good response for hydrogen sulfide (H_2S) reported by many researchers.

Biosensors: Biosensors have been extensively researched and developed as a tool for the medical, food, pharmaceutical and environmental field. Biosensors are gadgets that consolidate a biological component to identify an analyte and a physicochemical segment to create a quantifiable signal. The biosensors are intended to create a computerized electronic signal which is depending upon the concentration of a particular biochemical or a bunch of biochemicals within the sight of a few meddling species [136].

Humidity sensors: The estimation of humidity has gotten incredible consideration because of the perceived significance of water vapour concentration in several areas for instance medicine, meteorology, industry, and agriculture. The humidity sensors must have required properties like good chemical, and thermal stability, high sensitivity, and short response time. The relative humidity (RH)is ratio of actual vapor pressure to the saturated vapor pressure at a given temperature, is frequently used parameter to specify humidity.

Microwave or high-frequency devices: In alternating magnetic field ferrites are used, as a result of their high value electrical resistivity and less eddy current losses, and in high frequency they also have outstanding magnetic property than the magnetic metallic materials. Ferrites are magnetic iron oxides, and gives the best obtainable combination of magnetic materials and electrical insulators and provides the extraordinary flexibility to control the magnetic, conductive properties with crystal lattice parameters. At high frequency magnetic materials properties are considerably different from low-frequency behavior, which leads to a number of microwave phenomena. At higher frequencies self biased hexaferriteis is used to reduce the requirement of a huge external bias magnet [137]. Ferrites have applications in magnetostatic (MSW) devices, offering a better alternative to surface acoustic wave (SAW) circuits. Nonlinear ferrite devices uses the high-power instabilities of ferrites. Dispersed ferrites are used in thin EM wave absorbers [138].

LTCC (low temperature co-fired ceramic) devices: Low temperature co-fired ceramic technology is perhaps the most encouraging incorporated innovations. In this innovation passive components, for instances capacitors, inductors, and filters, are used to integrate the substrate of multilayer LTCC. The telecommunication industry nowadays requires high-volume and low-cost circuit fabrication, whereas at the same time it also require reliability, circuit miniaturization, surface mounting methods and exceptional electrical performance. LTCCs for example NiZnCu ferrite [139] and M-type barium ferrite [78] have been studied intensively for such practical applications.

Supercapacitor applications: Swiftly increasing in requirement for energy conversion devices, energy-storage devices have become significantly attractive in electronic devices. Certainly, supercapacitors are well thought-out mainly capable energy storage devices, due to their outstanding reversibility, high power density, swift charge/discharge. Supercapacitor also have cyclic stability and long-life than the analogous (electrochemical) energy storage devices. Many ferrites for instances Co ferrites, Sn ferrites etc. are largely used for energy storage application because of their electrochemical activity. Various morphologies influence the surface area resulting to the different supercapacitance values. The higher the surface area higher is the storage capacity of charge. Normally, supercapacitors have three basic types, pseudocapacitors, battery-like, and electrochemical double-layer capacitors (EDLC). The fabricated nanoneedles showed narrow pore size distribution, high surface area, and mesoporous nanostructure. $MnZnFe_2O_4$ [140] nanoneedles-based supercapacitorselectrodes display high value of specific capacitance (783 F g−1) in 0.5 M H2SO4 and high columbic efficiency through the stability test on the bases of pseudocapacitance faradaic mechanism.

5.3 Environmental applications

Catalytic applications: Ferrites can convey lattice oxygen to fuel, in some extend they recover it within the air. Copper ferrite seems to possess the best performance, since it can transport the highest amount of lattice O, than the other materials tested, is very active in CH4 oxidation with initial conversion greater than 99% and its selectivity towards CO and H2 is very low. In addition, it is quite stable, and ranked first, still after five redox cycles, and therefore it is appropriate as an oxygen carrier[141-143]

Photocatalytic applications: Photocatalysts are important materials employing which solar energy is used in oxidation and reduction processes. Photocatalysis is apt in numerous areas including the eradication of contaminants from water and air, odor control, bacterial inactivation, production of H_2 by water splitting, inactivation of cancer cells, and many others [144-148]. Organic compounds that are introduced in environment because contaminants are of major concern and the contaminants can be degraded photocatalytically, ideally to CO_2 and water, with no additional waste or by-products. Photocatalysts utilize light energy (hv) to carry out oxidation and reduction reactions. In the photocatalyst an electron (e) get excited from the valence band (VB) to the conduction band (C_B) when exposed to light energy, create a photogenerated hole (h^+). The total solar energy that arrives at the Earth's surface is ultraviolet (UV), visible, and infrared (IR) irradiation. It can be observed that visible light accounts for 46% while UV only accounts for 5% of the total energy from the sun, with the remaining portion belongs to the infrared region. Many of the photocatalysts that are generally used have large band gaps (>3.1 eV) and are only able to utilize a small portion of sunlight. It is ideal to use visible solar energy because the large amount of this light that is received by Earth's surface annually, which is approximately 10,000 times more than the current yearly energy consumption. for that reason, development of photocatalysts which can be capable for using protected and economical solar energy successfully and productively is significant. The development of such material which have a narrow bandgap less than 3.1 eV, which allows for photocatalytic activity in visible light exposer. Ferrites nanomaterial such as $CaFe_2O_4$ [104], $MgFe_2O_4$ [105], $ZnFe_2O_4$ [106], can be efficiently used in visible light because of small band gaps **(Table 4).**

Wastewater treatment: Water pollution is mainly caused by releasing of a heavy number of different category of harmful and toxic metals in the water stream beyond their acceptable limits. The occurrence of such heavy metals in water can create a serious danger for the environment because they are non-degradable and toxic in nature [149-151]. Of the available technologies, SFNPs are applied in water treatment industries are becoming extremely important for the immediate removal of contaminants through the adsorption or degradation process. For example, SFNPs are used in industries for wastewater treatment to remove dyes, phenols, and toxic trace metals. By the external magnetic field they can be recovered easily from their reaction mixture and the chance of several times reuse are other additional advantage of SFNPs [121, 152]

Radar absorbing devices: The emission of radiation from radar increases pollution of electromagnetic radiation in surrounding environment and reduces the effectiveness and performance rate of the electronic instruments, therefore decrease their life span and safety [153]. Radar absorbing material RAM has the property to change electromagnetic radiation energy by thermal energy. These materials are used in numerous areas, for instances in cellular telephones and (reception/transmission) antennas in telecommunication. The most commonly used ferrites spinels ferrites used for microwave absorption. Among the various ferrites CoZn [63], MnZn(2–18 GHz) [154], and NiZn-ferrites [155] are appropriate in the high frequency range (3–30 GHz) microwave absorption.

Conclusions

The synthesis and characterization of FNPs are becoming a popular research subject. This is primarily due to the high demand for high-quality FNPs for biomedical applications and appropriate materials for electronic devices. FNPs generated using excellent synthesis methods can hold up the conditions under which they are synthesized and perform better. The most frequently used synthesis techniques with their advantages and some limitations are also discussed. One important field that needs further attention is to make use of FNPs in electronic materials as sensors and biosensors. As a consequence of doping ferrites with transition metals or rare earth metals, many novel FNPs with various and advanced properties are emerging, as evidenced by recent research studies. The capability to prepare new and improved FNPs by doping Mn, Ni, Co, or Zn into Fe_3O_4 has been established in research. Researchers have not yet completely exploited the rare earth metals doping of ferrite, but progress is being made and results are expected soon. Doping of first-row transition metals for instances manganese, iron, cobalt, and copper appears to be one of the main active research fields, according to the most recent published literature. They are broadly studied for their structural, electrical, magnetic, adsorptive, and catalytic properties to achieve the best quality ferrites for biomedical, electronics devices, and in environmental remediation applications. This chapter briefly summarizes types of ferrites, different synthesis and characterization techniques with their respective applications.

Conflict of Interest

The authors declare that they have no conflict of interest.

Acknowledgment

The authors are grateful to the research and scientific community who have made relentless efforts to bring novel data in the fields and topics of research and development discussed in this review. This work did not receive any funding. The authors are also thankful to the RightsLink® service of the Copyright Clearance Center for providing the permissions and licenses for the reuse and inclusion of the figures in this review (for both the print and online versions).

References

[1] J. Silver, Chemistry of iron, Springer, 1993. https://doi.org/10.1007/978-94-011-2140-8

[2] M. Sugimoto, The past, present, and future of ferrites, Journal of the American Ceramic Society, 82 (1999) 269-280. https://doi.org/10.1111/j.1551-2916.1999.tb20058.x

[3] S.H.a.N.N. Ghosh, Preparation of nanoferrites and their applications, Journal of Nanoscience and Nanotechnology, 14 (2014) 1983-2000. https://doi.org/10.1166/jnn.2014.8745

[4] D.S. Mathew, R.-S. Juang, An overview of the structure and magnetism of spinel ferrite nanoparticles and their synthesis in microemulsions, Chemical engineering journal, 129 (2007) 51-65. https://doi.org/10.1016/j.cej.2006.11.001

[5] S.B. Narang, K. Pubby, Nickel Spinel Ferrites: A Review, Journal of Magnetism and Magnetic Materials, (2020) 167163. https://doi.org/10.1016/j.jmmm.2020.167163

[6] K.K. Kefeni, T.A. Msagati, B.B. Mamba, Ferrite nanoparticles: synthesis, characterisation and applications in electronic device, Materials Science and Engineering B, 215 (2017) 37-55. https://doi.org/10.1016/j.mseb.2016.11.002

[7] R. Srivastava, B. Yadav, Ferrite materials: introduction, synthesis techniques, and applications as sensors, International Journal of Green Nanotechnology, 4 (2012) 141-154. https://doi.org/10.1080/19430892.2012.676918

[8] S.P. John, J.J.J.o.M. Mathew, Determination of ferromagnetic, superparamagnetic and paramagnetic components of magnetization and the effect of magnesium substitution on structural, magnetic and hyperfine properties of zinc ferrite nanoparticles, Journal of Magnetism and Magnetic Materials, 475 (2019) 160-170. https://doi.org/10.1016/j.jmmm.2018.11.030

[9] J. Chatterjee, Y. Haik, C.-J. Chen, M. Materials, Size dependent magnetic properties of iron oxide nanoparticles, Journal of Magnetism, 257 (2003) 113-118. https://doi.org/10.1016/S0304-8853(02)01066-1

[10] P. Dhiman, N. Dhiman, A. Kumar, G. Sharma, M. Naushad, A.A. Ghfar, Solar active nano-$Zn_{1-x}Mg_xFe_2O_4$ as a magnetically separable sustainable photocatalyst for degradation of sulfadiazine antibiotic, Journal of Molecular Liquids, 294 (2019) 111574. https://doi.org/10.1016/j.molliq.2019.111574

[11] E.J.C.S.R. Roduner, Size matters: why nanomaterials are different, 35 (2006) 583-592. https://doi.org/10.1039/b502142c

[12] C. Heck, Magnetic materials and their applications, Elsevier, 2013.

[13] A.M. Ealias, M. Saravanakumar, A review on the classification, characterisation, synthesis of nanoparticles and their application, in: IOP Conf. Ser. Mater. Sci. Eng, 2017, pp. 032019. https://doi.org/10.1088/1757-899X/263/3/032019

[14] W.S. Galvão, D. Neto, R.M. Freire, P.B. Fechine, Super-paramagnetic nanoparticles with spinel structure: a review of synthesis and biomedical applications, in: solid state

phenomena, Trans Tech Publ, 2016, pp. 139-176. https://doi.org/10.4028/www.scientific.net/SSP.241.139

[15] O. Masala, D. Hoffman, N. Sundaram, K. Page, T. Proffen, G. Lawes, R. Seshadri, Preparation of magnetic spinel ferrite core/shell nanoparticles: Soft ferrites on hard ferrites and vice versa, Solid state sciences, 8 (2006) 1015-1022. https://doi.org/10.1016/j.solidstatesciences.2006.04.014

[16] M.N. Akhtar, M.A. Khan, M. Ahmad, G. Murtaza, R. Raza, S. Shaukat, M. Asif, N. Nasir, G. Abbas, M. Nazir, $Y_3Fe_5O_{12}$ nanoparticulate garnet ferrites: Comprehensive study on the synthesis and characterization fabricated by various routes, Journal of magnetism and magnetic materials, 368 (2014) 393-400. https://doi.org/10.1016/j.jmmm.2014.06.004

[17] E. Pollert, Crystal chemistry of magnetic oxides part 2: Hexagonal ferrites, Progress in crystal growth and characterization, 11 (1985) 155-205. https://doi.org/10.1016/0146-3535(85)90033-4

[18] R.C. Pullar, Hexagonal ferrites: a review of the synthesis, properties and applications of hexaferrite ceramics, Progress in Materials Science, 57 (2012) 1191-1334. https://doi.org/10.1016/j.pmatsci.2012.04.001

[19] A.A.S. Hassan, W. Khan, S. Husain, P. Dhiman, M. Singh, M.M. Alanazi, Influence of Ni doping on physical properties of $La_{0.7}Sr_{0.3}FeO_3$ synthesized by reverse micelle technique, Journal of Materials Science: Materials in Electronics, 32 (2021) 3753-3765. https://doi.org/10.1007/s10854-020-05120-w

[20] R. Abazari, S. Sanati, Perovskite $LaFeO_3$ nanoparticles synthesized by the reverse microemulsion nanoreactors in the presence of aerosol-OT: morphology, crystal structure, and their optical properties, Superlattices and Microstructures, 64 (2013) 148-157. https://doi.org/10.1016/j.spmi.2013.09.017

[21] O.B. Pavlovska, L.O. Vasylechko, I.V. Lutsyuk, N.M. Koval, Y.A. Zhydachevskii, A. Pieniążek, Structure Peculiarities of Micro- and Nanocrystalline Perovskite Ferrites $La_{1-x}SmxFeO_3$, Nanoscale Research Letters, 12 (2017) 153. https://doi.org/10.1186/s11671-017-1946-7

[22] S.K. Srivastav, N.J.J.o.t.A.C.S. S. Gajbhiye, Low temperature synthesis, structural, optical and magnetic properties of bismuth ferrite nanoparticles, 95 (2012) 3678-3682. https://doi.org/10.1111/j.1551-2916.2012.05411.x

[23] T.P. Yadav, R.M. Yadav, D.P. Singh, Mechanical milling: a top down approach for the synthesis of nanomaterials and nanocomposites, Nanoscience and Nanotechnology, 2 (2012) 22-48. https://doi.org/10.5923/j.nn.20120203.01

[24] Z. Zhang, G. Yao, X. Zhang, J. Ma, H. Lin, Synthesis and characterization of nickel ferrite nanoparticles via planetary ball milling assisted solid-state reaction, Ceramics International, 41 (2015) 4523-4530. https://doi.org/10.1016/j.ceramint.2014.11.147

[25] Z. Zhang, Y. Liu, G. Yao, G. Zu, Y. Hao, Synthesis and Characterization of $NiFe_2O_4$ Nanoparticles via Solid-State Reaction, International Journal of Applied Ceramic Technology, 10 (2013) 142-149. https://doi.org/10.1111/j.1744-7402.2011.02719.x

[26] E. Manova, T. Tsoncheva, D. Paneva, M. Popova, N. Velinov, B. Kunev, K. Tenchev, I. Mitov, Nanosized copper ferrite materials: mechanochemical synthesis and characterization, Journal of Solid State Chemistry, 184 (2011) 1153-1158. https://doi.org/10.1016/j.jssc.2011.03.035

[27] A. Javidan, S. Rafizadeh, S.M. Hosseinpour-Mashkani, Strontium ferrite nanoparticle study: thermal decomposition synthesis, characterization, and optical and magnetic properties, Materials science in semiconductor processing, 27 (2014) 468-473. https://doi.org/10.1016/j.mssp.2014.07.024

[28] G. Singh, S. Chandra, Electrochemical performance of $MnFe_2O_4$ nano-ferrites synthesized using thermal decomposition method, International Journal of Hydrogen Energy, 43 (2018) 4058-4066. https://doi.org/10.1016/j.ijhydene.2017.08.181

[29] A.P. Herrera, L. Polo-Corrales, E. Chavez, J. Cabarcas-Bolivar, O.N. Uwakweh, C. Rinaldi, Influence of aging time of oleate precursor on the magnetic relaxation of cobalt ferrite nanoparticles synthesized by the thermal decomposition method, Journal of magnetism and magnetic materials, 328 (2013) 41-52. https://doi.org/10.1016/j.jmmm.2012.09.069

[30] K.K. Jaiswal, D. Manikandan, R. Murugan, A.P. Ramaswamy, Microwave-assisted rapid synthesis of Fe_3O_4/poly (styrene-divinylbenzene-acrylic acid) polymeric magnetic composites and investigation of their structural and magnetic properties, European Polymer Journal, 98 (2018) 177-190. https://doi.org/10.1016/j.eurpolymj.2017.11.005

[31] H. Ashassi-Sorkhabi, B. Rezaei-moghadam, R. Bagheri, L. Abdoli, E. Asghari, Synthesis of Au nanoparticles by thermal, sonochemical and electrochemical methods: optimization and characterization, Physical Chemistry Research, 3 (2015) 24-34.

[32] D.D. Andhare, S.R. Patade, J.S. Kounsalye, K. Jadhav, Effect of Zn doping on structural, magnetic and optical properties of cobalt ferrite nanoparticles synthesized via. Co-precipitation method, Physica B: Condensed Matter, 583 (2020) 412051. https://doi.org/10.1016/j.physb.2020.412051

[33] M.S. Darwish, H. Kim, H. Lee, C. Ryu, J.Y. Lee, J. Yoon, Synthesis of magnetic ferrite nanoparticles with high hyperthermia performance via a controlled co-precipitation method, Nanomaterials, 9 (2019) 1176. https://doi.org/10.3390/nano9081176

[34] H.J.P.t. Shokrollahi, Magnetic, electrical and structural characterization of $BiFeO_3$ nanoparticles synthesized by co-precipitation, 235 (2013) 953-958. https://doi.org/10.1016/j.powtec.2012.12.008

[35] M. Houshiar, F. Zebhi, Z.J. Razi, A. Alidoust, Z. Askari, Synthesis of cobalt ferrite ($CoFe_2O_4$) nanoparticles using combustion, coprecipitation, and precipitation methods: A comparison study of size, structural, and magnetic properties, Journal of Magnetism and Magnetic Materials, 371 (2014) 43-48. https://doi.org/10.1016/j.jmmm.2014.06.059

[36] R.R. Shahraki, M. Ebrahimi, S.S. Ebrahimi, S.J.J.o.M. Masoudpanah, M. Materials, Structural characterization and magnetic properties of superparamagnetic zinc ferrite nanoparticles synthesized by the coprecipitation method, 324 (2012) 3762-3765.

[37] A. Mazrouei, A. Saidi, Microstructure and magnetic properties of cobalt ferrite nano powder prepared by solution combustion synthesis, Materials Chemistry and Physics, 209 (2018) 152-158. https://doi.org/10.1016/j.matchemphys.2018.01.075

[38] P. Dhiman, S. Sharma, A. Kumar, M. Shekh, G. Sharma, M. Naushad, Rapid visible and solar photocatalytic Cr(VI) reduction and electrochemical sensing of dopamine using solution combustion synthesized ZnO–Fe_2O_3 nano heterojunctions: Mechanism Elucidation, Ceramics International, 46 (2020) 12255-12268. https://doi.org/10.1016/j.ceramint.2020.01.275

[39] V.J. Angadi, B. Rudraswamy, K. Sadhana, K.J.J.o.M. Praveena, M. Materials, Structural and magnetic properties of manganese zinc ferrite nanoparticles prepared by solution combustion method using mixture of fuels, 409 (2016) 111-115. https://doi.org/10.1016/j.jmmm.2016.02.096

[40] V. Sudheesh, N. Thomas, N. Roona, H. Choudhary, B. Sahoo, N. Lakshmi, V. Sebastian, Synthesis of nanocrystalline spinel ferrite (MFe_2O_4, M= Zn and Mg) by solution combustion method: influence of fuel to oxidizer ratio, Journal of Alloys and Compounds, 742 (2018) 577-586. https://doi.org/10.1016/j.jallcom.2018.01.266

[41] N. Thomas, T. Shimna, P. Jithin, V. Sudheesh, H.K. Choudhary, B. Sahoo, S.S. Nair, N. Lakshmi, V. Sebastian, Comparative study of the structural and magnetic properties of alpha and beta phases of lithium ferrite nanoparticles synthesized by solution combustion method, Journal of Magnetism and Magnetic Materials, 462 (2018) 136-143. https://doi.org/10.1016/j.jmmm.2018.05.010

[42] G. Kumar, S. Sharma, R. Kotnala, J. Shah, S.E. Shirsath, K.M. Batoo, M. Singh, Electric, dielectric and ac electrical conductivity study of nanocrystalline cobalt substituted Mg–Mn ferrites synthesized via solution combustion technique, Journal of Molecular Structure, 1051 (2013) 336-344. https://doi.org/10.1016/j.molstruc.2013.08.019

[43] L. Zhan, L. Jiang, Y. Zhang, B. Gao, Z. Xu, Reduction, detoxification and recycling of solid waste by hydrothermal technology: A review, Chemical Engineering Journal, (2020) 124651. https://doi.org/10.1016/j.cej.2020.124651

[44] M. Li, X. Liu, T. Xu, Y. Nie, H. Li, C. Zhang, Synthesis and characterization of nanosized MnZn ferrites via a modified hydrothermal method, Journal of Magnetism and Magnetic Materials, 439 (2017) 228-235. https://doi.org/10.1016/j.jmmm.2017.04.015

[45] J. Li, Q. Wu, J.J.H.o.N. Wu, Synthesis of nanoparticles via solvothermal and hydrothermal methods, (2015) 1-28. https://doi.org/10.1007/978-3-319-13188-7_17-1

[46] S. Komarneni, M.C. D'Arrigo, C. Leonelli, G.C. Pellacani, H. Katsuki, Microwave-hydrothermal synthesis of nanophase ferrites, Journal of the American Ceramic Society, 81 (1998) 3041-3043. https://doi.org/10.1111/j.1151-2916.1998.tb02738.x

[47] A. Xia, C. Zuo, L. Chen, C. Jin, Y. Lv, Hexagonal SrFe12O19 ferrites: Hydrothermal synthesis and their sintering properties, Journal of magnetism and magnetic materials, 332 (2013) 186-191. https://doi.org/10.1016/j.jmmm.2012.12.035

[48] K. Nejati, R. Zabihi, Preparation and magnetic properties of nano size nickel ferrite particles using hydrothermal method, Chemistry Central Journal, 6 (2012) 23. https://doi.org/10.1186/1752-153X-6-23

[49] A. Gatelytė, D. Jasaitis, A. Beganskienė, A. Kareiva, Sol-gel synthesis and characterization of selected transition metal nano-ferrites, Materials science, 17 (2011) 302-307. https://doi.org/10.5755/j01.ms.17.3.598

[50] P. Dhiman, J. Chand, A. Kumar, R.K. Kotnala, K.M. Batoo, M. Singh, Synthesis and characterization of novel Fe@ZnO nanosystem, Journal of Alloys and Compounds, 578 (2013) 235-241. https://doi.org/10.1016/j.jallcom.2013.05.015

[51] P. Hankare, K. Sanadi, K. Garadkar, D. Patil, I. Mulla, Synthesis and characterization of nickel substituted cobalt ferrite nanoparticles by sol–gel auto-combustion method, Journal of Alloys and Compounds, 553 (2013) 383-388. https://doi.org/10.1016/j.jallcom.2012.11.181

[52] S. Nasir, M. Anis-ur-Rehman, Structural, electrical and magnetic studies of nickel–zinc nanoferrites prepared by simplified sol–gel and co-precipitation methods, Physica Scripta, 84 (2011) 025603. https://doi.org/10.1088/0031-8949/84/02/025603

[53] M.N. Akhtar, M. Saleem, M.A. Khan, Al doped spinel and garnet nanostructured ferrites for microwave frequency C and X-band applications, Journal of Physics and Chemistry of Solids, 123 (2018) 260-265. https://doi.org/10.1016/j.jpcs.2018.08.007

[54] C. Sujatha, K.V. Reddy, K.S. Babu, A.R. Reddy, K. Rao, Effect of sintering temperature on electromagnetic properties of NiCuZn ferrite, Ceramics International, 39 (2013) 3077-3086. https://doi.org/10.1016/j.ceramint.2012.09.087

[55] M.A. Malik, M.Y. Wani, M.A.J.A.j.o.C. Hashim, Microemulsion method: A novel route to synthesize organic and inorganic nanomaterials: 1st Nano Update, 5 (2012) 397-417. https://doi.org/10.1016/j.arabjc.2010.09.027

[56] D. Makovec, A. Kodre, I. Arčon, M. Drofenik, The structure of compositionally constrained zinc-ferrite spinel nanoparticles, Journal of Nanoparticle Research, 13 (2011) 1781-1790. https://doi.org/10.1007/s11051-010-9929-y

[57] M.M. Naik, H.B. Naik, G. Nagaraju, M. Vinuth, H.R. Naika, K. Vinu, Green synthesis of zinc ferrite nanoparticles in Limonia acidissima juice: characterization and their application as photocatalytic and antibacterial activities, Microchemical Journal, 146 (2019) 1227-1235. https://doi.org/10.1016/j.microc.2019.02.059

[58] A. Singh, P.K. Gautam, A. Verma, V. Singh, P.M. Shivapriya, S. Shivalkar, A.K. Sahoo, S.K. Samanta, Green synthesis of metallic nanoparticles as effective alternatives to treat antibiotics resistant bacterial infections: A review, Biotechnology Reports, 25 (2020) e00427. https://doi.org/10.1016/j.btre.2020.e00427

[59] W. Abbas, I. Ahmad, M. Kanwal, G. Murtaza, I. Ali, M.A. Khan, M.N. Akhtar, M. Ahmad, Structural and magnetic behavior of Pr-substituted M-type hexagonal ferrites synthesized by sol–gel autocombustion for a variety of applications, Journal of Magnetism and Magnetic Materials, 374 (2015) 187-191. https://doi.org/10.1016/j.jmmm.2014.08.029

[60] C.-C. Huang, A.-H. Jiang, C.-H. Liou, Y.-C. Wang, C.-P. Lee, T.-Y. Hung, C.-C. Shaw, Y.-H. Hung, M.-F. Kuo, C.-H. Cheng, Magnetic property enhancement of

cobalt-free M-type strontium hexagonal ferrites by $CaCO_3$ and SiO_2 addition, Intermetallics, 89 (2017) 111-117. https://doi.org/10.1016/j.intermet.2017.06.001

[61] K.M. Batoo, Study of dielectric and impedance properties of Mn ferrites, Physica B: Condensed Matter, 406 (2011) 382-387. https://doi.org/10.1016/j.physb.2010.10.075

[62] R. Nongjai, S. Khan, K. Asokan, H. Ahmed, I. Khan, Magnetic and electrical properties of In doped cobalt ferrite nanoparticles, Journal of Applied Physics, 112 (2012) 084321. https://doi.org/10.1063/1.4759436

[63] X. Huang, J. Zhang, S. Xiao, G. Chen, The cobalt zinc spinel ferrite nanofiber: lightweight and efficient microwave absorber, Journal of the American Ceramic Society, 97 (2014) 1363-1366. https://doi.org/10.1111/jace.12909

[64] C. Liu, B. Zou, A.J. Rondinone, Z.J. Zhang, Reverse micelle synthesis and characterization of superparamagnetic MnFe2O4 spinel ferrite nanocrystallites, The Journal of Physical Chemistry B, 104 (2000) 1141-1145. https://doi.org/10.1021/jp993552g

[65] R.H. Vignesh, K.V. Sankar, S. Amaresh, Y.S. Lee, R.K.J.S. Selvan, Synthesis and characterization of $MnFe_2O_4$ nanoparticles for impedometric ammonia gas sensor, Sensors and Actuators B: Chemical, 220 (2015) 50-58. https://doi.org/10.1016/j.snb.2015.04.115

[66] K. Praveena, K. Sadhana, S. Bharadwaj, S. Murthy, Development of nanocrystalline Mn–Zn ferrites for high frequency transformer applications, Journal of Magnetism and Magnetic Materials, 321 (2009) 2433-2437. https://doi.org/10.1016/j.jmmm.2009.02.138

[67] H.-J. Cui, J.-W. Shi, B. Yuan, M.-L. Fu, Synthesis of porous magnetic ferrite nanowires containing Mn and their application in water treatment, Journal of Materials Chemistry A, 1 (2013) 5902-5907. https://doi.org/10.1039/c3ta01692g

[68] K. Maaz, S. Karim, A. Mumtaz, S. Hasanain, J. Liu, J. Duan, Synthesis and magnetic characterization of nickel ferrite nanoparticles prepared by co-precipitation route, Journal of Magnetism and Magnetic Materials, 321 (2009) 1838-1842. https://doi.org/10.1016/j.jmmm.2008.11.098

[69] G. Nabiyouni, M.J. Fesharaki, M. Mozafari, C.P.L. Amighian, Characterization and magnetic properties of nickel ferrite nanoparticles prepared by ball milling technique, Chinese Physics Letters, 27 (2010) 126401. https://doi.org/10.1088/0256-307X/27/12/126401

[70] M. Amiri, M. Salavati-Niasari, A. Akbari, Magnetic nanocarriers: evolution of spinel ferrites for medical applications, Advances in Colloid and Interface Science, 265 (2019) 29-44. https://doi.org/10.1016/j.cis.2019.01.003

[71] A. Dumitrescu, G. Lisa, A. Iordan, F. Tudorache, I. Petrila, A. Borhan, M. Palamaru, C. Mihailescu, L. Leontie, C. Munteanu, Ni ferrite highly organized as humidity sensors, Materials Chemistry and Physics, 156 (2015) 170-179. https://doi.org/10.1016/j.matchemphys.2015.02.044

[72] M. Rashad, O. Fouad, Synthesis and characterization of nano-sized nickel ferrites from fly ash for catalytic oxidation of CO, Materials Chemistry and Physics, 94 (2005) 365-370. https://doi.org/10.1016/j.matchemphys.2005.05.028

[73] O. Karaagac, B.B. Yildiz, H. Köçkar, The influence of synthesis parameters on one-step synthesized superparamagnetic cobalt ferrite nanoparticles with high saturation magnetization, Journal of Magnetism and Magnetic Materials, 473 (2019) 262-267. https://doi.org/10.1016/j.jmmm.2018.10.063

[74] E. Swatsitang, S. Phokha, S. Hunpratub, B. Usher, A. Bootchanont, S. Maensiri, P. Chindaprasirt, Characterization and magnetic properties of cobalt ferrite nanoparticles, Journal of Alloys and Compounds, 664 (2016) 792-797. https://doi.org/10.1016/j.jallcom.2015.12.230

[75] G. Lal, K. Punia, S.N. Dolia, P. Alvi, S. Dalela, S. Kumar, Rietveld refinement, Raman, optical, dielectric, Mössbauer and magnetic characterization of superparamagnetic fcc-$CaFe_2O_4$ nanoparticles, Ceramics International, 45 (2019) 5837-5847. https://doi.org/10.1016/j.ceramint.2018.12.050

[76] N. Sulaiman, M. Ghazali, B. Majlis, J. Yunas, M. Razali, Superparamagnetic calcium ferrite nanoparticles synthesized using a simple sol-gel method for targeted drug delivery, Bio-medical materials and engineering, 26 (2015) S103-S110. https://doi.org/10.3233/BME-151295

[77] W.-z. Lv, B. Liu, Z.-k. Luo, X.-z. Ren, P.-x. Zhang, XRD studies on the nanosized copper ferrite powders synthesized by sonochemical method, Journal of Alloys and Compounds, 465 (2008) 261-264. https://doi.org/10.1016/j.jallcom.2007.10.049

[78] G. Yang, S.-J. Park, Conventional and microwave hydrothermal synthesis and application of functional materials: A review, Materials, 12 (2019) 1177. https://doi.org/10.3390/ma12071177

[79] A. Fernández-Ropero, J. Porras-Vázquez, A. Cabeza, P. Slater, D. Marrero-López, E. Losilla, High valence transition metal doped strontium ferrites for electrode

materials in symmetrical SOFCs, Journal of Power Sources, 249 (2014) 405-413. https://doi.org/10.1016/j.jpowsour.2013.10.118

[80] M.M. Naik, H.B. Naik, G. Nagaraju, M. Vinuth, K. Vinu, R.J.N.-S. Viswanath, Green synthesis of zinc doped cobalt ferrite nanoparticles: Structural, optical, photocatalytic and antibacterial studies, Nano-Structures and Nano-Objects, 19 (2019) 100322. https://doi.org/10.1016/j.nanoso.2019.100322

[81] S.M. Hoque, M.S. Hossain, S. Choudhury, S. Akhter, F. Hyder, Synthesis and characterization of $ZnFe_2O_4$ nanoparticles and its biomedical applications, Materials letters, 162 (2016) 60-63. https://doi.org/10.1016/j.matlet.2015.09.066

[82] V. Rathod, A. Anupama, R.V. Kumar, V. Jali, B. Sahoo, Correlated vibrations of the tetrahedral and octahedral complexes and splitting of the absorption bands in FTIR spectra of Li-Zn ferrites, Vibrational Spectroscopy, 92 (2017) 267-272. https://doi.org/10.1016/j.vibspec.2017.08.008

[83] Z.K. Heiba, M. Sanad, M.B. Mohamed, Influence of Mg-deficiency on the functional properties of magnesium ferrite anode material, Solid State Ionics, 341 (2019) 115042. https://doi.org/10.1016/j.ssi.2019.115042

[84] M.A. Albalah, Y.A. Alsabah, D.E. Mustafa, Characteristics of co-precipitation synthesized cobalt nanoferrites and their potential in industrial wastewater treatment, SN Applied Sciences, 2 (2020) 1-9. https://doi.org/10.1007/s42452-020-2586-6

[85] B.J. Rani, G. Ravi, R. Yuvakkumar, V. Ganesh, S. Ravichandran, M. Thambidurai, A. Rajalakshmi, A. Sakunthala, Pure and cobalt-substituted zinc-ferrite magnetic ceramics for supercapacitor applications, Applied Physics A, 124 (2018) 511. https://doi.org/10.1007/s00339-018-1936-3

[86] A.J. Rondinone, A.C. Samia, Z.J. Zhang, Characterizing the magnetic anisotropy constant of spinel cobalt ferrite nanoparticles, Applied Physics Letters, 76 (2000) 3624-3626. https://doi.org/10.1063/1.126727

[87] S. Mourdikoudis, R.M. Pallares, N.T. Thanh, Characterization techniques for nanoparticles: comparison and complementarity upon studying nanoparticle properties, J Nanoscale, 10 (2018) 12871-12934. https://doi.org/10.1039/C8NR02278J

[88] K. Maaz, S. Karim, A. Mashiatullah, J. Liu, M. Hou, Y. Sun, J. Duan, H. Yao, D. Mo, Y. Chen, Structural analysis of nickel doped cobalt ferrite nanoparticles prepared by coprecipitation route, Physica B: Condensed Matter, 404 (2009) 3947-3951. https://doi.org/10.1016/j.physb.2009.07.134

[89] M. Amer, M.J.J.o.m. El Hiti, m. materials, Mössbauer and X-ray studies for $Ni_{0.2}Zn_xMg_{0.8-x}Fe_2O_4$ ferrites, 234 (2001) 118-125. https://doi.org/10.1016/S0304-8853(00)01406-2

[90] K. Mohammed, A. Al-Rawas, A. Gismelseed, A. Sellai, H. Widatallah, A. Yousif, M. Elzain, M. Shongwe, Infrared and structural studies of $Mg_{1-x}Zn_xFe_2O_4$ ferrites, Physica B: Condensed Matter, 407 (2012) 795-804. https://doi.org/10.1016/j.physb.2011.12.097

[91] R. Melo, P. Banerjee, A.J.J.o.M.S.M.i.E. Franco, Hydrothermal synthesis of nickel doped cobalt ferrite nanoparticles: optical and magnetic properties, 29 (2018) 14657-14667. https://doi.org/10.1007/s10854-018-9602-2

[92] S. Shenoy, P. Joy, M. Anantharaman, Effect of mechanical milling on the structural, magnetic and dielectric properties of coprecipitated ultrafine zinc ferrite, Journal of magnetism and magnetic materials, 269 (2004) 217-226. https://doi.org/10.1016/S0304-8853(03)00596-1

[93] T. Tatarchuk, M. Bououdina, N. Paliychuk, I. Yaremiy, V. Moklyak, Structural characterization and antistructure modeling of cobalt-substituted zinc ferrites, Journal of Alloys and Compounds, 694 (2017) 777-791. https://doi.org/10.1016/j.jallcom.2016.10.067

[94] I. Gul, A. Maqsood, Structural, magnetic and electrical properties of cobalt ferrites prepared by the sol–gel route, Journal of Alloys and Compounds, 465 (2008) 227-231. https://doi.org/10.1016/j.jallcom.2007.11.006

[95] R. Qindeel, N.H. Alonizan, Structural, dielectric and magnetic properties of cobalt based spinel ferrites, Current Applied Physics, 18 (2018) 519-525. https://doi.org/10.1016/j.cap.2018.03.004

[96] R. Ahmad, I.H. Gul, M. Zarrar, H. Anwar, M.B. Khan Niazi, A. Khan, Improved electrical properties of cadmium substituted cobalt ferrites nano-particles for microwave application, Journal of magnetism and magnetic materials, 405 (2016) 28-35. https://doi.org/10.1016/j.jmmm.2015.12.019

[97] R. Mahajan, K. Patankar, M. Kothale, S. Patil, Conductivity, dielectric behaviour and magnetoelectric effect in copper ferrite-barium titanate composites, Bulletin of Materials Science, 23 (2000) 273-279. https://doi.org/10.1007/BF02720082

[98] A. Ashok, L.J. Kennedy, J.J. Vijaya, Structural, optical and magnetic properties of $Zn_{1-x}Mn_xFe_2O_4$ ($0 \leq x \leq 0.5$) spinel nano particles for transesterification of used

cooking oil, Journal of Alloys and Compounds, 780 (2019) 816-828. https://doi.org/10.1016/j.jallcom.2018.11.390

[99] S. Joshi, M. Kumar, S. Chhoker, G. Srivastava, M. Jewariya, V. Singh, Structural, magnetic, dielectric and optical properties of nickel ferrite nanoparticles synthesized by co-precipitation method, Journal of Molecular structure, 1076 (2014) 55-62. https://doi.org/10.1016/j.molstruc.2014.07.048

[100] K.A. McDonnell, N. Wadnerkar, N.J. English, M. Rahman, D. Dowling, Photo-active and optical properties of bismuth ferrite ($BiFeO_3$): an experimental and theoretical study, Chemical Physics Letters, 572 (2013) 78-84. https://doi.org/10.1016/j.cplett.2013.04.024

[101] V.S. Bushkova, I.P. Yaremiy, Magnetic, electric, mechanical, and optical properties of $NiCr_xFe_{2-x}O_4$ ferrites, Journal of Magnetism and Magnetic Materials, 461 (2018) 37-47. https://doi.org/10.1016/j.jmmm.2018.04.025

[102] T.R. Tatarchuk, N.D. Paliychuk, M. Bououdina, B. Al-Najar, M. Pacia, W. Macyk, A. Shyichuk, Effect of cobalt substitution on structural, elastic, magnetic and optical properties of zinc ferrite nanoparticles, Journal of Alloys and Compounds, 731 (2018) 1256-1266. https://doi.org/10.1016/j.jallcom.2017.10.103

[103] S. Güner, M. Amir, M. Geleri, M. Sertkol, A. Baykal, Magneto-optical properties of Mn^{3+} substituted Fe3O4 nanoparticles, Ceramics International, 41 (2015) 10915-10922. https://doi.org/10.1016/j.ceramint.2015.05.034

[104] Z. Zhang, W. Wang, Solution combustion synthesis of $CaFe_2O_4$ nanocrystal as a magnetically separable photocatalyst, Materials letters 133 (2014) 212-215. https://doi.org/10.1016/j.matlet.2014.07.050

[105] M. Shahid, L. Jingling, Z. Ali, I. Shakir, M.F. Warsi, R. Parveen, M. Nadeem, Photocatalytic degradation of methylene blue on magnetically separable $MgFe_2O_4$ under visible light irradiation, Materials Chemistry and Physics, 139 (2013) 566-571. https://doi.org/10.1016/j.matchemphys.2013.01.058

[106] M. Valenzuela, P. Bosch, J. Jiménez-Becerrill, O. Quiroz, A. Páez, Preparation, characterization and photocatalytic activity of ZnO, Fe_2O_3 and $ZnFe_2O_4$, Journal of Photochemistry and photobiology A: Chemistry, 148 (2002) 177-182. https://doi.org/10.1016/S1010-6030(02)00040-0

[107] A.B. Harris, C. Kallin, A.J. Berlinsky, Possible Néel orderings of the Kagomé antiferromagnet, Physical Review B, 45 (1992) 2899. https://doi.org/10.1103/PhysRevB.45.2899

[108] S. Chikazumi, C.D. Graham, Physics of Ferromagnetism 2e, Oxford University Press on Demand, 2009.

[109] Y. Yafet, C. Kittel, Antiferromagnetic arrangements in ferrites, Physical Review, 87 (1952) 290. https://doi.org/10.1103/PhysRev.87.290

[110] E. Manova, B. Kunev, D. Paneva, I. Mitov, L. Petrov, C. Estournès, C. D'Orléan, J.-L. Rehspringer, M.J.C.o.m. Kurmoo, Mechano-synthesis, characterization, and magnetic properties of nanoparticles of cobalt ferrite, CoFe2O4, 16 (2004) 5689-5696. https://doi.org/10.1021/cm049189u

[111] M. Buzinaro, N. Ferreira, F. Cunha, M. Macêdo, Hopkinson effect, structural and magnetic properties of M-type Sm^{3+}-doped $SrFe_{12}O_{19}$ nanoparticles produced by a proteic sol–gel process, Ceramics International, 42 (2016) 5865-5872. https://doi.org/10.1016/j.ceramint.2015.12.130

[112] D.S. Nikam, S.V. Jadhav, V.M. Khot, R. Bohara, C.K. Hong, S.S. Mali, S. Pawar, Cation distribution, structural, morphological and magnetic properties of Co_{1-x} Zn x Fe_2O_4 (x= $_{0–1}$) nanoparticles, RSC Advances, 5 (2015) 2338-2345. https://doi.org/10.1039/C4RA08342C

[113] L. Chauhan, A. Shukla, K. Sreenivas, Dielectric and magnetic properties of Nickel ferrite ceramics using crystalline powders derived from DL alanine fuel in sol–gel auto-combustion, Ceramics International, 41 (2015) 8341-8351. https://doi.org/10.1016/j.ceramint.2015.03.014

[114] M. Naseri, Optical and magnetic properties of monophasic cadmium ferrite ($CdFe_2O_4$) nanostructure prepared by thermal treatment method, Journal of magnetism and magnetic materials, 392 (2015) 107-113. https://doi.org/10.1016/j.jmmm.2015.05.026

[115] E. Hema, A. Manikandan, M. Gayathri, M. Durka, S.A. Antony, B. Venkatraman, The role of Mn^{2+}-doping on structural, morphological, optical, magnetic and catalytic properties of spinel $ZnFe_2O_4$ nanoparticles, Journal of nanoscience and nanotechnology, 16 (2016) 5929-5943. https://doi.org/10.1166/jnn.2016.11037

[116] P. Chand, S. Vaish, P. Kumar, Structural, optical and dielectric properties of transition metal (MFe_2O_4; M= Co, Ni and Zn) nanoferrites, Physica B: Condensed Matter, 524 (2017) 53-63. https://doi.org/10.1016/j.physb.2017.08.060

[117] R.S. Yadav, I. Kuřitka, J. Vilcakova, P. Urbánek, M. Machovsky, M. Masař, M. Holek, Structural, magnetic, optical, dielectric, electrical and modulus spectroscopic characteristics of $ZnFe_2O_4$ spinel ferrite nanoparticles synthesized via honey-mediated

sol-gel combustion method, Journal of Physics and Chemistry of Solids, 110 (2017) 87-99. https://doi.org/10.1016/j.jpcs.2017.05.029

[118] S. Gowreesan, A.R. Kumar, Synthesis, structural, dielectric and magnetic properties of spinel structure of Ca2+ substitute in Cobalt ferrites ($Co_{1-x}CaxFe_2O_4$), Chinese journal of physics, 56 (2018) 1262-1272. https://doi.org/10.1016/j.cjph.2018.02.014

[119] S. Iftikhar, M.F. Warsi, S. Haider, S. Musaddiq, I. Shakir, M. Shahid, The impact of carbon nanotubes on the optical, electrical, and magnetic parameters of Ni^{2+} and Co^{2+} based spinel ferrites, Ceramics International, 45 (2019) 21150-21161. https://doi.org/10.1016/j.ceramint.2019.07.092

[120] M.F. Warsi, A. Iftikhar, M.A. Yousuf, M.I. Sarwar, S. Yousaf, S. Haider, M.F.A. Aboud, I. Shakir, S. Zulfiqar, Erbium substituted nickel–cobalt spinel ferrite nanoparticles: Tailoring the structural, magnetic and electrical parameters, Ceramics International, 46 (2020) 24194-24203. https://doi.org/10.1016/j.ceramint.2020.06.199

[121] K.K. Kefeni, B.B. Mamba, T.A.J.S. Msagati, P. Technology, Application of spinel ferrite nanoparticles in water and wastewater treatment: a review, 188 (2017) 399-422. https://doi.org/10.1016/j.seppur.2017.07.015

[122] R. Kappiyoor, M. Liangruksa, R. Ganguly, I.K. Puri, The effects of magnetic nanoparticle properties on magnetic fluid hyperthermia, Journal of Applied Physics, 108 (2010) 094702. https://doi.org/10.1063/1.3500337

[123] Z. Hedayatnasab, F. Abnisa, W.M.A.W. Daud, Review on magnetic nanoparticles for magnetic nanofluid hyperthermia application, Materials and design, 123 (2017) 174-196. https://doi.org/10.1016/j.matdes.2017.03.036

[124] S. Laurent, S. Dutz, U.O. Häfeli, M. Mahmoudi, Magnetic fluid hyperthermia: focus on superparamagnetic iron oxide nanoparticles, Advances in colloid and interface science, 166 (2011) 8-23. https://doi.org/10.1016/j.cis.2011.04.003

[125] I. Sharifi, H. Shokrollahi, S.J.J.o.m. Amiri, m. materials, Ferrite-based magnetic nanofluids used in hyperthermia applications, 324 (2012) 903-915. https://doi.org/10.1016/j.jmmm.2011.10.017

[126] H.B. Na, I.C. Song, T. Hyeon, Inorganic nanoparticles for MRI contrast agents, Advanced materials, 21 (2009) 2133-2148. https://doi.org/10.1002/adma.200802366

[127] C. Bárcena, A.K. Sra, G.S. Chaubey, C. Khemtong, J.P. Liu, J.J.C.C. Gao, Zinc ferrite nanoparticles as MRI contrast agents, Chemical Communications, (2008) 2224-2226. https://doi.org/10.1039/b801041b

[128] S. Rana, A. Gallo, R. Srivastava, R. Misra, On the suitability of nanocrystalline ferrites as a magnetic carrier for drug delivery: functionalization, conjugation and drug release kinetics, Acta Biomaterialia, 3 (2007) 233-242. https://doi.org/10.1016/j.actbio.2006.10.006

[129] M. Ansari, A. Bigham, S. Hassanzadeh-Tabrizi, H.A. Ahangar, Synthesis and characterization of $Cu_{0.3}Zn_{0.5}Mg_{0.2}Fe_2O_4$ nanoparticles as a magnetic drug delivery system, Journal of Magnetism and Magnetic Materials, 439 (2017) 67-75. https://doi.org/10.1016/j.jmmm.2017.04.084

[130] E.R. Kumar, R. Jayaprakash, G.S. Devi, P.S.P. Reddy, Magnetic, dielectric and sensing properties of manganese substituted copper ferrite nanoparticles, Journal of magnetism and magnetic materials, 355 (2014) 87-92. https://doi.org/10.1016/j.jmmm.2013.11.051

[131] S. Joshi, V.B. Kamble, M. Kumar, A.M. Umarji, G. Srivastava, Nickel substitution induced effects on gas sensing properties of cobalt ferrite nanoparticles, Journal of Alloys and Compounds, 654 (2016) 460-466. https://doi.org/10.1016/j.jallcom.2015.09.119

[132] Y.-L. Liu, Z.-M. Liu, Y. Yang, H.-F. Yang, G.-L. Shen, R.-Q. Yu, Simple synthesis of $MgFe_2O_4$ nanoparticles as gas sensing materials, Sensors and Actuators B: Chemical, 107 (2005) 600-604. https://doi.org/10.1016/j.snb.2004.11.026

[133] Z. Sun, L. Liu, D. zeng Jia, W. Pan, Simple synthesis of $CuFe_2O_4$ nanoparticles as gas-sensing materials, Sensors and Actuators B: Chemical, 125 (2007) 144-148. https://doi.org/10.1016/j.snb.2007.01.050

[134] H.-J. Zhang, F.-N. Meng, L.-Z. Liu, Y.-J. Chen, P.-J.J.J.o.A. Wang, Highly sensitive H2S sensor based on solvothermally prepared spinel $ZnFe_2O_4$ nanoparticles, Journal of Alloys and Compounds, 764 (2018) 147-154. https://doi.org/10.1016/j.jallcom.2018.06.052

[135] K. Wu, J. Li, C. Zhang, Zinc ferrite based gas sensors: A review, Ceramics International, 45 (2019) 11143-11157. https://doi.org/10.1016/j.ceramint.2019.03.086

[136] E.O. Blair, D.K. Corrigan, A review of microfabricated electrochemical biosensors for DNA detection, Biosensors and Bioelectronics, 134 (2019) 57-67. https://doi.org/10.1016/j.bios.2019.03.055

[137] G. Asghar, M. Anis-ur-Rehman, Structural, dielectric and magnetic properties of Cr–Zn doped strontium hexa-ferrites for high frequency applications, Journal of alloys and compounds, 526 (2012) 85-90. https://doi.org/10.1016/j.jallcom.2012.02.086

[138] M. Pardavi-Horvath, Microwave applications of soft ferrites, Journal of Magnetism and Magnetic Materials, 215 (2000) 171-183. https://doi.org/10.1016/S0304-8853(00)00106-2

[139] P. Guzdek, J. Kulawik, K. Zaraska, A. Bieńkowski, NiZnCu ferrite applied for LTCC microinductor, Journal of Magnetism and Magnetic Materials, 322 (2010) 2897-2901. https://doi.org/10.1016/j.jmmm.2010.05.001

[140] F.M. Ismail, M. Ramadan, A.M. Abdellah, I. Ismail, N.K. Allam, Mesoporous spinel manganese zinc ferrite for high-performance supercapacitors, Journal of Electroanalytical Chemistry, 817 (2018) 111-117. https://doi.org/10.1016/j.jelechem.2018.04.002

[141] B.I. Kharisov, H.R. Dias, O.V. Kharissova, Mini-review: ferrite nanoparticles in the catalysis, Arabian Journal of Chemistry, 12 (2019) 1234-1246. https://doi.org/10.1016/j.arabjc.2014.10.049

[142] A. Evdou, V. Zaspalis, L.J.F. Nalbandian, Ferrites as redox catalysts for chemical looping processes, 165 (2016) 367-378. https://doi.org/10.1016/j.fuel.2015.10.049

[143] A. Kumar, A. Kumari, G. Sharma, B. Du, M. Naushad, F.J. Stadler, Carbon quantum dots and reduced graphene oxide modified self-assembled S@ C_3N_4/B@ C_3N_4 metal-free nano-photocatalyst for high performance degradation of chloramphenicol, Journal of Molecular Liquids, 300 (2020) 112356. https://doi.org/10.1016/j.molliq.2019.112356

[144] E. Casbeer, V.K. Sharma, X.-Z. Li, Synthesis and photocatalytic activity of ferrites under visible light: a review, Separation and Purification Technology, 87 (2012) 1-14. https://doi.org/10.1016/j.seppur.2011.11.034

[145] M. Kaur, N. Kaur, Ferrites: synthesis and applications for environmental remediation, in: Ferrites and Ferrates: Chemistry and Applications in Sustainable Energy and Environmental Remediation, ACS Publications, 2016, pp. 113-136. https://doi.org/10.1021/bk-2016-1238.ch004

[146] A. Kumar, A. Kumar, G. Sharma, M. Naushad, F.J. Stadler, A.A. Ghfar, P. Dhiman, R.V. Saini, Sustainable nano-hybrids of magnetic biochar supported g-C3N4/FeVO4 for solar powered degradation of noxious pollutants- Synergism of adsorption, photocatalysis & photo-ozonation, Journal of Cleaner Production, 165 (2017) 431-451. https://doi.org/10.1016/j.jclepro.2017.07.117

[147] A. Kumar, A. Rana, C. Guo, G. Sharma, K.M. M Katubi, F.M. Alzahrani, M. Naushad, M. Sillanpää, P. Dhiman, F.J. Stadler, Acceleration of photo-reduction and

oxidation capabilities of $B_{i4}O_5I_2$/SPION@calcium alginate by metallic Ag: Wide spectral removal of nitrate and azithromycin, Chemical Engineering Journal, 423 (2021) 130173. https://doi.org/10.1016/j.cej.2021.130173

[148] A. Kumar, S.K. Sharma, G. Sharma, A.a.H. Al-Muhtaseb, M. Naushad, A.A. Ghfar, F.J. Stadler, Wide spectral degradation of Norfloxacin by Ag@$BiPO_4$/BiOBr/$BiFeO_3$ nano-assembly: Elucidating the photocatalytic mechanism under different light sources, Journal of Hazardous Materials, 364 (2019) 429-440. https://doi.org/10.1016/j.jhazmat.2018.10.060

[149] V. Srivastava, Y. Sharma, M. Sillanpää, Application of nano-magnesso ferrite (n-$MgFe_2O_4$) for the removal of Co^{2+} ions from synthetic wastewater: Kinetic, equilibrium and thermodynamic studies, Applied Surface Science, 338 (2015) 42-54. https://doi.org/10.1016/j.apsusc.2015.02.072

[150] P. Dhiman, M. Patial, A. Kumar, M. Alam, M. Naushad, G. Sharma, D.-V.N. Vo, R. Kumar, Environmental friendly and robust $Mg_{0.5-x}CuxZn_{0.5}Fe_2O_4$ spinel nanoparticles for visible light driven degradation of Carbamazepine: Band shift driven by dopants, Materials Letters, 284 (2021) 129005. https://doi.org/10.1016/j.matlet.2020.129005

[151] G. Sharma, A. Kumar, S. Sharma, M. Naushad, P. Dhiman, D.-V.N. Vo, F.J. Stadler, Fe_3O_4/ZnO/Si_3N_4 nanocomposite based photocatalyst for the degradation of dyes from aqueous solution, Materials Letters, 278 (2020) 128359. https://doi.org/10.1016/j.matlet.2020.128359

[152] P. Dhiman, A. Kumar, M. Shekh, G. Sharma, G. Rana, D.-V.N. Vo, N. AlMasoud, M. Naushad, Z.A. Alothman, Robust magnetic ZnO-Fe_2O_3 Z-scheme hetereojunctions with in-built metal-redox for high performance photo-degradation of sulfamethoxazole and electrochemical dopamine detection, Environmental Research, 197 (2021) 111074. https://doi.org/10.1016/j.envres.2021.111074

[153] X.C. Tong, Advanced materials and design for electromagnetic interference shielding, CRC press, 2016. https://doi.org/10.1201/9781420073591

[154] A.M. Gama, M.C. Rezende, C.C. Dantas, Dependence of microwave absorption properties on ferrite volume fraction in MnZn ferrite/rubber radar absorbing materials, Journal of Magnetism and Magnetic Materials, 323 (2011) 2782-2785. https://doi.org/10.1016/j.jmmm.2011.05.052

[155] U. Lima, M. Nasar, R. Nasar, M. Rezende, J.J.J.o.M. Araújo, Ni–Zn nanoferrite for radar-absorbing material, Journal of Magnetism and Magnetic Materials, 320 (2008) 1666-1670. https://doi.org/10.1016/j.jmmm.2008.01.022

Ferrite - Nanostructures with Tunable Properties and Diverse Applications Materials Research Forum LLC
Materials Research Foundations **112** (2021) 62-120 https://doi.org/10.21741/9781644901595-2

Chapter 2

Recent Advances in Processing, Characterizations and Biomedical Applications of Spinel Ferrite Nanoparticles

Gaurav Katoch[1], Garima Rana[2], M. Singh[3], Alberto García-Peñas[6], Sumit Bhardwaj[1], Indu Sharma[4], Pankaj Sharma[5], Gagan Kumar[1*]

[1]Department of Physics, Chandigarh University, Gharuan, Mohali, Punjab, India

[2]School of Physics and Materials Science, Shoolini University, Bajhol, Solan, India

[3]Department of Physics, Himachal Pradesh University, Shimla, India

[4]Department of Physics, Career Point University, Hamirpur, HP, India

[5]Department of Applied Sciences, National Institute of Technical Teachers Training and Research, Sector 26, Chandigarh, 160019, India

[6]Departamento de Ciencia e Ingeniería de Materiales e Ingeniería Química (IAAB), Universidad Carlos III de Madrid, 28911, Leganés, Madrid, Spain

*physics.bhargava@gmail.com

Abstract

Many researchers are interested in investigating ceramic materials because of the potential for their use in nanotechnology. Spinel ferrites are a diverse group of materials with many applications. Electronic devices such as inductors, power, information storage, microwave, and induction tuners are only a few examples. As ferrite materials exhibit super-paramagnetic activity, their potential for biological applications such as drug delivery, hyperthermia, and resonance magnetic imaging. As a result, super-paramagnetism is a highly desirable property in spinel ferrites. Due to the size dependence, the methodologies used to synthesis of these materials have emerged as a critical step in achieving the desired properties. Many synthesis strategies have been developed in this regard such as sol-gel, co-precipitation, solid-state, solution combustion method and so on. As a result, this study provides a historical overview of spinel ferrites, as well as key principles for comprehending their various characterization techniques and properties. Recent developments in the synthesis and applications of spinel ferrites are also discussed.

Keywords

Spinel Structure, Synthesis Routes, Characterization, Biomedical Applications

Contents

Materials Research Foundations **112** (2021) 62-120 https://doi.org/10.21741/9781644901595-2

1. Introduction

Thanks to their unique design and promising applications such as high-quality data storage, catalysts, gas sensors, rechargeable lithium batteries, data storage systems, magnetic bulk cores, magnetic fluids, microwave absorbers, and medical and therapeutic diagnostics, spinel ferrite nanomaterials (SFNMs) have received a lot of attention in recent years from a fundamental science perspective [1-5]. Besides, spinel ferrite-based NPs have potential antimicrobial properties against certain microorganisms that cause disease[6]. Hilpert synthesized diverse spinel ferrites at the start of the century and attributed their chemical composition to their magnetic houses [7]. Kato and Takei [8], two Japanese scholars, made significant contributions to the field of ferrite in 1932. The first discovery was a low magnetism in Fe_3O_4 and $CoFe_2O_4$ solid solutions at 300 degrees Celsius. At room temperature, the same sample showed strong magnetization [8]. Another breakthrough was the development of industrial magnetic core materials. They combined two spinel ferrites, one inverse and the other normal, one with high magnetization and the other with almost no magnetization. The permeability of the composite was establish to multiply by more than ten times [9]. These significant discoveries prepared the path for the industrialization of these fabricated samples, exposing the electronic industry's potential for ferrites-based materials. Apart from significant commercial progress, the fundamental knowledge about the magnetic behavior of spinel ferrites was unknown. To demonstrate this, in 1945, Neél [10] employed exchange interactions and cation distribution to introduce the idea of ferrimagnetism into the scientific community. The advancement of electronic items such as television, computers, and electronic devices were aided by a greater accepting of the magnetic properties of ferrites [11]. So, the fundamental information connected with the market for electronic goods, the researchers decided to concentrate on ferrites-based electronic devices and circuits. As a result, the ferrite industry grew around the world [12-16]. On the other side, biochemists demonstrated why Fe_3O_4 was found within the

Materials Research Foundations **112** (2021) 62-120 https://doi.org/10.21741/9781644901595-2

bodies of certain species, such as pigeons [17], honey bees [18], and bacteria [19]. This showed that this material has a lot of promise in biomedical applications including drug delivery [20], magnetic hyperthermia [21], and diagnostics [22].

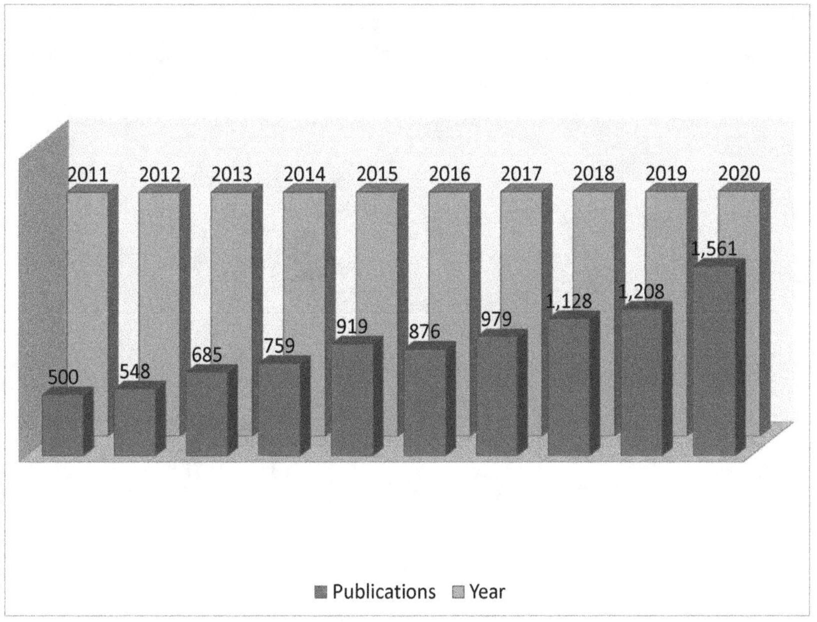

Figure 1: As per Scopus data, the number of publications for the keyword "Spinel ferrites" from 2011-2020.

Scale, shape, synthesis process, and dopant amount all affect these desirable properties. The surface area to volume ratio of nanoparticles increases dramatically as particle size decreases. For a particle with a diameter of 1 μm, 6 nm, and 1.6 nm, approximately 0.15, 20, and 60% of the atoms in the composition are found on the surface, respectively [23, 24]. As compared to their bulk counterparts, these size differences result in specific physical, chemical, and mechanical properties. Among these unique properties of spinel ferrite nanoparticles, and negligible inter-particle interactions are important attributes of SFNPs [25, 26] and to make use of such unique properties, spinel ferrite nanoparticles have been briefly studied by several researchers to address and solve the existing health and environmental problems. The most interesting discoveries and novel applications over time are perfectly described in **Figure 2**. Low toxicity, super-paramagnetism, and high magnetization are only a few of the requirements for using magnetic nanoparticles in biomedical applications [27]. Synthetic processes are essential in this observation because they can regulate the magnetic, structural, and morphological properties of MNPs. We will study the basic principles to better understand the properties of spinel ferrites and their methods in this paper. Finally, MNPs will be addressed concerning MRI, hyperthermia, drug delivery, and other applications. **Figure1** depicts the number of research paper publication by the research community for spinel ferrite from 2011 to 2020.

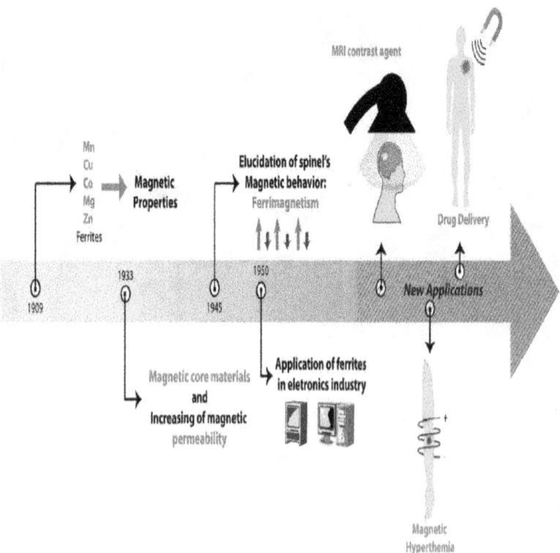

Figure 2: Time-line of main findings and new applications of spinel ferrite nanoparticles(Reproduced with permission from reference [9]).

2. Structure of spinel ferrite nanoparticles

Spinel ferrites are compliant with the common chemical formula AB2O4, where A and B are metals in two different crystallographic, tetrahedral (A sites) and octahedral (B sites), and one of the key components in their formation is Fe. Three potential properties of spinel ferrite are known in the formula AFe2O4, depending on the site selection site A (II) and Fe (III). They are known as normal, inverse, and mixed structures. A (II) is present at tetrahedral sites in a normal spinel structure, while Fe(III) is present at octahedral sites. Fe (III) is uniformly scattered at both sites in an inverse spinel structure, while A (II) occupies only the octahedral sites. Both ions occupy the tetrahedral and octahedral sites at random in a mixed spinel structure. Normal spinel is $ZnFe_2O_4$ as an example. Inverse spinel ferrites including Fe_3O_4 and $NiFe_2O_4$ are a popular example, while $Mn_{0.8}Fe_{0.2}$ $(Mn_{0.2}Fe_{1.8})O_4$ are a good example of mixed spinel ferrite. **Figure 3** shows the structure of spinel ferrite. Spinel ferrites are a large class of materials that have the same structure as natural spinel i.e. $MgAl_2O_4$. According to the literature [28], there were already over 140 oxides and 80 sulphides, and their physicochemical properties were investigated after they were synthesized. Spinel ferrites have an extensive range of properties due to their ability to integrate cations of various charges into their structure. To balance the charge of the anions, the total positive charge should not be greater than 8. The cation radii are also a necessity. The values must be between 0.4 and 0.9 Å. The general formula for magnetic spinel is $A^{2+}Fe_2O_4$ (or $AO.Fe_2O_3$), where the divalent cation can be Mn, Ni, Fe, Co, Zn, Mg, and so on. Natural Fe_3O_4 (or $FeO.Fe_2O_3$) is the most essential and abundant [29]. In 1915, Bragg [30] was the first to discover the crystalline structure of spinel ferrites. The space group of the spinel ferrite is Fd3m, and the unit cell comprises eight units of AB_2O_4, a total of 56 ions (16 in A sites, 32 in B sites, and 32 oxygen directly bonded to the cations) are present. The lattice parameter variation based on the cations in A or B sites, and it can be determined theoretically using equation 1.

$$a_{th} = \frac{8}{3\sqrt{3}}[(r_a+r_O) + \sqrt{3}\,(r_b + r_O)] \qquad (1)$$

where r_o is the oxygen anions of ionic radii, and r_a and r_b are the A and B sites ionic radii, respectively. Each spinel ferrites has a unique physicochemical property.

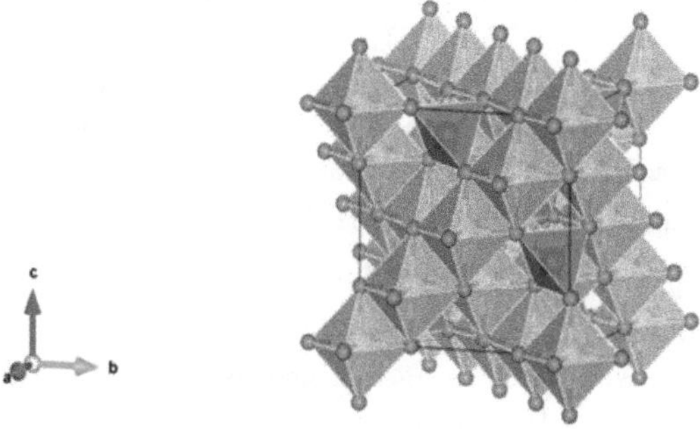

Figure 3: Spinel structure of ferrite nanoparticles.

For example, inverse spinel ferrites are ferrimagnetic, while normal spinel ferrites are paramagnetic [31]. As a result, it's critical to comprehend which factors influence site occupancy, as well as the cationic distribution on structural and magnetic properties.
For example, the average cation radii at the tetrahedral and octahedral sites, r_A and r_B, for the $Zn_{1-x}Co_xFe_2O_4$ system [32], were determined using the following cation distribution:

$$(Zn^{2+}_{1-x}Fe^{3+}_{x+y}Co^{2+}_{y})_A[Co^{2+}_{x-y}Fe^{3+}_{2-x-y}]_B\,(O^{2-}_{4})_O \quad (2)$$

$$r_A = C(Zn^{2+}_{A}).r(Zn^{2+}_{A}) + C(Fe^{3+}_{A}).r(Fe^{3+}_{A}) + C(Co^{2+}_{A}).r(Co^{2+}_{A}) \quad (3)$$

$$r_B = C(Co^{2+}_{B}).r(Co^{2+}_{B}) + C(Fe^{3+}_{B}).r(Fe^{3+}_{B})\,/\,2 \quad (4)$$

where $r(Zn^{2+}_{A})$, $r(Fe^{3+}_{A})$, and $r(Co^{2+}_{A})$ denote the ionic concentration at tetrahedral and octahedral sites, respectively; C denotes the ionic concentration. $r(Co^{2+}_{B})$,$r(Fe^{3+}_{B})$ are the ionic radii of **Co^{2+} (0.745 Å)** and **Fe^{3+} (0.645 Å)** ions in the octahedral sites and **Zn^{2+} (0.60Å),Fe^{3+} (0.49 Å),** and **Co^{2+} (0.58 Å)** ions in the tetrahedral sites, respectively[32].With decrease the ionic radii r(A) at tetrahedral site, while r(B) increases with increasing Co^{2+}concentration because Zn^{2+} ions (0.60 Å) with larger ionic radius are replaced by Co^{2+} ions (0.58 Å) with smaller ionic radius[33]. Kane et al. prepared $Zn_{0.85-x}Ni_xMg_{0.05}Cu_{0.1}Fe_2O_4$ ferrites by sol-gel method and studied the cation distribution. Since there is a clear correlation between the magnetic properties and the cation distribution of spinel ferrites, cationic distribution will help us better understand the magnetic interactions of these elements [34, 35]. Table 1 shows the chemical composition and chemical parameters of several ferrites in a cubic spinel structure. The constant lattice decrease in spinel ferrite systems can be explained in several ways: a decrease in ionic radius due to low ionic surface of the doping cation relative to that of the parent cation

(i) a potential redistribution of Me^{2+} and Me^{3+} ions within the tetrahedral/octahedral ionic sites, due to substantial changes in magnetic properties

(ii) a fraction of Me^{2+} ions occupying the octahedral sites, driving Me^{3+} to the tetrahedral sites against their chemical preferences [34].

Table 1: Composition and crystal-chemical parameters for several ferrites with a spinel structure.

S.No.	Composition	Synthesis Methods	Size(nm)	Lattice Constant (Å)	Strain (ε)	References
1	$ZnFe_2O_4$	microwave combustion	5–15	8.4410	0.034	[34]
2	$Co_{1-x}Zn_xFe_2O_4$	microwave combustion	45–32	8.3799–8.3959	0.035 – 0.043	[35]
3	$Mg_xZn_{1-x}Fe_2O_4$ (x = 0.5, 0.6, 0.7)	Co-precipitation	6.3–14.6	8.422–8.438	~0.005	[36]
4	$Zn_{1-x}Mg_xFe_2O_4$ (x=0.0, 0.1, 0.2, 0.3 0.4, 0.5, 0.6, 0.7, and 0.8)	microwave combustion	41.20–15.87	8.443–8.427	0.067 – 0.047	[37]
5	$Zn_{1-x}Cu_xFe_2O_4$(x = 0.0, 0.1, 0.2, 0.3, 0.4, and 0.5)	microwave combustion	43–54	8.443–8.413	0.067 – 0.056	[38]
6	$Mn_{0.5}Zn_{0.5}Fe_2O_4$	Sol–gel	25	8.4339	0.0009	[39]
7	$Mn_{1-x}Ni_xFe_2O_4$(x = 0.0, 0.1, 0.2, 0.3, 0.4, 0.5)	microwave combustion	-	8.477–8.419	0.412 – 0.200	[40]
8	$Co_{1-x}Ca_xFe_2O_4$(x=0.00, 0.01, 0.015, 0.02, 0.05, 0.1, and 0.15)	sol–gel	22-9	8.376-8.399	1.26-9.6	[41]
9	$NiGd_xFe_{2-x}O_4$(x=0,0.025,0.050,0.075and 0.1)	Sol-gel	25.02-11.1	8.303-8.309	0.00148-0.00262	[42]
10	$Al_xZnFe_{2-x}O_4$, (x = 0.0, 0.04, 0.08, 0.12, 0.16, 0.20)	co-precipitation	36.1-31.4	8.44-8.425	0.00117-0.000121	[43]
11	$CoFe_{1.95}Ho_{0.05}O_4$	mechanical alloyed	50	8.3745	0.463	[44]

12	$Ni_xZn_{1-x}Fe_2O_4$(x=0.5,0.6 and 0.7)	Coprecipitated	10.6-12.3	8.381-8.346	-	[45]
13	$Co_{0.5}Zn_{0.5}Fe_2O_4$	sol–gel	21.68	8.411	0.577	[46]
14	$Co_{0.5}Zn_{0.5}Fe_2O_4$	ball-mill	18.99	8.409	0.702	[46]
15	$MgFe_2O_4$	sol–gel auto combustion	17.5	8.301	-	[47]
16	$ZnFe_2O_4$	sol–gel auto combustion	21.3	8.366	-	[47]
17	$MnFe_2O_4$	sol–gel auto combustion	23.3	8.434	-	[47]
18	$Ni_xZn_{1-x}Fe_2O_4$ (x=0.1, 0.3, 0.5)	co-precipitation	11.85-16.76	8.4271-8.4093	0.003539-0.002166	[48]
19	$NiCe_xFe_{2-x}O_4$(x=0,0.025,0.050, 0.075 and 1.0)	Sol-gel	35.02-15.33	8.303-8.319	0.00124-0.00293	[49]
20	$Ni_{0.9}Cd_{0.1}Gd_xFe_{2-x}O_4$ (x = 0, 0.1, 0.2 and 0.3)	sol–gel	17.5 ± 1.2-19.3 ± 0.8	0.8402-0.8363	-	[50]
21	$Cu_{0.5}Ni_{0.5}In_xFe_{2-x}O_4$ (x = 0.00–0.32)	sol–gel	33.21–17.78	8.3349 to 8.3723	0.054 - 0.094	[51]

3. Synthesis methods of spinel ferrite nanoparticles

To synthesize nanoparticles there is two techniques: top-down and bottom-up shown in **Figure 4**. In the first technique, mass material is separated to get nano estimated particles. This technique has many breaking points at particles like commonly metal oxides are utilized, a prerequisite of extremely high temperature for the response, items are inhomogeneous, presence of pollutions, gem abandons, expansive size appropriation and defect in surface design. In the second technique, nanoparticles are produced from small atomic building blocks. To synthesize nanoparticles there are various methods to produce the product uniform and pure. Tremendous procedures are utilized to plan nano ferrites, for example, solution combustion method, co-precipitation method, sol-gel method, microemulsions method, solid-state reaction method, thermal method. Top-down synthesis techniques include solid-state and other are bottom-up techniques various properties improved with various procedures.

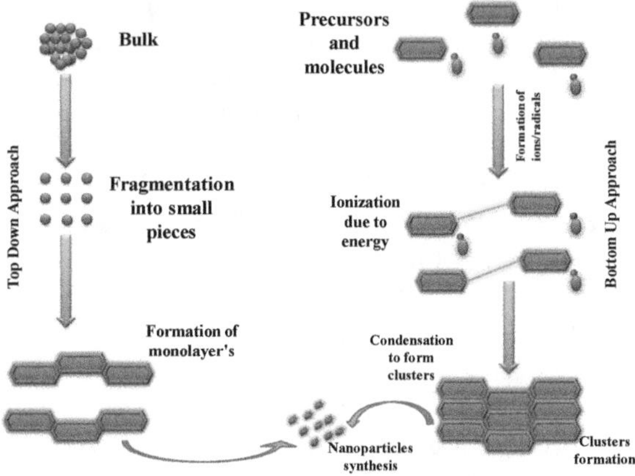

Figure 4: shows the top-down and bottom-up approach spinel ferrite nanoparticles.

3.1 Co-precipitation method

This method is easy to integrate nanomaterials. To get homogenous construction and small size this method is used. The ferrites arranged utilizing this technique are of controlled size, profoundly unadulterated, and have a homogeneous construction. For making a homogenous solution different materials are dissolved in H_2O and may be other medium and here basic material that is used like (chloride, sulphate, nitrate, etc.) as the initial material. Set the pH range 7-10 to get nanoparticles. In the end, materials are together after precipitation. After the precipitation, the material should be calcined. Thakur et al. [52] utilized this technique to prepare MnZn ferrite. For starting materials $FeCl_3$, $MgCl_3$ and $ZnCl_2$ were utilized. In 60 ml of refined D.I. water the 3M arrangement was prepared. Then this arrangement was filled with a bubbling sodium hydroxide (NaOH) arrangement although mixing for 60 min at different temperatures 355-359 K, keeping up pH somewhere in the range of 10 and 11. For getting the nanoparticles to settle down the stirring process is allowed to precipitates and after this, with the help of distilled water, the final sample was washed many times. Using mortar pestle the crushing process was started while before this the sample washed properly and dried in a hot air oven. Anwar et al. [53] more over prepared MnZn ferrites, $Mn(NO3)_2.4H_2O$, $Zn(NO_3)_2H_2O$, and $Fe_2(NO_3)_3.9H_2O$ as the initial materials. At 355K homogenous solution prepared when these were assorted. CPT has successfully synthesized MFe_2O_4 (M=Fe, Co, Mn) NPs with sizes ranging from 4 to 12 nm in high yields using alkanolamines instead of sodium hydroxide, ammonia, and tetra alkyl-ammonium

Materials Research Foundations **112** (2021) 62-120 https://doi.org/10.21741/9781644901595-2

hydroxides. When compared to NaOH-prepared nanoparticles, this method produced nanoparticles with higher colloidal stability, smaller particle sizes, and improved magnetic properties [54]. The CPT method was used to make spinel-type $CuFeMnO_4$ with an average size of 100 nm, which was achieved by calcinating the precipitate at 500–900 °C. Cu^+ ions at tetrahedral sites and Mn^{4+} at octahedral sites turned into Cu^{2+} and Mn^{3+} during calcinations, according to X-ray photoelectron spectroscopy (XPS) study [55]. $Mn_{1-x}Zn_xFe_2O_4$ [56], $Ni_{1-x}Zn_xFe_2O$ [57,58], $MnFe_2O_4$ [59], $CoFe_2O_4$ [60], $Ni_{0.53}Cu_{0.12}Zn_{0.35}Fe_2O_4$ and $Mg_{0.2}Cu_{0.3}Zn_{0.5}Fe_2O_4$ [61] has been prepared by CPT method.

3.2 Sol-gel method

It is a method for producing solid materials from small particles. In this compound methodology, a "sol" (a colloidal arrangement) is shaped and also, gelation of the sol in a nonstop fluid stage (gel) to shape a three-dimensional organization structure. For this method, metal ferrite is prepared by a sol-gel method like nickel ferrite, cobalt ferrite, zinc ferrite. The salt precursors (oxidizer) and complexing agents (reductant) are added to form a gel in these processes. Citric acid is widely used as a chelating agent and a reductant in the auto combustion sol-gel process, where it is oxidised by nitrate ions and thus acts as a fuel [62]. To make ferrite powder, the solution is heated and self-ignited [63]. After that, the ferrite is annealed at temperatures ranging from 400 to 1000 degrees Celsius. The sol-gel method [64] allows for more precise control of elemental composition, structure, homogeneity, and powder morphology. The key drawback of sol-gel techniques is that they are expensive and produce poor yields [65].

It has been possible to monitor the stability, crystallinity, particle size, and electromagnetic properties of SF products by selecting an acceptable complexing agent, the heating source, and the atmosphere [66]. The sol-gel method was used to synthesize $NiFe_2O_4$NPs [67]. The chelating agent was polyacrylic acid. The average particle size ranged from 5 to 30 nm. Particle sizes decreased and crystallinity increased as the molar ratio of PAA to total metal ions was increased. The findings revealed that PAA not only acts as a chelating agent but also aids in increasing the crystallinity of nanoparticles through its combustion heat. R.H. Kadam et al. [68] prepared Gd3 + doped Mn-Cr nanoferrites in the form of sol-gel auto combining and then investigate the structural properties and magnetic fields.

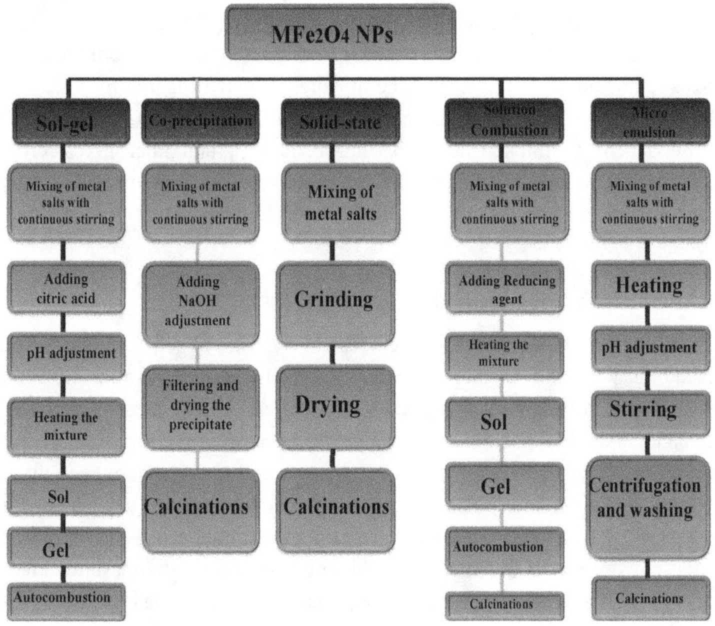

Figure 5: Flow chart for the synthesis of spinel ferrite nanoparticles by different methods.

3.3 Microemulsion method

Microemulsions are macroscopically homogeneous and thermodynamically stable solutions made up of at least three components: a polar phase (typically water), a nonpolar phase (typically oil), and a surfactant [69]. Oil-in-water (O/W) (normal micelle), water-in-oil (W/O) (reverse micelle), and water-in-supercritical CO_2 (W/sc-CO_2) are all examples of microemulsion systems. For the most part, the reverse micelle process is used to make ferrite nanoparticles. A surfactant is dissolved in organic solvents, and tiny droplets of water are encircled by surfactant molecules, creating so-called "water ponds." These reverse micelle water pools serve as nanoreactors, allowing inorganic and organic materials to precipitate [69, 70]. This method produces nanoparticles that are uniform and homogeneous, with a small size distribution. Small modifications to the synthesis will easily regulate the size. On the other hand, the process is costly and low yield. Furthermore, significant amounts of organic solvents are used, which is not environmentally friendly [58, 71]. In a W/O microemulsion device [water–

isooctane–AOT (sodium di 2ethylhexylsulfosuccinate)] [72], nanocrystalline Mn-Zn ferrite was synthesized. The particles ranged in size from 20 to 80 nanometres. The effect of synthesis temperature on the size of the micelles was investigated in this analysis. At 15°C, the smallest particles were collected. $CoFe_2O_4$ [73], $MgFe_2O_4$ [74], $ZnFe_2O_4$ [75], $Co_{0.5}Ni_{0.5}Fe_2O_4$ [76], CoeSn substituted barium ferrite [77], and other mixed ferrites have all been synthesised using the microemulsion technique. As previously mentioned, the method of preparation used may affect the structural and physical properties of ferrite nanoparticles [78].

3.4 Solution combustion method

SCS is a thrilling phenomenon, which includes engendering of self-supported exothermic responses along with fluid or sol-gel media. This method is very simple and synthesizes nanomaterial especially complexing oxides, alloys, etc. It is a method that is used to prepared nanocrystalline powder viz ceramic oxides, which are extremely pure and uniform powders [79]. In SCS there are numerous initial materials are used viz nitrates, metal sulfates, and carbonates, as oxidants and reducing reagents, fuels such as glycine, sucrose, urea [80, 81]. During combustion reaction, the nitrate acts as an oxidizer fuel. In the end, the powder that is obtained from this method of a single-phase meanwhile is a combination of metal oxides and required heat treatment to form a single phase. SCS is a technique dependent on the rule that once a response is started under warming, an exothermic response happens that gets self-supporting inside a specific period, bringing about a powder as the result. The exothermic response starts at the start temperature and creates a specific measure of warmth that is shown in the greatest temperature or temperature of burning. Arrangement ignition blend has the benefit of quickly creating the powder homogenous. Since it is an exothermic, auto-spread cycle, and with a high warmth discharge rate, it tends to be hazardous and ought to be attempted with additional safeguards. Arrangement burning blend is a speedy and simple cycle, with as primary favorable circumstances the saving of time and energy.

3.5 Solid-state method

Grinding of iron and metal salts is accompanied by a direct reaction at high temperatures in the solid-state synthesis of the ceramic process. The ceramic technique can be influenced by the particle size of the reactants, the gas environment, and international additives [82]. The procedure is straightforward and low-cost, with high efficiency and selectivity [83]. Since the reaction takes place without the use of a solvent, contamination from solvents is avoided. It's difficult to keep track of the reaction's progress and find the right conditions for the reaction to finish. Furthermore, it is difficult to separate the target product from a mixture of reactants and products. Furthermore, even though the reaction continues almost to completion, the solid-state reaction makes it impossible to obtain a homogeneous pure product [84]. Nb^{5+}/Ti^{3+}doped $NiFe_2O_4$, $MnFe_2O_4$, and $NiZnFe_2O_4$ have been synthesized using the solid-state reaction [85]. As the concentration of dopant ions increased, the particle diameters shrank. The ferrite grain size had a big impact on saturation magnetization and coercive area. Zinc ferrite [86], spinel cobalt ferrite [87], and LiZn ferrite ($Li_{0.4}Zn_{0.2}Fe_{2.4}O_4$) [88] have all been synthesized using the solid-state method.

There are the following steps we followed after fabrication of samples by different methods:

a) Calcination

Calcination enables the neighbouring grains to react by inter-diffusion of their ions in the furnace, to obtain a new stable process the diffusion must be reduced during sintering. Controlling the shrinkage of the sample during the sintering of the calcination is an important process. Calcination is a process to heat the solid material in the absence of oxygen and air. The powder to be calcined is not normally pressed. Escaping gases often produces extra porosity, and the reactions do not go to completion partly owing to small contact areas at interfaces between grains. Repeated mixing, pressing, and grinding are needed to form a homogeneous product.

b) Pelletization

The simplest way to shape a ceramic is using pelletization. The sample is ground uniformly by mixing with a suitable binder, and then the powder is pressed by hydraulic pressing which is used for the shaping of the materials. Unidirectional pressing is carried out in a die with movable top and bottom punches. The bottom punch is used for free-flowing of powder in which cavity is formed and the other side top punch is used to compress the powder and determine the volume of the sample. These punches are moved upwards when the top punch is clear from the powder and the bottom punch level with the die. The bottom punch is cleared when the compact is removed. Under this method, the dry powder is pressed in a metal die at a sufficiently high pressure of 10 tons to form a dense, strong piece. Pressing ensures a large area of surface contact between the grains. This increases the reaction rates, density, and strength of final products.

Macroscopic pressure gradients that develop across the pellet during pressurization can result in cracking and therefore they should be minimized.

i) Friction at die wall

ii) Die surfaces are not flat, and

iii) Die edges are not perfect

After pressing, pellets need to be handled carefully to avoid mechanical stress. A metal die of 1.5 cm. diameter opening is used for pelletization purposes. Pressure usually used was nearly equal to 10 tons.

c) Sintering

It is interaction as the expulsion of the pores between the underlying fine particles joined by shrinkage of the calcined item joined with development together and arrangement of solid connections between adjacent particles.

In ceramics, this increases mechanical strength, density, and thermal conductivity. These are correlated to the grain size, grain shape, pore size, and pore shape. During heat treatment, the average grain size increases due to grain growth in this process. Some grain ought to shrink and disappear when the average grain size increases. While overpowering is the driving force in all systems, the main difference in the behavior of different types of systems must be related to different asset conversion processes. Some of these processes are evaporation, condensing, visible flow, expansion of the earth, and the deterioration of plastic.

Material transformation by evaporation and condensation arises from the difference in surface curvature and consequently of the vapour pressure at the various parts of the system. (Vapour phase material requires heating of the material to a temperature sufficiently high for vapour pressure to be appreciable). The diffusion can occur in many ways, e.g. at surfaces, grain boundaries, or in bulk involving migration of vacancies. The sintering temperature is chosen so that solid-state diffusion predominates over vapour transport.

To carry out sintering in the present case, a high-temperature programmable furnace (Heraeus Make) is used with a chromal-alumal thermocouple. For this furnace, the upper-temperature limit was 1400°C. High-temperature alumina boats were used to place the powder/ samples in the furnace.

d) Finishing

Sintered samples are then lapped using micro grit powder to reduce the thickness of the pellet and finally silver metallization is done for electrical and magnetic characterization. **Figure 5** shows the flow chart for the synthesis of spinel ferrite nanoparticles by different methods. **Table 2** summarizes the composition of various spinel ferrites synthesized with different methods and study their magnetic and dielectric properties. Their magnetic properties, as well as the ease with which they can be modified and regulated in terms of size, shape, and composition, make them an appealing medium for biomedical applications [89, 90]. Their high saturation magnetization, high coercivity and low curie temperature are ideal for a variety of biomedical applications.

4. Characterizations of spinel ferrite nanoparticles

The nano ferrites are characterized using a variety of techniques. These techniques provide high-quality details about the sample. The structure, morphological, magnetic, and dielectric properties of spinel ferrite nanoparticles have been defined using a variety of characterization methods. The following are some of them:

4.1 X-ray diffraction

For characterizing crystalline materials, X-ray diffraction (XRD) is an effective non-destructive technique. Structures, phases, desired crystal orientations (texture), and other structural parameters such as average grain size, crystallinity, strain, and crystal defects, provided all the information. The constructive distortion of the monochromatic X-ray beam is dispersed at certain angles from each set of lattice planes in the sample producing X-ray diffraction peaks. The distribution of atoms within the lattice determines the maximum force. For extracting XRD patterns from samples, XRD used CuK as a radiation source. The parameters of the site radii structure, bond hopping length, shared and unallocated edges are all determined using distribution X-ray data. X-ray beam with a length of λ = 1.5406Å is directed at an angle to the sample in this process. The X-ray power shown is recorded as the sample and detector rotate.The Debye-Scherrer equation can be used to determine the crystallite size [91]:

$$D = \frac{0.9\lambda}{\beta cos\theta} \quad (5)$$

where D is the crystallite size in nm, λ is the wavelength, β is the highest Bragg's peak, and θ is the diffraction angle.

The lattice parameter is determined using the equation:

$$a = d\sqrt{h^2 + k^2 + k^2} \quad (6)$$

where d is the interplanar distance and the miller indices are “h,” “k,” and “l.”

The Williamson–Hall method [92] was used to measure the induced strain

$$\beta cos\theta = 4\varepsilon sin\theta \frac{\lambda}{D} \tag{7}$$

Where D is the particle size, X-ray wavelength is λ, FWHM is measured in radians, induced lattice strain is ε, and Bragg Angle is θ. The above-mentioned formula yielded an induced lattice strain.

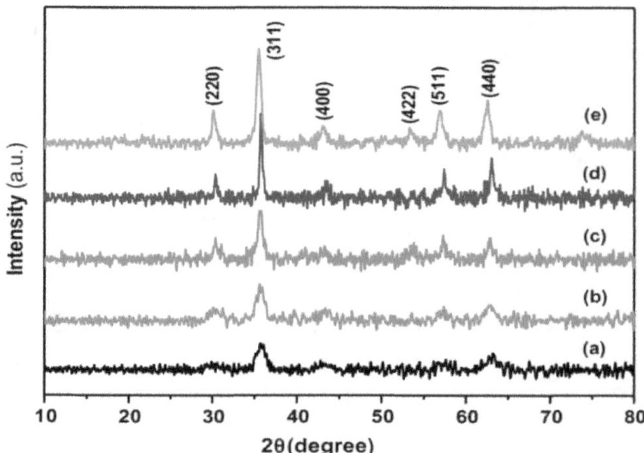

Figure 6: shows the XRD pattern of the prepared ferrite sample (a) $NiFe_2O_4$, (b) $CoFe_2O_4$, (c) $MnFe_2O_4$, (d) $MgFe_2O_4$ and (e) $ZnFe_2O_4$ [93] (copyright permission with license number:5056370206855).

Darshane et al. published an XRD pattern that fits well with JCPDC data (card no. 10-0325) for $NiFe_2O_4$ calcined at 700°C, with an average crystallite size of 20 nm, and crystal studies show the formation of a large number of distinct uniform nanoparticles, confirming the position of NaCl as a growth inhibitor [94]. As grain size increases from nano to micro-scale during annealing at temperatures ranging from 900°C to 1300°C, Sutka et al. reported big differences in the impact of grain size on electrical resistivity between p-type and n-type materials [95]. Wang et al. found that the XRD pattern of cubic $NiFe_2O_4$ (JCPDS file no. 44-1485) has no other iron or nickel oxide phases and that the formation of spherical nanoparticles is dependent on the precursor concentration. Ethylene glycol is used as a solvent, reducing agent, complex agent, and stabilizing agent, according to Wang et al. [96]. Figure 6 shows that the XRD results show that powders made with AFe2O4 have a spinel structure without any phase impurities [93].

Materials Research Foundations **112** (2021) 62-120 https://doi.org/10.21741/9781644901595-2

4.2 Morphological study

Scanning electron microscopy and transmission electron microscopy were used to examine the surface morphology of spinel ferrite nanoparticles. These methods are used on a wide scale to calculate the diameter of ferrite nanoparticles and to visualize their morphology. TEM is a more useful technique for determining the morphology, structure, and crystallite of ferrite nanoparticles. The TEM has a better resolution down to the atomic level. Scanning electron microscopy (SEM) is a method for imaging surfaces with high sensitivity. Its use of the electrons generated by the interaction of the beam with the sample surface. It's widely used to investigate nanoparticles aggregation and surface morphology. SEM is more commonly used than TEM, but SEM resolution is lower than TEM. Zhu et al. documented a porous brick-like structure of nickel-ferrite with numerous nanoparticles agglomerated on it [96]. Zhang et al. discovered a rod-like morphology with an average diameter of 100 nm and a 5 mm average length. Amorphous precursor rods have a very smooth surface [97]. Wang et al. described the elemental composition of aggregated nickel ferrite nanoparticles. Wang et al. published aggregated spheres of nickel ferrite nanoparticles, and EDS analysis revealed that the products only contain O, Fe, and Ni, with no contamination elements found. According to a TEM analysis, the average particle size is 90 nm [98]. **Figure 7** shows the TEM and SEM images of nickel spinel ferrite nanoparticles [99].

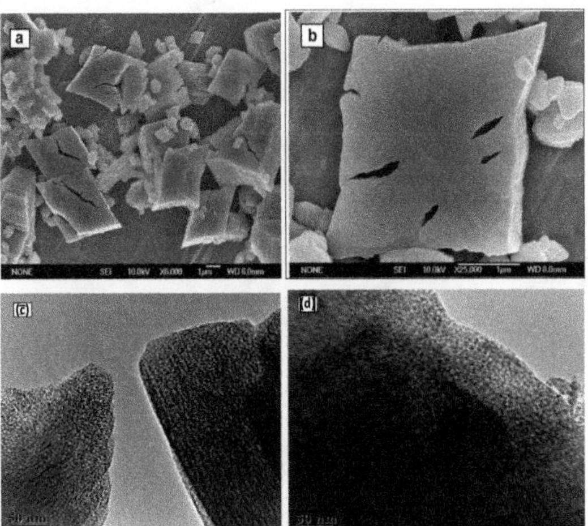

Figure 7: SEM(a,b) and TEM(c,d) image of NFSF NPs [99] (copyright permission with license number:5056401432599).

4.3 Fourier transform infrared study

FTIR analysis has been used by several researchers to determine the structure of ferrite nanoparticles. The FTIR spectra reveal the ions' locations in the crystal structure and their vibration modes, which reflect the effects of different ordering positions on the structural properties of synthesized compounds. Zhu et al. found that calcination temperature affects the FTIR and that all of the bands that appeared in the spectra of the $NiFe_2O_4$ samples gradually became weaker as the calcination temperature of the samples increases [99]. By removing the presence of unwanted ions, the FTIR spectrum is an effective tool for determining the temperature of nickel ferrite [100]. FTIR spectroscopic evidence published by Baykal et al. suggests that magnetic particle migration from aqueous to organic phase is followed by CTAB, which acts as a transfer agent [101]. Waldron suggested [102] that the $Fe^{3+}O^{2-}$ complex at the tetrahedral site is connected to the vibrational frequency (1) about 600–500 cm–1 (A). **Figure 8** shows FTIR peaks of LiZn ferrite nanoparticles [103].

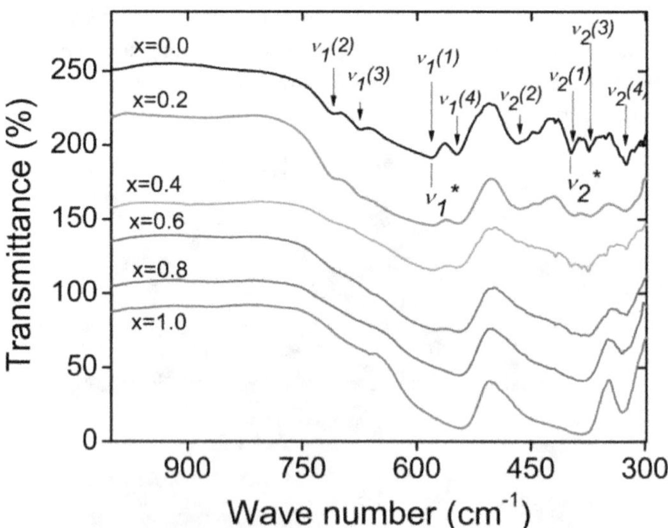

Figure 8: FTIR spectra of $Li_{(0.5-0.5x)}Zn_x\,Fe_{(2.5-0.5x)}O_4$ samples (x= 0.0, 0.2, 0.4, 0.6, 0.8 and 1.0). For clarity, the transmittance values of samples with x = 0.0, 0.2, 0.4, 0.6 and 0.8 are shifted by 200, 140, 75, 35 and 30 units, respectively [103] (copyright permission with license number:5056430338199).

4.4 Magnetic properties

The investigation and analysis of magnetic properties were carried out using standard methods i.e. SQUID and VSM. The SQUID was employed where highly sensitive magnetisation measurements are required, and VSM was employed where the sensitivity maximum is approximately 10^{-6} emu [104]. The Saturation Magnetization (M_s), the Squareness ratio (SQR), coercivity (H_c), remanence (M_r) investigated in the VSM research. SQUID can calculate crystals, powders, thin films, gases, and liquids. The samples that are being investigated are packed into capsules made of gelatine. It is possible to calculate Ms, Hs, and residual magnetisation of ferrite nanoparticles using the SQUID technique when an external magnetic field and temperature are applied.

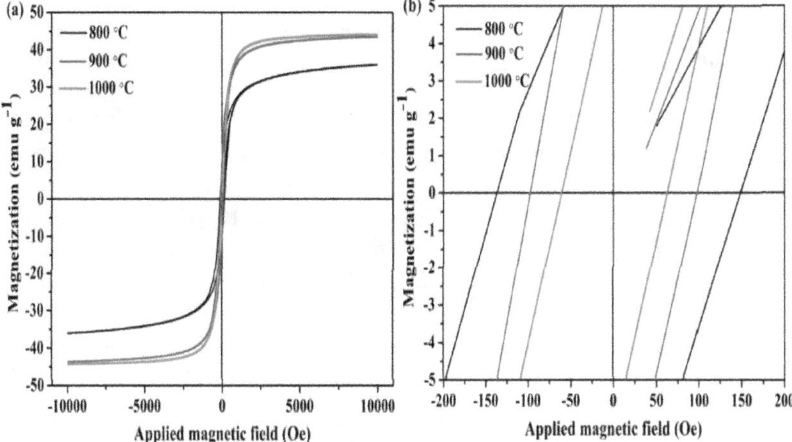

Figure 9: The magnetic hysteresis loops of the as-prepared (Ni, Cu, Co)Fe_2O_4 calcined at different temperatures and (b) H_C values of the products [105] (copyright permission with license number: 5056880383838).

Figure 9 shows the hysteresis loops of the prepared sample and it observes that the value of magnetic saturation was increased with increasing the temperature [105]. According to Naseri et al. the temperature of the calcinations is one of the most important factors that can increase or decrease the saturation of nickel ferrite when the sample is prepared by the heat treatment process. If the heating rate is slow, the crystallization and magnetic phases will emerge, resulting in higher saturation magnetization [100]. According to Maaz et al. the coercivity of nickel ferrite increases with particle size (for small particles) due to the transition from a superparamagnetic to a blocked/unblocked condition. This happens when thermal energy triumphs over volume-dependent anisotropy in small particles [106].

4.5 Dielectric properties

One of the most important characteristics of ferrites is the dielectric behavior, which is highly dependent on the preparation conditions, such as sintering time and temperature, as well as the type and quantity of additives. Analysis of dielectric structures reveals useful information on the behavior of energy-carrying companies, leading to a better understanding of the dielectric separation of ferrite samples under investigation. The dielectric properties of spinel ferrites were investigated as a function of temperature, frequency, and composition [107]. While dielectric loss decreased as frequency increased, hopping resonance was discovered at higher frequencies and this research indicates that these soft ferrite nanoparticles may be used in a variety of advanced technical devices [108]. Dar et al. prepared $Cu_{1-x}Co_xFe_2O_4$($x = 0.0, 0.1, 0.2, 0.4, 0.6, 1.0$) nanoparticles and reported that with increasing frequency, both the dielectric constant and the dielectric loss decrease. Furthermore, it is observe that with increase the concentration of Co^{2+}, dielectric constant also increases shown in **Figure 10** [109]. The dielectric constant in MnZn ferrite is found to decrease with the composition of Mn^{2+} using sol-gel auto combustion and co-precipitation techniques. The dependency of boundary charges located in the shell region of nanoparticles, which causes surface charge polarisation, is thought to be the cause of the dielectric constant variation. In the current series of samples, the hopping mechanism for electrical conduction is very different, and it is determined by the presence of a specific metal ion at B–site in the spinel structure. In this analysis, the method of preparation is important for achieving spinel structure [110]. The high dc resistivity of Gd–Mg ferrite makes it suitable for high frequency applications where eddy current losses are important. At 3 MHz, the dielectric loss in the currently studied ferrite is just 0.003 at room temperature. This ferrite's low dielectric constant and dielectric losses mean that it may be useful in microwave communications [83].

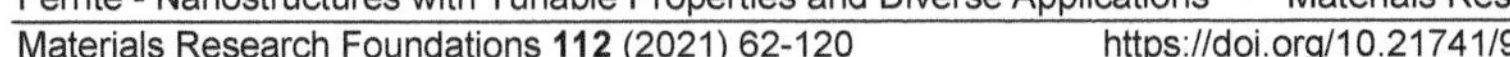

Materials Research Foundations **112** (2021) 62-120 https://doi.org/10.21741/9781644901595-2

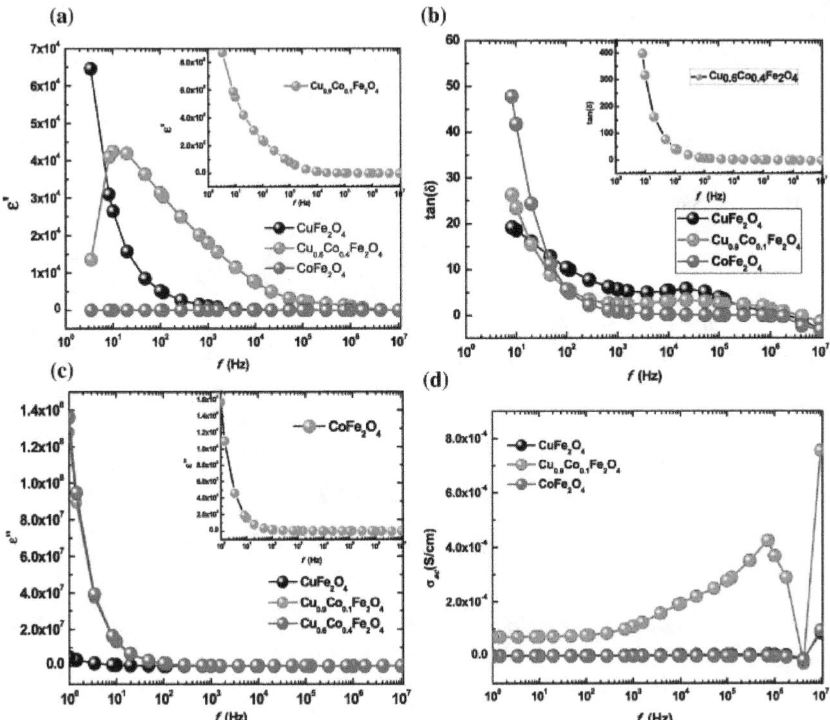

Figure10: Room temperature variation of (a) real part of dielectric constant (ε') (b) imaginary part of dielectric constant (c) dielectric Loss (d) ac conductivity with applied frequency (f) for different Co^{+2} substitution in $Cu_{1-x}Co_xFe_2O_4$ (x = 0.0 – 1.0) ferrite samples [109] (copyright permission with license number: 5057020073404).

Table2: Magnetic and electrical properties of spinel ferrites with different methods.

S. No.	Composition	Methods	Size (nm)	Calcinations Temperature (°)	Magnetic Properties	Dielectric Properties	Refs.
1	$Ni_{0.65}Zn_{0.35}Fe_{2-x}Mn_xO_4$ (x= 0,0.05,0.10,0.15,0.20,0.25,0.30 and 0.40)	Conventional ceramic processing	-	-	M_r = 3475-3180 emu/g M_s = 5287-4476 emu/g H_c= 1.1 Oe	ρ = 1.7-1.9 tan δ = $<10^{-3}$	[111]
2	$NiCr_{0.2}Ga_xFe_{1.8-x}O_4$ (x= 0.00, 0.02, 0.04, 0.06, 0.08)	Co precipitation	23.01-17.48	800-7h	M_r = 7.41-0.16 emu/g M_s = 27.83-7.01 emu/g H_c= 162.7-41.9 Oe	-	[112]
3	$Ni_xZn_{1-x}Fe_2O_4$(x = 0.1, 0.2, 0.3, 0.4 and 0.5)	Combustion	71-53	-	M_s= 2.90-67.62 emu/g H_c=14.61-81.09 Oe	ρ = 9.56-10.62 10x (Ώcm)	[113]
4	$Co_{1.2-x}Mn_xFe_{1.8}O_4$ (0 ≤ x ≥ 0.4)	Chemical Autocombustion	-	700-4h	M_r=93.02-108.17 emu/g M_s=106.50-97.78 emu/g H_c=1520.82 - 1165.36 Oe	-	[114]
5	$NiHo_xFe_{2-x}O_4$(x=0.0-0.15)	sol–gel auto-combustion	600-4h	14-25 ± 2	M_s= 21.23-11.57 M_r=8.718-2.567 H_c=253.47-707	ῥ(Ωcm^{-1})=8.9 *10^7 ἑ=7.748	[115]
6	$NiFe_{1.5}Sm_{0.5}O_4$	Sol-gel	67.43	900-2h	-	σ_{AC}=5.203 × 10^{-7}	[116]
7	$Ni_{0.5}Co_{0.5-x}Cd_xFe_{1.78}Nd_{0.02}O_4$ (x ≤ 0.25)	Electro spinning	18.5-10.5	600-2h	-	ἑ = 0.360eV	[117]

8	$NiFe_{1.925}Sm_{0.075}O_4$	Solid-state	-	1250-12h	M_s =42.5 emu/g	ἐ=4.9× 10^4	[118]
9	$NiFe_2O_4$	Sol-gel	35	800-2h	M_s=45.902 emu/g M_r=1.506 emu/g H_c = 27.374 Oe	σ_{dc} = $7.7\times10^{-9}\Omega$ cm^{-1}	[119]
10	 $CoFe_2O_4$ $NiFe_2O_4$ $ZnFe_2O_4$	Co-precipitation method	 ~14 ~ 9 ~12	 600-4h	-	Resistivity (Ώ)= 0.2344×10^6 0.1153×10^6 0.2063×10^6	[120]
11	$ZnFe_2O_4$	Sol-gel	16	500	M_s=12.01emu/g M_r=0.345 emu/g H_c=20.83Oe	σ_{ac}=0.42×10^{-9} Scm^{-1}	[121]
12	$Ni_{0.4}Co_{0.6}ErFe_2O_4$	Co-precipitation	 14.8336	600-5h	M_s=35.99emu/g M_r=12.21 emu/g H_c= 0.16 Oe	Resistivity(Ώ)= 6.20 $\times10^7$	[122]
13	$Ni_{0.5}Zn_{0.3}Cd_{0.2}La_yFe_{2-y}O_4$(y = 0.00–0.21)	Sol-gel	54.05-38.69	1023 K for 6 h	M_s=40.31-34.1 emu/g M_r=5.2-3.25 emu/g H_c= 71.85-54.37 Oe	-	[123]
14	$Co_{1-x}Zr_xFe_2O_4$ (x = 0, 0.25, 0.5, 0.75 and 1)	Co-precipitation	 17-26	 700-4h	M_s= 47.09 - 0.33 emu/g M_r= 18.22 - 0.06 emu/g H_c=1144.10 - 1225.34 Oe	-	[124]
15	$CoFe_2O_4$ $CoFe_{1.9}Nd_{0.1}O_4$	Co-precipitatio	 38.59	 600-2h	M_s=73.58 emu/g	(Ω^{-1} cm^{-1}) (10^{-6})=	[125]

	$CoFe_{1.9}Sm_{0.1}O_4$ $CoFe_{1.9}Gd_{0.1}O_4$	n	34.23 32.43 31.15		M_s=48.75 emu/g M_s=54.36 emu/g M_s=61.64 emu/g	5.095 (Ω^{-1} cm^{-1}) (10^{-6})=5.885 (Ω^{-1} cm^{-1}) (10^{-6})=5.678 (Ω^{-1} cm^{-1}) (10^{-6})=6.016	
16	$Cu_{0.8}Zn_{0.2}Cr_xFe_{2-x}O_4$; ($0 \leq x \leq 0.1$)	Citrate–nitrate combustion	37.7-47	-	M_s=33.22-30.31 emu/g M_r=6.3-4.07 emu/g H_c=265.23-158.58 Oe	-	[126]
17	$Ni_{0.6}Mg_{0.4}Fe_2O_4$	Sol-gel	-	1050 - 48 h	M_s = 41 emu/g M_r = 1.9 emu/g H_c= 18Oe	Rg (Ω)= 2249	[127]
18	$Co_{0.7}Zn_{0.3}La_xFe_{2-x}O_4$(x = 0, 0.025, 0.05, 0.075 and 0.1)	Sol-gel	20-29	450-4h	M_s=63-51 emu/g H_c=481-420 Oe M_r=24-15emu/g	-	[128]
19	$Mg_{1+x}Mn_xFe_{2-2x}O_4$ ($0 \leq x \leq 0.9$)	Ceramic technique	-	1050-24h	M_s= 17.99-8.08 emu/g	-	[129]
20	$Ni_{0.2}Mg_{0.8-x}Zn_xFe_2O_4$ (x = 0.0–0.8)	Sol-gel	8.6-48.6	400-5h	M_s = 25.6 - 45.5 emu/g	ε' (1 MHz)= 18.1 - 47.4 σ_{ac} (1 MHz) = 2.57E-6 - 1.11E-5 ε''(1 MHz) = 4.62 - 20.1	[130]

5. Biomedical applications of SFNPs

In the last few decades, MNPs have been thoroughly studied. Their magnetic properties have always attracted the attention of scientists all over the world. Catalysis [131], information technology [132], biosensor [133, 134], and medicine [135], have produced a wide range of applications. SFNPs are used in a variety of biomedical applications, the most popular of which are disease detection, cancer treatment, and drug delivery. Non-toxic Superparamagnetic NPs with high Ms, high coercivity, and low Curie temperature are ideal for a variety of biomedical applications [136]. Spinel ferrite NPs with all these properties at room temperature, are clearly supported and are actually one of the priority areas of research to address and solve some of the health issues. Early identification of diseases before they cause significant harm and systematic delivery of medication at the targeted area that need surgent attention.

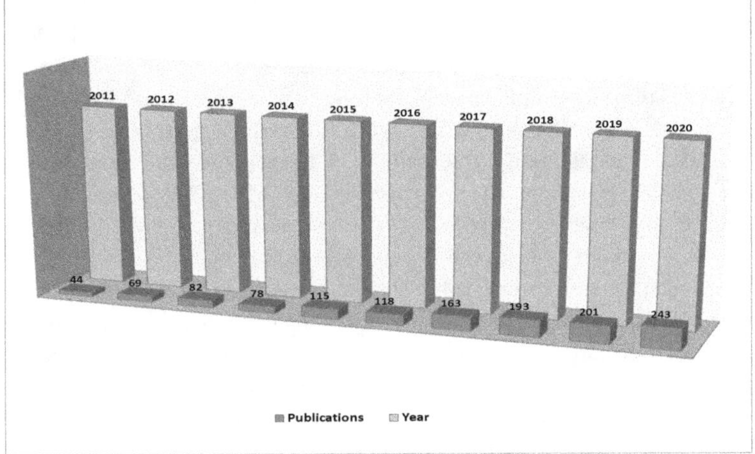

Figure 11: As per Scopus data, the number of publication for the keyword "Biomedical applications of spinel ferrites" from 2011-2020.

Advanced diagnostics and imaging are required for the former, while a targeted drug delivery system is required for the latter. All of these possibilities are available with SFNPs if they are coated with biocompatible polymeric chemicals. They have innovative and impressive uses in magnetically driven drug delivery, hyperthermia, and imaging techniques [137-142]. Traditional diagnostic and imaging methods are unable to identify tumour cells in the early stages of development and have little ability to distinguish between normal and malignant cells [142]. Researchers from all over the world are interested in the special properties of SFNPs for resolving health issues, as well as their excellent properties in imaging, cancer care, and drug delivery. As a result, this section will provide a summary of the evolution of spinel ferrite biomedical applications. **Figure 11** shows the number of published papers in the field of biomedical application of spinel ferrite nanoparticles from 2011 to 2020.

5.1 Magnetic resonance imaging (MRI)

One of the most effective medical imaging methods for obtaining diagnosis for clinical purposes is magnetic resonance imaging (MRI). It's a non-invasive technique that relies on the interaction of protons & biological tissues in the presence of a magnetic field. Briefly, ^{1}H, ^{11}B, ^{13}C, ^{19}F, or ^{31}P can be used as MRI signal sources because of its sensitivity and abundance in biological tissues, the water proton (^{1}H) is the most commonly used for clinical purposes [143, 144]. As the proton nuclear spin is affected by an external magnetic field, longitudinal magnetization can be observed for imaging purposes. Radiofrequency pulses are used to excite the spin. The longitudinal magnetization decreases in this process, while a transverse magnetization is created. The pulse is removed, and the spin emits energy to return to the ground state. Relaxation is the term for this operation, which is recorded by MRI. The relaxation times T1 (longitudinal: spin–lattice relaxation) and T2 (transverse: spin–spin relaxation), which enable protons to return to their original state, can be used to create an MRI image. The complementation of the NP's magnetic field and its magnetic field and the protons from water molecules are required for T1 and T2. Reducing the resting time limits of water by following tissue separation, using different speeds of tissue regeneration and comparison agents [145]. Presence of cobalt nanoparticles Ms. (1422 emu cm^{-3}) has a much higher at room temperature than iron oxides (395 emu cm^{-3}), which could result in a stronger proton relaxation effect. This improves MRI contrast and allows for the use of smaller magnetic nanoparticles core without sacrificing sensitivity. Water-soluble Cobalt nanoparticles are difficult to make since [145] they are resistant to oxidation.

Materials Research Foundations **112** (2021) 62-120 https://doi.org/10.21741/9781644901595-2

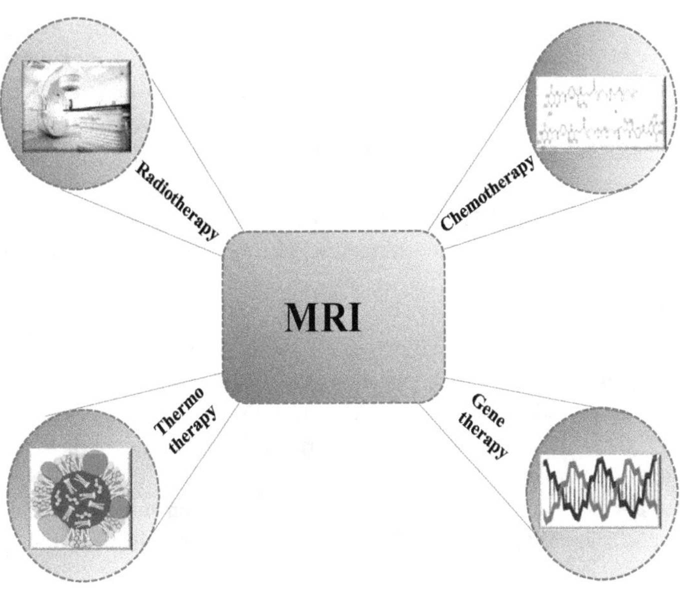

Figure12: Schematic representation of multifunctional MRI-contrast agents and their applications.

Larger magnetic particles have a higher r_1 relaxivity than shorter size. Field strength and particle size have little impact on r_2 relaxivity, but the relatively high value of r_2 makes Co nanoparticles useful as a negative contrast agent. Other metallic nanoparticles (Fe, Fe\\Ni, Fe\\Co and Fe\\Pt), in addition to Co, can be used as MRI contrast agents. At the same particle size, the r_2 and r_2* relaxivity of Fe is significantly higher than that of iron oxide [146]. The r_2/r_1 relaxivity ratio of FePt magnetic nanoparticles is 3–14 times greater than that of standard iron-oxide dependent contrast agents [147]. The lack of specificity in MRI contrast agents is still an issue. Their low sensitivity makes successful identification of tiny targets like tumours in the early stages of cancer [148]. MNPs tend to be an outstanding candidate in this regard. Their magnetic properties, as well as the ease with which they can be modified and regulated in terms of scale, shape, and composition, make them an appealing medium for MRI contrast agents [89, 90]. MNPs have proven to be effective contrast agents. The accumulation of MNPs in the target tissue will alter the local physiochemical properties. They looked at the magnetic properties of a series of spinel structures with the general formula MFe_2O_4, where M can be Mn^{2+}, Fe^{2+}, Co^{2+}, or Ni^{2+}, as well as their effect on MRI signals. In the range 85–110 emu.g^{-1} (Fe + M), there is high saturation magnetization, which has a big impact on the T2 relaxation time. $MnFe_2O_4$ outperformed as compared to other ferrites in terms of MR imaging sensitivity, being able to unambiguously detect small size cancers (~50 mg) in vivo documented by Jae-Hyun et al. [149]. Zhenghuan et al. [150] used octapod and spherical magnetic NPs for high-performance T2 contrast agents to successfully detect liver cancer in vivo. Both MNPs were able to detect liver cancer in vivo, but the octapod magnetic NPs were clearly better at distinguishing between healthy and diseased tissue.

5.2 Drug delivery

Because of their efficacy, simplicity, ease of preparation, and ability to adapt their properties for particular biological applications, the use of spinel ferrite nanoparticles as a drug delivery agent under the control of an external magnetic field has gotten a lot of attention [151-154]. About a century ago, German scientist Paul Ehrlich (1854-1915) proposed the concept of a "magic bullet" that would destroy only diseased tissue [155]. In 1950s the modern history of drug delivery can be work on micro-encapsulated drug particles [156]. Since then, the number of publications in this biomedical area has increased dramatically. Many analyses and research papers involving nanoparticles derived from silica [157-159], gold [160-162], polymers [163, 164], and other materials are currently available in the literature [157, 165-168]. As a result, it's easy to see that magnetism isn't a necessary feature when designing a drug delivery nano system, since a wide range of nano materials can be used. However, in recent decades, the use of MNPs in drug delivery has gotten a lot of attention [167]. The traditional method of drug delivery from nonspecific cell and tissue distributions with metabolic instability resulting in whole body toxicity and reduced therapeutic efficacy [169]. Another intriguing feature of SFNPs is their ability to encapsulate cytotoxic drugs within the polymer matrix and deliver them to cells [170-173]. SFNPs can hold drugs and circulate without spilling, and with the help of an external magnetic field, they can easily travel to the target tumour site and assist in delivering successful therapeutic care to cancer cells by bypassing normal cells [174]. The medication will be released and have therapeutic effects until it reaches the site of action [151]. Multifunctional NPs/NCs, which are made up of spinel ferrite nanoparticles, anti-cancer drugs, semiconductors (for cell imaging), and biocompatible coating agents, are particularly valuable for cancer therapy. Shen et al. [175] used in situ assembly to create multifunctional chitosan-based magnetic hybrid NCs with folate-conjugated tetrapeptide composite.

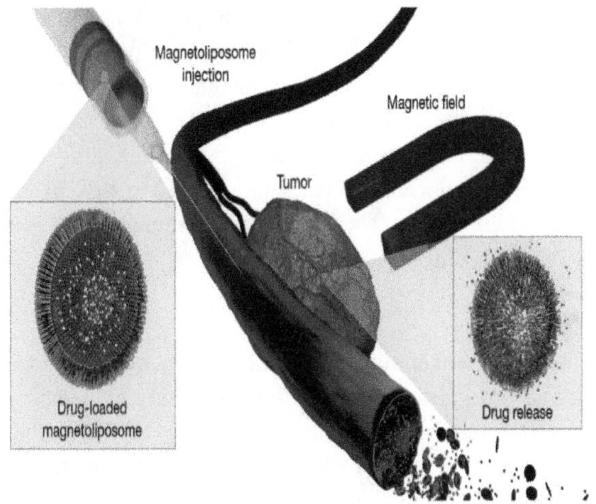

Figure13: Schematic representation of a SPION-liposome hybrid drug delivery system specifically designed for the triggered release of an encapsulated hydrophilic drug (Reproduced with permission from reference [176] Copyright 2014).

Chitosan, CdTe quantum dots, SPM Fe_3O_4, anticancer agent camptothecin, and folate are the key components of synthesised NCs.$MnFe_2O_4$@graphene oxide NC with SPM properties was recently synthesised using a sono-chemical process, and the NC has a high DOX loading potential [177]. The higher DOX loading ability has been due to hydrogen bonding and - interactions between graphene oxide and DOX, according to the author's observations. In addition, under acidic rather than alkaline conditions, DOX sensitivity and release were found to be higher. As a result, the $MnFe_2O_4$@GO is a promising candidate for DOX drug delivery to the tumour site and due to the acidic microenvironment of cancer tissue, the drug could be easily released at the tumour site. It has also been documented that lauric acid capped $CoFe_2O_4$ nanoparticles could be used as a promising drug delivery agent with pH sensitive release [178]. Furthermore, capping or coating SFNPs with biocompatible materials may help to improve their stability and reduce their toxicity in cells. To be used for magnetically driven drug delivery, SFNPs must have no residual magnetisation after the external magnet field has been removed, which means they must be SPM. This also allows them to preserve colloidal stability and prevent aggregation, allowing them to be used in biomedical applications [179]. Magnetic attraction between the particles is one of the predicted reasons for SFNP agglomeration [180]. Targeted drug delivery with SFNPs has a number of benefits, including minimising drug waste, reducing drug administration frequency, reducing side effects, removing side effects on other organs due to site-specific drug delivery, improving treatment efficacy in a limited amount of time, and being safe and efficient due to drug delivery to preferred target organs. Hyperthermia, drug delivery and release at the target organs, where an increase in temperature to 42 to 46°C has no direct negative impact on the body **(Figure 13).** The reader is recommended to read the following reviews [89, 139, 151, 181-185] for more detail.

5.3 Hyperthermia

The basic principle of hyperthermia treatment is to generate heat locally by using an external system to pass energy to the tissue, thus increasing the temperature of the local atmosphere. Observation of certain patients with high fevers led to the development of this physical therapy. Surprisingly, the cancer cells were found to be significantly reduced or completely destroyed. Furthermore, other experiments using radiation-resistant cancer cells revealed a promising treatment combination capable of increasing the radiotherapy's efficacy [186]. As a result, hyperthermia will damage, destroy, or make cancerous cells more vulnerable to the effects of radiation and anticancer drugs [187]. To induce heat, various techniques such as ultrasound, microwave, radiofrequency, and others are currently available. The key issue with these thermal methodologies [188] is the absence of homogeneous heat distribution and the depth of therapeutic temperatures (45°C). In this regard, a thermal process based on MNPs is an excellent candidate for addressing these issues. As previously mentioned, their scale, magnetism, and ease of tuning surface properties to accumulate MNPs in tumour tissue are significant advantages. Both of these factors may contribute to the development of a cancer-treatment thermal therapy. Gordon et al. [189]were the first to describe a *"modern multidisciplinary "intracellular" biophysical treatment of cancer"* in 1979.

After an external high frequency or pulsed electromagnetic field was applied to colloidal magnetic submicron particles phagocytized by cancer cells, local heat was generated. As a result, a highly specific intracellular heat would be produced. Chan and his colleagues published Chan et al. in 1993. As a result, a highly specific intracellular heat would be produced. Chan et al. [155] created an aqueous magnetic fluid with dextran-modified SPM iron oxides in 1993. It was investigated how well they could absorb energy from an external magnetic field of varying intensity and frequency and convert it to heat. Jung-tak et al. [160] investigated the use of mixed ferrites (metal-doped MNPs). The authors investigated the magnetic influence of Zn^{2+} in the spinel structure of Fe_3O_4 and $MnFe_2O_4$ in their research. In addition, the Zn doped-MNPs $((Zn_xMn_{1-x})$ Fe_2O_4 and (ZnxFe1-

Materials Research Foundations **112** (2021) 62-120 https://doi.org/10.21741/9781644901595-2

x)Fe_2O_4, x ranging from 0.0 to 0.8) were tested for MRI and hyperthermia treatment. Since high values of saturation magnetization were observed for x equal to 0.4, it was possible to observe the mostly positive effect of Zn in the spinel structure of the MNPs studied. For $(Zn_{0.4}Fe_{0.6})$ Fe_2O_4 and $(Zn_{0.4}Mn_{0.6})Fe_2O_4$, the values were 161 and 175 emu.g^{-1}, respectively. HeLa cells were used in an vitro trial, and 84.4 percent of those treated with $(Zn_{0.4}Mn_{0.6})Fe_2O_4$ died after 10 minutes, compared to just 13.5 percent of those treated with Feridex®. With all of these findings, it's easy to see how hyperthermia for cancer care has progressed rapidly over time. Jae-Hyun et al. [136] developed a core-shell structure compound consisting of $MnFe_2O_4$ (Shell) and $CoFe_2O_4$ (core) for in vivo hyperthermia treatment. In comparison to common antidrug cancers like doxorubicin, the antitumor study in mice showed higher efficacy. Tanmoy et al. [165] investigated magnetic hyperthermia therapy. According to the findings, hyperthermia was found to have an important in vivo inhibitory effect on tumour development. Noriyasu et al. [159] investigated the anticancer effects of Fe_3O_4 cationic liposomes (MCLs) in syngeneic rats with prostate cancer. The findings in this study strongly suggest that MCLs may be a useful tool in the treatment of prostate cancer. Not only do the MCLs appear to destroy cancerous cells by heating them, but they also appear to cause an immune response.

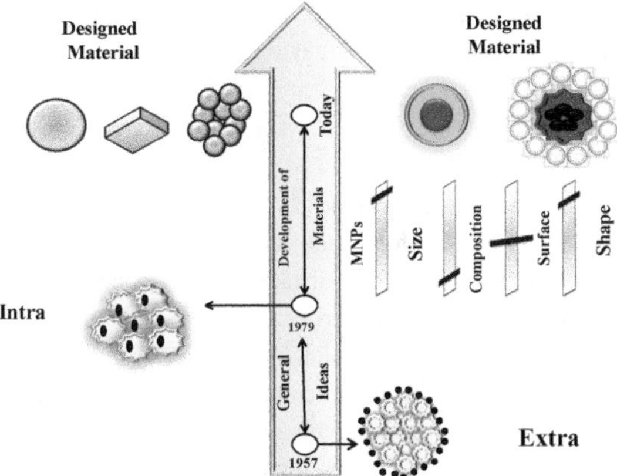

Figure14: Timeline of the development of hyperthermia using magnetic nanoparticles.

5.4 Other applications

5.4.1 Inductors and transformers for small-signal applications

A major use of high quality pot-core inductors is in combination with capacitors in filter circuits. They were extensively used in telephone systems but solid state switched systems have replaced them to a large extent. The value of Snoek's constant for cubic (standard) ferrites is around 7 GHz for a high-quality NiZn ferrite and around 4 GHz for a MnZn ferrite. If capacitance and line-length effects are ignored, most ferrites for EMC filters or transformers have similar HF efficiency [190]. The soft MnZn and NiZn ferrite systems are dominant for pot-core manufacture, although metal 'dust cores' are used for certain applications.

5.4.2 Transformers for power applications

There are many high-frequency (typically 1 kHz to 1 MHz) applications where a high saturation flux density is the prime requirement. Examples are power transformers to transmit a single frequency, those required to transmit over a narrow frequency band, as in the power unit of an ultrasonic generator and wide-band matching transformers to feed transmitting aerials. Television line output transformers are, commercially speaking, very important examples of power transformers and although not a transformer, the beam deflection yoke falls into the same category. At 300 kHz/70 mT, the total loss of the Mn Zn Co Fe O ferrite core did not exceed 420 mW/cm. This shows that it has a lot of scope for use in high-frequency power electronics circuits [191].

5.4.3 Information storage

Computer memory systems are classified as 'volatile' or 'non-volatile', the latter are of the type which retains the stored information if the power supply fails. Non-volatility is, of course, a very desirable feature, but other factors such as size, speed and cost all have to be considered. Two decades ago ferrite cores were dominant in random access memories; since then they have been almost completely superseded in commercial computers by the inherently volatile semiconductor memory elements. Sharma et al. reported that the moderate coercivity (maximum coercivity is 125.58 Oe) and saturation magnetization (decreases from 57.84 emu/gm to 25.30 emu/gm as Ni^{2+} content increases) suggest that these materials may be used in high density information storage devices [192].

5.4.4 Ferrite core memory

A material with a 'square' hysteresis loop has the important property that its remanent state is very little altered by the application or removal of small fields such as +1/2 Hm. If the field greater than +Hc are applied, the remanent state switches to the opposite sense. This is the basis of operation of a ferrite core memory, where the +Br states represent the digits 0 and 1.

5.4.4.1 Magnetic bubble memory

Magnetic bubble memories are based on the mobility and stability of magnetic domains under certain conditions. If a thin transparent layer of anisotropic magnetic garnet is viewed through a crossed polarizer and analyzer system in a microscope.

5.4.5 Microwave devices

The elementary description of microwave propagation in magnetically saturated ferrites provides a basis for an understanding of the principles of operation of devices such as circulators, isolators and phase shifters, all of which find extensive use in microwave engineering. For example, a circulator might be used in a radar system where a single antenna is employed for both transmitting microwave power and receiving a reflected signal. Because of its high specific resistance (10^8-10^{10} Ωcm), Mg ferrite is well suited for microwave applications [193].

Faraday rotation isolator

Essential parts of a Faraday rotation isolator. The electric field of the forward wave is perpendicular to the plane of the resistive card and so it passes it with little attenuation.

Y-junction circulators

The most frequently used ferrite device is a three port circulator in waveguide or strip line configuration. In the waveguide configuration a ferrite cylindrical disc is located symmetrically with respect to three ports. In the strip line configuration two ferrite discs are placed on each side of the strips. In both cases a magnetic field is applied perpendicular to the planes of the discs.

Phase shifters

The phase of a microwave passing along a guide can be shifted by inserting pieces of ferrite at appropriate positions. For example, in a rectangular guide along which the simplest microwave field configuration is propagating, the magnetic field component is circularly polarized in the plane of the broad face of the waveguide at a distance approximately a quarter of the way across the guide; on the opposite side of the guide the magnetic field component is circularly polarized in the opposite sense.

Microwave Ferrite Types

Most other examples that can be cited of materials exploitation, designing a ferrite for a particular microwave application is a compromise: a desire value for a particular parameter is generally achieved at the expense of the optimum values of other parameters. Nevertheless, a select number of spinel and garnet compositions have become favoured for microwave applications.

Permanent Magnets

Permanent magnet materials are required to give a maximum flux in the surrounding medium for a minimum volume of material and to be resistant to demagnetization. A material's suitability for a particular application is judged by its Br, Hc and (BH)max values. The hexaferrites have the advantage of lower-cost raw materials their metal counterparts; both cobalt and the rare earths are expensive. They also have higher coercive field relative to those for the mass-produced metal magnet types and this allows them to be made into shapes involving high demagnetizing fields. According to the hysteresis loop, $Co_{0.5}Nd_{0.5}Fe_2O_4$ has a maximum energy product (BH) max of 0.431 MG Oe (3.434 KJ/m^3) at room temperature, which is a 7 percent increase above pure cobalt ferrites nanoparticles, confirming the material's performance in permanent magnet applications [194].

Wastewater treatment

Colours are distinguished as most noteworthy chemical contaminants that cause wastewater contamination [195, 196, 197, 198]. There has been developing exploration interests on photocatalytic decomposition using magnetic nanoparticles for environmental remediation because of the capability of these materials for enormous-scope application and compatibility with current innovations [199, 200, 201]. Magnetic ferrites and spinel structure have a restricted band gap (~1.9eV) and that's why they are able for the engrossing visible light [202]. Photocatalytic degradation of colors in wastewater utilizing zinc ferrite and its spinel are exceptionally alluring as it is a zero waste procedure, and the impetuses can be reused [203-207]. The dye decomposition efficiency and kinetics of the pristine catalyst under direct and diffused sunlight was studied. Mandal S. et al. reported that direct sunlight works better than diffused light dye rot and more than 95%, can be found in the concentration of the first dye under the sun [206]. A second pseudonym for kinetics. AgBr / ZF / NG, AgBr / CF / NG and NG demonstrated significant adsorption potential for MO and MG elimination while AgBr, ZF and CF showed optimal image performance under 12 % and 10%, respectively. The adsorption

Materials Research Foundations **112** (2021) 62-120 https://doi.org/10.21741/9781644901595-2

strength of AgBr / CF / NG, AgBr / ZF / NG, and NG of MG was recorded as 64.00, 62.56 and 63.00 mg / g, respectively. 65.12, 66.00, and 67.10 mg / g of MO were extracted from AgBr / CF / NG, AgBr / ZF / NG, and NG, respectively. Claiming to reuse experiments, AgBr / ZF / NG and AgBr / CF / NG achieved 85% and 83% degradation efficacy, respectively, after ten consecutive cycles [207].

6. Future prospects

The use of SFNPs in biomedicine for diagnostics, hyperthermia, cancer gene therapy, and selective drug delivery and release has piqued the scientific community's interest. Despite the availability of several synthesis methods, the existing methods for the preparation of pure SFNPs with all of the required properties and in the desired quantities still need improvement. As a result, finding simple standard synthesis methods for SFNPs with the necessary scale, shape, and properties requires advanced technology that improves SFNP quality at a scalable production level. The ability to prepare new and improved SFNPs by doping or substituting transition or rare earth metals into SFNPs has been demonstrated in research. Doping of transition metals such as manganese, iron, cobalt, nickel, copper, and zinc has been found to be one of the most active research areas right now. On the other hand, the impact of incorporating rare earth metals into first-row transition metal ferrites hasn't been thoroughly investigated. As a result, we conclude that selective doping of low-concentration rare earth metals would most likely aid in the development of new SFNPs that have the potential to solve current health problems. To modified the structure and properties of SFNPs can be changing their size, synthesis methods, and doping them with transition metals or rare earth metals. As it has been observed in recent studies, doping of metals into ferrites has resulted in the variety of new SFNPs with various useful properties. SFNPs' multipurpose applications in biomedical fields like drug delivery, drug release, MRI, and cancer therapy are very promising. However, efficiency of synthesised SFNPs, optimisation of heating properties, concentrations of anticancer agents and spinel ferrite NPs, reproducibility of the synthesis system, and process parameters are all important. More research and attention are still required. Furthermore, the optimum size of SFNPs that can generate optimal heat and the alternative magnetic field needed to produce optimal heating are areas that should be investigated further.

Conclusions

Spinel ferrites have emerged as a promising material for the advancement of nanotechnology and nanomedicine. Spinel ferrite NPs have a variety of medical applications, including cancer diagnostics, cancer gene therapy, and drug administration. In biomedical applications, the efficacy of spinel ferrite NPs is more dependent on synthesis procedures; good synthesis methods provide SFNPs that can endure the circumstances for which they are synthesised while also boosting their efficiency. The ease with which they may be manipulated and managed in terms of size, form, and composition, as well as their magnetic properties (high Ms and coercivity), make them an interesting medium for biomedical applications. In general, the excellent properties of SFNPs observed, as well as the ease with which they can be functionalized by coating their surfaces with non-toxic chemicals, provide an exciting opportunity for future generations to solve environmental and medical problems. SFNPs, in particular, are highly beneficial in the treatment of cancer. Recent conventional methodologies and approaches to spinel ferrites synthesis are discussed in this study. In addition, an overview of the issue was attempted, highlighting significant applications and discussing their characterisation methodologies, which provide high-quality details on the sample.

Conflict of Interest

The authors declare that they have no conflict of interest.

Acknowledgment

The authors are grateful to the research and scientific community who have made relentless efforts to bring novel data in the fields and topics of research and development discussed in this review. This work did not receive any funding.

References

[1] Z. Yan, J. Gao, Y. Li, M. Zhang, M. Guo, Hydrothermal synthesis and structure evolution of metal-doped magnesium ferrite from saprolite laterite, RSC Advances, 5 (2015) 92778-92787. https://doi.org/10.1039/C5RA17145H

[2] R.T. Olsson, G. Salazar-Alvarez, M.S. Hedenqvist, U.W. Gedde, F. Lindberg, S.J. Savage, Controlled synthesis of near-stoichiometric cobalt ferrite nanoparticles, Chemistry of Materials, 17 (2005) 5109-5118. https://doi.org/10.1021/cm0501665

[3] A.H. Latham, M.E. Williams, Controlling transport and chemical functionality of magnetic nanoparticles, Accounts of chemical research, 41 (2008) 411-420. https://doi.org/10.1021/ar700183b

[4] H. El Moussaoui, T. Mahfoud, S. Habouti, K. El Maalam, M.B. Ali, M. Hamedoun, O. Mounkachi, R. Masrour, E. Hlil, A. Benyoussef, Synthesis and magnetic properties of tin spinel ferrites doped manganese, Journal of magnetism and magnetic materials, 405 (2016) 181-186. https://doi.org/10.1016/j.jmmm.2015.12.059

[5] D.S. Mathew, R.-S. Juang, An overview of the structure and magnetism of spinel ferrite nanoparticles and their synthesis in microemulsions, Chemical engineering journal, 129 (2007) 51-65. https://doi.org/10.1016/j.cej.2006.11.001

[6] M.A. Maksoud, G.S. El-Sayyad, A. Ashour, A.I. El-Batal, M.S. Abd-Elmonem, H.A. Hendawy, E. Abdel-Khalek, S. Labib, E. Abdeltwab, M. El-Okr, Synthesis and characterization of metals-substituted cobalt ferrite [Mx Co $_{(1-x)}$ Fe_2O_4;(M= Zn, Cu and Mn; x= 0 and 0.5)] nanoparticles as antimicrobial agents and sensors for Anagrelide determination in biological samples, Materials Science and Engineering: C, 92 (2018) 644-656. https://doi.org/10.1016/j.msec.2018.07.007

[7] S. Hilpert, V. Verf, Genetische und konstitutive Zusammenhänge in den magnetischen Eigenschaften bei Ferriten und Eisenoxyden, Berichte der deutschen chemischen Gesellschaft, 42 (1909) 2248-2261. https://doi.org/10.1002/cber.190904202121

[8] Y. Kato, T. Takei, Characteristics of metallic oxide magnetic, Journal of the Institute Eletronic Engineering of Japan, 53 (1933) 408-412.

[9] W.S. Galvão, D. Neto, R.M. Freire, P.B. Fechine, Super-paramagnetic nanoparticles with spinel structure: a review of synthesis and biomedical applications, in: solid state phenomena, Trans Tech Publ, 2016, pp. 139-176. https://doi.org/10.4028/www.scientific.net/SSP.241.139

[10] L. Néel, Théorie du traînage magnétique des ferromagnétiques en grains fins avec applications aux terres cuites, Ann. géophys., 5 (1949) 99-136.

[11] M. Sugimoto, The past, present, and future of ferrites, Journal of the American Ceramic Society, 82 (1999) 269-280. https://doi.org/10.1111/j.1551-2916.1999.tb20058.x

[12] E. Albers-Schoenberg, Ferrites for microwave circuits and digital computers, Journal of Applied Physics, 25 (1954) 152-154. https://doi.org/10.1063/1.1721594

[13] A. Bobeck, Properties and device applications of magnetic domains in orthoferrites, The bell system technical journal, 46 (1967) 1901-1925. https://doi.org/10.1002/j.1538-7305.1967.tb03177.x

[14] A. Bobeck, R. Fischer, A. Perneski, J. Remeika, L. Van Uitert, Application of orthoferrites to domain-wall devices, IEEE Transactions on Magnetics, 5 (1969) 544-553. https://doi.org/10.1109/TMAG.1969.1066480

[15] J. Dillon Jr, E. Gyorgy, J. Remeika, Photoinduced magnetic anisotropy and optical dichroism in silicon-doped yttrium iron garnet, Physical Review Letters, 22 (1969) 643. https://doi.org/10.1103/PhysRevLett.22.643

[16] C.L. Hogan, The ferromagnetic Faraday effect at microwave frequencies and its applications: the microwave gyrator, The Bell System Technical Journal, 31 (1952) 1-31. https://doi.org/10.1002/j.1538-7305.1952.tb01374.x

[17] C. Walcott, J.L. Gould, J. Kirschvink, Pigeons have magnets, Science, 205 (1979) 1027-1029. https://doi.org/10.1126/science.472725

[18] J.L. Gould, J. Kirschvink, K. Deffeyes, Bees have magnetic remanence, Science, 201 (1978) 1026-1028. https://doi.org/10.1126/science.201.4360.1026

[19] R.B. Frankel, R.P. Blakemore, R.S. Wolfe, Magnetite in freshwater magnetotactic bacteria, Science, 203 (1979) 1355-1356. https://doi.org/10.1126/science.203.4387.1355

[20] K.J. Widder, A.E. Senyei, D.G. Scarpelli, Magnetic microspheres: a model system for site specific drug delivery in vivo, Proceedings of the Society for Experimental Biology and Medicine, 158 (1978) 141-146. https://doi.org/10.3181/00379727-158-40158

[21] R. Gilchrist, R. Medal, W.D. Shorey, R.C. Hanselman, J.C. Parrott, C.B. Taylor, Selective inductive heating of lymph nodes, Annals of surgery, 146 (1957) 596. https://doi.org/10.1097/00000658-195710000-00007

[22] P. Fannin, Measurement of the Neel relaxation of magnetic particles in the frequency range 1 kHz to 160 MHz, Journal of Physics D: Applied Physics, 24 (1991) 76. https://doi.org/10.1088/0022-3727/24/1/013

[23] A.H. Lu, E.e.L. Salabas, F. Schüth, Magnetic nanoparticles: synthesis, protection, functionalization, and application, Angewandte Chemie International Edition, 46 (2007) 1222-1244. https://doi.org/10.1002/anie.200602866

[24] B. Issa, I.M. Obaidat, B.A. Albiss, Y. Haik, Magnetic nanoparticles: surface effects and properties related to biomedicine applications, International journal of molecular sciences, 14 (2013) 21266-21305. https://doi.org/10.3390/ijms141121266

[25] K.L. Pisane, E.C. Despeaux, M.S. Seehra, Magnetic relaxation and correlating effective magnetic moment with particle size distribution in maghemite nanoparticles, Journal of Magnetism and Magnetic Materials, 384 (2015) 148-154. https://doi.org/10.1016/j.jmmm.2015.02.038

[26] S. Ruta, R. Chantrell, O. Hovorka, Unified model of hyperthermia via hysteresis heating in systems of interacting magnetic nanoparticles, Scientific reports, 5 (2015) 1-7. https://doi.org/10.1038/srep09090

[27] P. Tartaj, M. del Puerto Morales, S. Veintemillas-Verdaguer, T. González-Carreño, C.J. Serna, The preparation of magnetic nanoparticles for applications in biomedicine, Journal of physics D: Applied physics, 36 (2003) R182. https://doi.org/10.1088/0022-3727/36/13/202

[28] R.J. Hill, J.R. Craig, G. Gibbs, Systematics of the spinel structure type, Physics and chemistry of minerals, 4 (1979) 317-339. https://doi.org/10.1007/BF00307535

[29] V.G. Harris, Modern microwave ferrites, IEEE Transactions on Magnetics, 48 (2011) 1075-1104. https://doi.org/10.1109/TMAG.2011.2180732

[30] W. Bragg, The structure of magnetite and the spinels, Nature, 95 (1915) 561-561. https://doi.org/10.1038/095561a0

[31] B. Cullity, C. Graham, Introduction to Magnetic Materials, A John Wiley & Sons, Inc., Hoboken, New Jersey, (2009) 361.

[32] T. Tatarchuk, M. Bououdina, N. Paliychuk, I. Yaremiy, V. Moklyak, Structural characterization and antistructure modeling of cobalt-substituted zinc ferrites, Journal of Alloys and Compounds, 694 (2017) 777-791. https://doi.org/10.1016/j.jallcom.2016.10.067

[33] M.P. Leal, S. Rivera-Fernández, J.M. Franco, D. Pozo, M. Jesús, M.L. García-Martín, Long-circulating PEGylated manganese ferrite nanoparticles for MRI-based molecular imaging, Nanoscale, 7 (2015) 2050-2059. https://doi.org/10.1039/C4NR05781C

[34] G. Manjari, Green synthesis of silver and copper nanoparticles using Aglaia elaeagnoidea and its catalytic application on dye degradation, in, Department of Ecology and Environmental Sciences, Pondicherry University, 2018.

[35] M. Sundararajan, V. Sailaja, L.J. Kennedy, J.J. Vijaya, Photocatalytic degradation of rhodamine B under visible light using nanostructured zinc doped cobalt ferrite: kinetics and mechanism, Ceramics International, 43 (2017) 540-548. https://doi.org/10.1016/j.ceramint.2016.09.191

[36] S. Singh, C. Srinivas, B. Tirupanyam, C. Prajapat, M. Singh, S. Meena, P. Bhatt, S. Yusuf, D. Sastry, Structural, thermal and magnetic studies of $MgxZn_{1-x}Fe_2O_4$ nanoferrites: study of exchange interactions on magnetic anisotropy, Ceramics International, 42 (2016) 19179-19186. https://doi.org/10.1016/j.ceramint.2016.09.081

[37] A. Manikandan, L.J. Kennedy, M. Bououdina, J.J. Vijaya, Synthesis, optical and magnetic properties of pure and Co-doped $ZnFe_2O_4$ nanoparticles by microwave combustion method, Journal of magnetism and magnetic materials, 349 (2014) 249-258. https://doi.org/10.1016/j.jmmm.2013.09.013

[38] A. Manikandan, J.J. Vijaya, L.J. Kennedy, M. Bououdina, Structural, optical and magnetic properties of Zn1− xCuxFe$_2$O$_4$ nanoparticles prepared by microwave combustion method, Journal of molecular structure, 1035 (2013) 332-340. https://doi.org/10.1016/j.molstruc.2012.11.007

[39] V.J. Angadi, A. Anupama, H.K. Choudhary, R. Kumar, H. Somashekarappa, M. Mallappa, B. Rudraswamy, B. Sahoo, Mechanism of γ-irradiation induced phase transformations in nanocrystalline $Mn_{0.5}Zn_{0.5}Fe_2O_4$ ceramics, Journal of Solid State Chemistry, 246 (2017) 119-124. https://doi.org/10.1016/j.jssc.2016.11.017

[40] S. Jesudoss, J.J. Vijaya, L.J. Kennedy, P.I. Rajan, H.A. Al-Lohedan, R.J. Ramalingam, K. Kaviyarasu, M. Bououdina, Studies on the efficient dual performance of Mn1–xNixFe$_2$O$_4$ spinel nanoparticles in photodegradation and antibacterial activity, Journal of Photochemistry and Photobiology B: Biology, 165 (2016) 121-132. https://doi.org/10.1016/j.jphotobiol.2016.10.004

[41] R. Kumar, M. Kar, Correlation between lattice strain and magnetic behavior in non-magnetic Ca substituted nano-crystalline cobalt ferrite, Ceramics International, 42 (2016) 6640-6647. https://doi.org/10.1016/j.ceramint.2016.01.007

[42] M.M.L. Sonia, S. Anand, V.M. Vinosel, M.A. Janifer, S. Pauline, A. Manikandan, Effect of lattice strain on structure, morphology and magneto-dielectric properties of spinel $NiGd_xFe_{2-x}O_4$ ferrite nano-crystallites synthesized by sol-gel route, Journal of Magnetism and Magnetic Materials, 466 (2018) 238-251. https://doi.org/10.1016/j.jmmm.2018.07.017

[43] S. Gul, M.A. Yousuf, A. Anwar, M.F. Warsi, P.O. Agboola, I. Shakir, M. Shahid, Al-substituted zinc spinel ferrite nanoparticles: Preparation and evaluation of structural, electrical, magnetic and photocatalytic properties, Ceramics International, 46 (2020) 14195-14205. https://doi.org/10.1016/j.ceramint.2020.02.228

[44] I.P. Muthuselvam, R. Bhowmik, Mechanical alloyed Ho^{3+} doping in $CoFe_2O_4$ spinel ferrite and understanding of magnetic nanodomains, Journal of Magnetism and Magnetic Materials, 322 (2010) 767-776. https://doi.org/10.1016/j.jmmm.2009.10.057

[45] C. Srinivas, B. Tirupanyam, S. Meena, S. Yusuf, C.S. Babu, K. Ramakrishna, D. Potukuchi, D. Sastry, Structural and magnetic characterization of co-precipitated

$Ni_xZn_{1-x}Fe_2O_4$ ferrite nanoparticles, Journal of Magnetism and Magnetic Materials, 407 (2016) 135-141. https://doi.org/10.1016/j.jmmm.2016.01.060

[46] R.S. Yadav, J. Havlica, M. Hnatko, P. Šajgalík, C. Alexander, M. Palou, E. Bartoníčková, M. Boháč, F. Frajkorová, J. Masilko, Magnetic properties of $Co_{1-x}Zn_xFe_2O_4$ spinel ferrite nanoparticles synthesized by starch-assisted sol–gel autocombustion method and its ball milling, Journal of Magnetism and Magnetic Materials, 378 (2015) 190-199. https://doi.org/10.1016/j.jmmm.2014.11.027

[47] S. Prasad, M. Deepty, P. Ramesh, G. Prasad, K. Srinivasarao, C. Srinivas, K.V. Babu, E.R. Kumar, N.K. Mohan, D. Sastry, Synthesis of MFe_2O_4 (M= Mg^{2+}, Zn^{2+}, Mn^{2+}) spinel ferrites and their structural, elastic and electron magnetic resonance properties, Ceramics International, 44 (2018) 10517-10524. https://doi.org/10.1016/j.ceramint.2018.03.070

[48] M. Gabal, R.M. El-Shishtawy, Y. Al Angari, Structural and magnetic properties of nano-crystalline Ni–Zn ferrites synthesized using egg-white precursor, Journal of Magnetism and Magnetic Materials, 324 (2012) 2258-2264. https://doi.org/10.1016/j.jmmm.2012.02.112

[49] M.M.L. Sonia, S. Anand, V.M. Vinosel, M.A. Janifer, S. Pauline, Effect of lattice strain on structural, magnetic and dielectric properties of sol–gel synthesized nanocrystalline Ce^{3+} substituted nickel ferrite, Journal of Materials Science: Materials in Electronics, 29 (2018) 15006-15021. https://doi.org/10.1007/s10854-018-9639-2

[50] B.P. Jacob, S. Thankachan, S. Xavier, E. Mohammed, Effect of Gd^{3+} doping on the structural and magnetic properties of nanocrystalline Ni–Cd mixed ferrite, Physica Scripta, 84 (2011) 045702. https://doi.org/10.1088/0031-8949/84/04/045702

[51] M. Junaid, M.A. Khan, M.N. Akhtar, A. Hussain, M.F. Warsi, Impact of indium substitution on dielectric and magnetic properties of $Cu_{0.5}Ni_{0.5}Fe_{2-x}O_4$ ferrite materials, Ceramics International, 45 (2019) 13431-13437. https://doi.org/10.1016/j.ceramint.2019.04.042

[52] A. Thakur, P. Thakur, J.-H. Hsu, Structural, Magnetic and Electromagnetic Characterization of In^{3+} Substituted Mn-Zn Nanoferrites, Zeitschrift für Physikalische Chemie, 228 (2014) 663-672. https://doi.org/10.1515/zpch-2014-0477

[53] H. Anwar, A. Maqsood, Effect of sintering temperature on structural, electrical and dielectric parameters of Mn-Zn nano ferrites, in: Key Engineering Materials, Trans Tech Publ, 2012, pp. 163-170. https://doi.org/10.4028/www.scientific.net/KEM.510-511.163

[54] C. Pereira, A.M. Pereira, C. Fernandes, M. Rocha, R. Mendes, M.P. Fernández-García, A. Guedes, P.B. Tavares, J.-M. Grenèche, J.o.P. Araújo, Superparamagnetic MFe_2O_4 (M= Fe, Co, Mn) nanoparticles: tuning the particle size and magnetic properties through a novel one-step coprecipitation route, Chemistry of Materials, 24 (2012) 1496-1504. https://doi.org/10.1021/cm300301c

[55] H. Ni, Z. Gao, X. Li, Y. Xiao, Y. Wang, Y. Zhang, Synthesis and characterization of $CuFeMnO_4$ prepared by co-precipitation method, Journal of Materials Science, 53 (2018) 3581-3589. https://doi.org/10.1007/s10853-017-1800-4

[56] K. Hoshi, H. Kato, T. Fukunaga, S. Furusawa, H. Sakurai, Synthesis of MnZn-Ferrite Using Coprecipitation Method, in: Key Engineering Materials, Trans Tech Publ, 2013, pp. 22-25. https://doi.org/10.4028/www.scientific.net/KEM.534.22

[57] K. Velmurugan, V.S.K. Venkatachalapathy, S. Sendhilnathan, Synthesis of nickel zinc iron nanoparticles by coprecipitation technique, Materials Research, 13 (2010) 299-303. https://doi.org/10.1590/S1516-14392010000300005

[58] N. Kaur, M. Kaur, Comparative studies on impact of synthesis methods on structural and magnetic properties of magnesium ferrite nanoparticles, Processing and Application of Ceramics, 8 (2014) 137-143. https://doi.org/10.2298/PAC1403137K

[59] A. Lungu, I. Malaescu, C. Marin, P. Vlazan, P. Sfirloaga, The electrical properties of manganese ferrite powders prepared by two different methods, Physica B: Condensed Matter, 462 (2015) 80-85. https://doi.org/10.1016/j.physb.2015.01.025

[60] M. Houshiar, F. Zebhi, Z.J. Razi, A. Alidoust, Z. Askari, Synthesis of cobalt ferrite ($CoFe_2O_4$) nanoparticles using combustion, coprecipitation, and precipitation methods: A comparison study of size, structural, and magnetic properties, Journal of Magnetism and Magnetic Materials, 371 (2014) 43-48. https://doi.org/10.1016/j.jmmm.2014.06.059

[61] Y. Kannan, R. Saravanan, N. Srinivasan, K. Praveena, K. Sadhana, Synthesis and characterization of some ferrite nanoparticles prepared by co-precipitation method, Journal of Materials Science: Materials in Electronics, 27 (2016) 12000-12008. https://doi.org/10.1007/s10854-016-5347-y

[62] P. Dhiman, J. Chand, A. Kumar, R.K. Kotnala, K.M. Batoo, M. Singh, Synthesis and characterization of novel Fe@ZnO nanosystem, Journal of Alloys and Compounds, 578 (2013) 235-241. https://doi.org/10.1016/j.jallcom.2013.05.015

[63] S. Bhukal, S. Mor, S. Bansal, J. Singh, S. Singhal, Influence of Cd_{2+} ions on the structural, electrical, optical and magnetic properties of Co–Zn nanoferrites prepared

by sol gel auto combustion method, Journal of Molecular Structure, 1071 (2014) 95-102. https://doi.org/10.1016/j.molstruc.2014.04.073

[64] N. Sanpo, J. Wang, C.C. Berndt, Influence of chelating agents on the microstructure and antibacterial property of cobalt ferrite nanopowders, Journal of the Australian Ceramic Society, 49 (2013) 84-91.

[65] G.J. Owens, R.K. Singh, F. Foroutan, M. Alqaysi, C.-M. Han, C. Mahapatra, H.-W. Kim, J.C. Knowles, Sol–gel based materials for biomedical applications, Progress in Materials Science, 77 (2016) 1-79. https://doi.org/10.1016/j.pmatsci.2015.12.001

[66] A. Sutka, G. Mezinskis, Sol-gel auto-combustion synthesis of spinel-type ferrite nanomaterials, Frontiers of Materials Science, 6 (2012) 128-141. https://doi.org/10.1007/s11706-012-0167-3

[67] D.-H. Chen, X.-R. He, Synthesis of nickel ferrite nanoparticles by sol-gel method, Materials Research Bulletin, 36 (2001) 1369-1377. https://doi.org/10.1016/S0025-5408(01)00620-1

[68] R. Kadam, K. Desai, V.S. Shinde, M. Hashim, S.E. Shirsath, Influence of Gd3+ ion substitution on the $MnCrFeO_4$ for their nanoparticle shape formation and magnetic properties, Journal of Alloys and Compounds, 657 (2016) 487-494. https://doi.org/10.1016/j.jallcom.2015.10.164

[69] M.A. Malik, M.Y. Wani, M.A. Hashim, Microemulsion method: A novel route to synthesize organic and inorganic nanomaterials: 1st Nano Update, Arabian journal of Chemistry, 5 (2012) 397-417. https://doi.org/10.1016/j.arabjc.2010.09.027

[70] A. Košak, D. Makovec, A. Žnidaršič, M. Drofenik, Preparation of MnZn-ferrite with microemulsion technique, Journal of the European Ceramic Society, 24 (2004) 959-962. https://doi.org/10.1016/S0955-2219(03)00524-7

[71] K. Pemartin, C. Solans, J. Alvarez-Quintana, M. Sanchez-Dominguez, Synthesis of Mn–Zn ferrite nanoparticles by the oil-in-water microemulsion reaction method, Colloids and Surfaces A: Physicochemical and Engineering Aspects, 451 (2014) 161-171. https://doi.org/10.1016/j.colsurfa.2014.03.036

[72] D.O. Yener, H. Giesche, Synthesis of Pure and Manganese-, Nickel-, and Zinc-Doped Ferrite Particles in Water-in-Oil Microemulsions, Journal of the American Ceramic Society, 84 (2001) 1987-1995. https://doi.org/10.1111/j.1151-2916.2001.tb00947.x

[73] V. Pillai, D. Shah, Synthesis of high-coercivity cobalt ferrite particles using water-in-oil microemulsions, Journal of Magnetism and Magnetic Materials, 163 (1996) 243-248. https://doi.org/10.1016/S0304-8853(96)00280-6

[74] P. Holec, J. Plocek, D. Nižňanský, J.P. Vejpravová, Preparation of $MgFe_2O_4$ nanoparticles by microemulsion method and their characterization, Journal of sol-gel science and technology, 51 (2009) 301-305. https://doi.org/10.1007/s10971-009-1962-x

[75] F. Grasset, N. Labhsetwar, D. Li, D. Park, N. Saito, H. Haneda, O. Cador, T. Roisnel, S. Mornet, E. Duguet, Synthesis and magnetic characterization of zinc ferrite nanoparticles with different environments: powder, colloidal solution, and zinc ferrite–silica core– shell nanoparticles, Langmuir, 18 (2002) 8209-8216. https://doi.org/10.1021/la020322b

[76] Y. Gao, Y. Zhao, Q. Jiao, H. Li, Microemulsion-based synthesis of porous Co–Ni ferrite nanorods and their magnetic properties, Journal of alloys and compounds, 555 (2013) 95-100. https://doi.org/10.1016/j.jallcom.2012.12.057

[77] X. Gao, Y. Du, X. Liu, P. Xu, X. Han, Synthesis and characterization of Co–Sn substituted barium ferrite particles by a reverse microemulsion technique, Materials Research Bulletin, 46 (2011) 643-648. https://doi.org/10.1016/j.materresbull.2011.02.002

[78] B.P. Jacob, A. Kumar, R. Pant, S. Singh, E. Mohammed, Influence of preparation method on structural and magnetic properties of nickel ferrite nanoparticles, Bulletin of Materials Science, 34 (2011) 1345-1350. https://doi.org/10.1007/s12034-011-0326-7

[79] P. Dhiman, J. Chand, S. Verma, Sarveena, M. Singh, Ni, Fe Co-doped ZnO nanoparticles synthesized by solution combustion method, AIP Conference Proceedings, 1591 (2014) 1443-1445. https://doi.org/10.1063/1.4872990

[80] P. Dhiman, M. Patial, A. Kumar, M. Alam, M. Naushad, G. Sharma, D.-V.N. Vo, R. Kumar, Environmental friendly and robust $Mg_{0.5-x}Cu_xZn_{0.5}Fe_2O_4$ spinel nanoparticles for visible light driven degradation of Carbamazepine: Band shift driven by dopants, Materials Letters, 284 (2021) 129005. https://doi.org/10.1016/j.matlet.2020.129005

[81] P. Dhiman, K.M. Batoo, R.K. Kotnala, J. Chand, M. Singh, Room temperature ferromagnetism and structural characterization of Fe,Ni co-doped ZnO nanocrystals, Applied Surface Science, 287 (2013) 287-292. https://doi.org/10.1016/j.apsusc.2013.09.144

[82] C. Rao, J. Gopalakrishnan, E. Banks, New Directions in Solid State Chemistry: Structures, Synthesis, Properties, Reactivity and Materials Design, Physics Today, 41 (1988) 112. https://doi.org/10.1063/1.2811679

[83] J. Chand, M. Singh, Electric and dielectric properties of $MgGd_{0.1}Fe_{1.9}O_4$ ferrite, Journal of alloys and compounds, 486 (2009) 376-379. https://doi.org/10.1016/j.jallcom.2009.06.150

[84] N. Chaibakhsh, Z. Moradi-Shoeili, Enzyme mimetic activities of spinel substituted nanoferrites (MFe_2O_4): A review of synthesis, mechanism and potential applications, Materials Science and Engineering: C, 99 (2019) 1424-1447. https://doi.org/10.1016/j.msec.2019.02.086

[85] T. Kundu, S. Mishra, Nanocrystalline spinel ferrites by solid state reaction route, Bulletin of Materials Science, 31 (2008) 507-510. https://doi.org/10.1007/s12034-008-0079-0

[86] S. Bera, A. Prince, S. Velmurugan, P. Raghavan, R. Gopalan, G. Panneerselvam, S. Narasimhan, Formation of zinc ferrite by solid-state reaction and its characterization by XRD and XPS, Journal of materials science, 36 (2001) 5379-5384. https://doi.org/10.1023/A:1012488422484

[87] V. Berbenni, C. Milanese, G. Bruni, A. Girella, A. Marini, Mechanothermal solid-state synthesis of cobalt (II) ferrite and determination of its heat capacity by MTDSC, Zeitschrift für Naturforschung B, 65 (2010) 1434-1438. https://doi.org/10.1515/znb-2010-1204

[88] E.N. Lysenko, E.V. Nikolaev, A.P. Surzhikov, TG study of the $Li_{0.4}Fe_{2.4}Zn_{0.2}O_4$ ferrite synthesis, in: IOP Conference Series: Materials Science and Engineering, IOP Publishing, 2016, pp. 012092. https://doi.org/10.1088/1757-899X/110/1/012092

[89] C. Sun, J.S. Lee, M. Zhang, Magnetic nanoparticles in MR imaging and drug delivery, Advanced drug delivery reviews, 60 (2008) 1252-1265. https://doi.org/10.1016/j.addr.2008.03.018

[90] J. Wang, Y. Huang, A. E David, B. Chertok, L. Zhang, F. Yu, V. C Yang, Magnetic nanoparticles for MRI of brain tumors, Current pharmaceutical biotechnology, 13 (2012) 2403-2416. https://doi.org/10.2174/138920112803341824

[91] A. Rana, V. Kumar, O. Thakur, R. Pant, Structural and Electrical Properties of Gd^{3+} Ion Substituted $CoGd_xFe_{2-x}O_4$ Nano-Ferrites, Journal of nanoscience and nanotechnology, 12 (2012) 6355-6358. https://doi.org/10.1166/jnn.2012.6418

[92] Y.-L. Liu, Z.-M. Liu, Y. Yang, H.-F. Yang, G.-L. Shen, R.-Q. Yu, Simple synthesis of $MgFe_2O_4$ nanoparticles as gas sensing materials, Sensors and Actuators B: Chemical, 107 (2005) 600-604. https://doi.org/10.1016/j.snb.2004.11.026

[93] S. Phumying, S. Labuayai, E. Swatsitang, V. Amornkitbamrung, S. Maensiri, Nanocrystalline spinel ferrite (MFe_2O_4, M= Ni, Co, Mn, Mg, Zn) powders prepared by a simple aloe vera plant-extracted solution hydrothermal route, Materials Research Bulletin, 48 (2013) 2060-2065. https://doi.org/10.1016/j.materresbull.2013.02.042

[94] N. Gupta, P. Jain, R. Rana, S. Shrivastava, Current development in synthesis and characterization of nickel ferrite nanoparticle, Materials Today: Proceedings, 4 (2017) 342-349. https://doi.org/10.1016/j.matpr.2017.01.031

[95] A. Šutka, R. Pärna, T. Käämbre, V. Kisand, Synthesis of p-type and n-type nickel ferrites and associated electrical properties, Physica B: Condensed Matter, 456 (2015) 232-236. https://doi.org/10.1016/j.physb.2014.09.013

[96] Z. Zhu, X. Li, Q. Zhao, H. Li, Y. Shen, G. Chen, Porous "brick-like" $NiFe_2O_4$ nanocrystals loaded with Ag species towards effective degradation of toluene, Chemical Engineering Journal, 165 (2010) 64-70. https://doi.org/10.1016/j.cej.2010.08.060

[97] D. Zhang, Z. Tong, G. Xu, S. Li, J. Ma, Templated fabrication of $NiFe_2O_4$ nanorods: characterization, magnetic and electrochemical properties, Solid State Sciences, 11 (2009) 113-117. https://doi.org/10.1016/j.solidstatesciences.2008.05.001

[98] J. Wang, F. Ren, R. Yi, A. Yan, G. Qiu, X. Liu, Solvothermal synthesis and magnetic properties of size-controlled nickel ferrite nanoparticles, Journal of Alloys and Compounds, 479 (2009) 791-796. https://doi.org/10.1016/j.jallcom.2009.01.059

[99] Z. Gao, F. Cui, S. Zeng, L. Guo, J. Shi, A high surface area superparamagnetic mesoporous spinel ferrite synthesized by a template-free approach and its adsorptive property, Microporous and mesoporous materials, 132 (2010) 188-195. https://doi.org/10.1016/j.micromeso.2010.02.019

[100] M.G. Naseri, E.B. Saion, H.A. Ahangar, M. Hashim, A.H. Shaari, Simple preparation and characterization of nickel ferrite nanocrystals by a thermal treatment method, Powder Technology, 212 (2011) 80-88. https://doi.org/10.1016/j.powtec.2011.04.033

[101] A. Baykal, N. Kasapoğlu, Y. Köseoğlu, M.S. Toprak, H. Bayrakdar, CTAB-assisted hydrothermal synthesis of $NiFe_2O_4$ and its magnetic characterization, Journal

of Alloys and Compounds, 464 (2008) 514-518. https://doi.org/10.1016/j.jallcom.2007.10.041

[102] R. Waldron, Infrared spectra of ferrites, Physical review, 99 (1955) 1727. https://doi.org/10.1103/PhysRev.99.1727

[103] V. Rathod, A. Anupama, R.V. Kumar, V. Jali, B. Sahoo, Correlated vibrations of the tetrahedral and octahedral complexes and splitting of the absorption bands in FTIR spectra of Li-Zn ferrites, Vibrational Spectroscopy, 92 (2017) 267-272. https://doi.org/10.1016/j.vibspec.2017.08.008

[104] D. Chaudhari, D. Choudhary, K. Rewatkar, Spinel Ferrite Nanoparticles: Synthesis, Characterization and Applications, (2020).

[105] Y.-j. Sun, Y.-f. Diao, H.-g. Wang, G. Chen, M. Zhang, M. Guo, Synthesis, structure and magnetic properties of spinel ferrite (Ni, Cu, Co) Fe_2O_4 from low nickel matte, Ceramics International, 43 (2017) 16474-16481. https://doi.org/10.1016/j.ceramint.2017.09.029

[106] K. Maaz, S. Karim, A. Mumtaz, S. Hasanain, J. Liu, J. Duan, Synthesis and magnetic characterization of nickel ferrite nanoparticles prepared by co-precipitation route, Journal of Magnetism and Magnetic Materials, 321 (2009) 1838-1842. https://doi.org/10.1016/j.jmmm.2008.11.098

[107] A. Abdeen, Dielectric behaviour in Ni–Zn ferrites, Journal of magnetism and magnetic materials, 192 (1999) 121-129. https://doi.org/10.1016/S0304-8853(98)00324-2

[108] M.A. Yousuf, M.M. Baig, N.F. Al-Khalli, M.A. Khan, M.F.A. Aboud, I. Shakir, M.F. Warsi, The impact of yttrium cations (Y^{3+}) on structural, spectral and dielectric properties of spinel manganese ferrite nanoparticles, Ceramics International, 45 (2019) 10936-10942. https://doi.org/10.1016/j.ceramint.2019.02.174

[109] M. Dar, D. Varshney, Effect of d-block element Co^{2+} substitution on structural, Mössbauer and dielectric properties of spinel copper ferrites, Journal of Magnetism and Magnetic Materials, 436 (2017) 101-112. https://doi.org/10.1016/j.jmmm.2017.04.046

[110] M. Deepty, S. Ch, P. Ramesh, N.K. Mohan, M.S. Singh, C. Prajapat, A. Verma, D. Sastry, Evaluation of structural and dielectric properties of Mn^{2+}-substituted Zn-spinel ferrite nanoparticles for gas sensor applications, Sensors and Actuators B: Chemical, 316 (2020) 128127. https://doi.org/10.1016/j.snb.2020.128127

[111] G.F. Dionne, R.G. West, Magnetic and dielectric properties of the spinel ferrite system $Ni_{0.65}Zn_{0.35}Fe_{2-x}Mn_xO_4$, Journal of Applied Physics, 61 (1987) 3868-3870. https://doi.org/10.1063/1.338623

[112] M. Ajmal, M.U. Islam, G.A. Ashraf, M.A. Nazir, M. Ghouri, The influence of Ga doping on structural magnetic and dielectric properties of $NiCr_{0.2}Fe_{1.8}O_4$ spinel ferrite, Physica B: Condensed Matter, 526 (2017) 149-154. https://doi.org/10.1016/j.physb.2017.05.044

[113] R. Kambale, N. Adhate, B. Chougule, Y. Kolekar, Magnetic and dielectric properties of mixed spinel Ni–Zn ferrites synthesized by citrate–nitrate combustion method, Journal of Alloys and Compounds, 491 (2010) 372-377. https://doi.org/10.1016/j.jallcom.2009.10.187

[114] R. Kambale, P. Shaikh, C. Bhosale, K. Rajpure, Y. Kolekar, The effect of Mn substitution on the magnetic and dielectric properties of cobalt ferrite synthesized by an autocombustion route, Smart Materials and structures, 18 (2009) 115028. https://doi.org/10.1088/0964-1726/18/11/115028

[115] E. Pervaiz, I. Gul, High frequency AC response, DC resistivity and magnetic studies of holmium substituted Ni-ferrite: a novel electromagnetic material, Journal of magnetism and magnetic materials, 349 (2014) 27-34. https://doi.org/10.1016/j.jmmm.2013.08.011

[116] M. Khan, S. Bisen, J. Shukla, A. Mishra, P. Sharma, Investigations on the Structural and Electrical Properties of Sm^{3+}-Doped Nickel Ferrite–Based Ceramics, Journal of Superconductivity and Novel Magnetism, (2020) 1-18. https://doi.org/10.1007/s10948-020-05754-1

[117] F. Alahmari, M. Almessiere, B. Ünal, Y. Slimani, A. Baykal, Electrical and optical properties of $Ni_{0.5}Co_{0.5-x}Cd_xNd_{0.02}Fe_{1.78}O_4$ ($x \leq 0.25$) spinel ferrite nanofibers, Ceramics International, 46 (2020) 24605-24614. https://doi.org/10.1016/j.ceramint.2020.06.249

[118] K.K. Bharathi, G. Markandeyulu, C. Ramana, Structural, magnetic, electrical, and magnetoelectric properties of Sm-and Ho-substituted nickel ferrites, The Journal of Physical Chemistry C, 115 (2011) 554-560. https://doi.org/10.1021/jp1060864

[119] L. Chauhan, A. Shukla, K. Sreenivas, Dielectric and magnetic properties of Nickel ferrite ceramics using crystalline powders derived from DL alanine fuel in sol–gel auto-combustion, Ceramics International, 41 (2015) 8341-8351. https://doi.org/10.1016/j.ceramint.2015.03.014

[120] P. Chand, S. Vaish, P. Kumar, Structural, optical and dielectric properties of transition metal (MFe_2O_4; M= Co, Ni and Zn) nanoferrites, Physica B: Condensed Matter, 524 (2017) 53-63. https://doi.org/10.1016/j.physb.2017.08.060

[121] R.S. Yadav, I. Kuřitka, J. Vilcakova, P. Urbánek, M. Machovsky, M. Masař, M. Holek, Structural, magnetic, optical, dielectric, electrical and modulus spectroscopic characteristics of $ZnFe_2O_4$ spinel ferrite nanoparticles synthesized via honey-mediated sol-gel combustion method, Journal of Physics and Chemistry of Solids, 110 (2017) 87-99. https://doi.org/10.1016/j.jpcs.2017.05.029

[122] M.F. Warsi, A. Iftikhar, M.A. Yousuf, M.I. Sarwar, S. Yousaf, S. Haider, M.F.A. Aboud, I. Shakir, S. Zulfiqar, Erbium substituted nickel–cobalt spinel ferrite nanoparticles: Tailoring the structural, magnetic and electrical parameters, Ceramics International, 46 (2020) 24194-24203. https://doi.org/10.1016/j.ceramint.2020.06.199

[123] S. Ikram, J. Jacob, M.I. Arshad, K. Mahmood, A. Ali, N. Sabir, N. Amin, S. Hussain, Tailoring the structural, magnetic and dielectric properties of Ni-Zn-$CdFe_2O_4$ spinel ferrites by the substitution of lanthanum ions, Ceramics International, 45 (2019) 3563-3569. https://doi.org/10.1016/j.ceramint.2018.11.015

[124] S. Kavitha, M. Kurian, Effect of zirconium doping in the microstructure, magnetic and dielectric properties of cobalt ferrite nanoparticles, Journal of Alloys and Compounds, 799 (2019) 147-159. https://doi.org/10.1016/j.jallcom.2019.05.183

[125] A. Nikumbh, R. Pawar, D. Nighot, G. Gugale, M. Sangale, M. Khanvilkar, A. Nagawade, Structural, electrical, magnetic and dielectric properties of rare-earth substituted cobalt ferrites nanoparticles synthesized by the co-precipitation method, Journal of Magnetism and Magnetic Materials, 355 (2014) 201-209. https://doi.org/10.1016/j.jmmm.2013.11.052

[126] S. Mansour, M. Abdo, F. Kzar, Effect of Cr dopant on the structural, magnetic and dielectric properties of Cu-Zn nanoferrites, Journal of Magnetism and Magnetic Materials, 465 (2018) 176-185. https://doi.org/10.1016/j.jmmm.2018.05.104

[127] N. Hamdaoui, Y. Azizian-Kalandaragh, M. Khlifi, L. Beji, Structural, magnetic and dielectric properties of $Ni_{0.6}Mg_{0.4}Fe_2O_4$ ferromagnetic ferrite prepared by sol gel method, Ceramics International, 45 (2019) 16458-16465. https://doi.org/10.1016/j.ceramint.2019.05.177

[128] A.B. Mugutkar, S.K. Gore, U.B. Tumberphale, V.V. Jadhav, R.S. Mane, S.M. Patange, S.F. Shaikh, M. Ubaidullah, A.M. Al-Enizi, S.S. Jadhav, The role of La^{3+} substitution in modification of the magnetic and dielectric properties of the

nanocrystalline Co-Zn ferrites, Journal of Magnetism and Magnetic Materials, 502 (2020) 166490. https://doi.org/10.1016/j.jmmm.2020.166490

[129] A. Pandit, A. Shitre, D. Shengule, K. Jadhav, Magnetic and dielectric properties of Mg $_{1+x}$ Mn $_x$, Fe $_{2-2x}$, O $_4$ ferrite system, Journal of materials science, 40 (2005) 423-428. https://doi.org/10.1007/s10853-005-6099-x

[130] T.V. Sagar, T.S. Rao, K.C.B. Naidu, Effect of calcination temperature on optical, magnetic and dielectric properties of sol-gel synthesized Ni0. 2Mg0. 8-xZnxFe$_2$O$_4$ (x= 0.0–0.8), Ceramics International, 46 (2020) 11515-11529. https://doi.org/10.1016/j.ceramint.2020.01.178

[131] N. Rezlescu, E. Rezlescu, L. Sachelarie, P. Popa, C. Doroftei, Structural and catalytic properties of mesoporous nanocrystalline mixed oxides containing magnesium, Catalysis Communications, 46 (2014) 51-56. https://doi.org/10.1016/j.catcom.2013.11.021

[132] X. Huang, J. Zhang, S. Xiao, T. Sang, G. Chen, Unique electromagnetic properties of the zinc ferrite nanofiber, Materials Letters, 124 (2014) 126-128. https://doi.org/10.1016/j.matlet.2014.03.049

[133] E.R. Kumar, R. Jayaprakash, G.S. Devi, P.S.P. Reddy, Synthesis of Mn substituted CuFe$_2$O$_4$ nanoparticles for liquefied petroleum gas sensor applications, Sensors and Actuators B: Chemical, 191 (2014) 186-191. https://doi.org/10.1016/j.snb.2013.09.108

[134] Y. Liu, M. Yuan, L. Qiao, R. Guo, An efficient colorimetric biosensor for glucose based on peroxidase-like protein-Fe$_3$O$_4$ and glucose oxidase nanocomposites, Biosensors and Bioelectronics, 52 (2014) 391-396. https://doi.org/10.1016/j.bios.2013.09.020

[135] D. Ling, N. Lee, T. Hyeon, Chemical synthesis and assembly of uniformly sized iron oxide nanoparticles for medical applications, Accounts of chemical research, 48 (2015) 1276-1285. https://doi.org/10.1021/acs.accounts.5b00038

[136] A. Yadollahpour, S. Rashidi, Magnetic nanoparticles: a review of chemical and physical characteristics important in medical applications, Orient J Chem, 31 (2015) 25-30. https://doi.org/10.13005/ojc/31.Special-Issue1.03

[137] A. Akbarzadeh, M. Samiei, S. Davaran, Magnetic nanoparticles: preparation, physical properties, and applications in biomedicine, Nanoscale research letters, 7 (2012) 1-13. https://doi.org/10.1186/1556-276X-7-144

[138] D. Psimadas, G. Baldi, C. Ravagli, M.C. Franchini, E. Locatelli, C. Innocenti, C. Sangregorio, G. Loudos, Comparison of the magnetic, radiolabeling, hyperthermic and

biodistribution properties of hybrid nanoparticles bearing $CoFe_2O_4$ and Fe_3O_4 metal cores, Nanotechnology, 25 (2013) 025101. https://doi.org/10.1088/0957-4484/25/2/025101

[139] M. Colombo, S. Carregal-Romero, M.F. Casula, L. Gutiérrez, M.P. Morales, I.B. Böhm, J.T. Heverhagen, D. Prosperi, W.J. Parak, Biological applications of magnetic nanoparticles, Chemical Society Reviews, 41 (2012) 4306-4334. https://doi.org/10.1039/c2cs15337h

[140] H. Ersoy, F.J. Rybicki, Biochemical safety profiles of gadolinium-based extracellular contrast agents and nephrogenic systemic fibrosis, Journal of Magnetic Resonance Imaging: An Official Journal of the International Society for Magnetic Resonance in Medicine, 26 (2007) 1190-1197. https://doi.org/10.1002/jmri.21135

[141] L. Huang, L. Ao, W. Wang, D. Hu, Z. Sheng, W. Su, Multifunctional magnetic silica nanotubes for MR imaging and targeted drug delivery, Chemical Communications, 51 (2015) 3923-3926. https://doi.org/10.1039/C4CC09382H

[142] W. Lin, Introduction: nanoparticles in medicine, Chemical reviews, 115 (2015) 10407-10409. https://doi.org/10.1021/acs.chemrev.5b00534

[143] C.P. Slichter, Principles of magnetic resonance, Springer Science & Business Media, 2013.

[144] N. Lee, T. Hyeon, Designed synthesis of uniformly sized iron oxide nanoparticles for efficient magnetic resonance imaging contrast agents, Chemical Society Reviews, 41 (2012) 2575-2589. https://doi.org/10.1039/C1CS15248C

[145] Q. Pankhurst, N. Thanh, S. Jones, J. Dobson, Progress in applications of magnetic nanoparticles in biomedicine, Journal of Physics D: Applied Physics, 42 (2009) 224001. https://doi.org/10.1088/0022-3727/42/22/224001

[146] C.G. Hadjipanayis, M.J. Bonder, S. Balakrishnan, X. Wang, H. Mao, G.C. Hadjipanayis, Metallic iron nanoparticles for MRI contrast enhancement and local hyperthermia, Small, 4 (2008) 1925-1929. https://doi.org/10.1002/smll.200800261

[147] S. Maenosono, T. Suzuki, S. Saita, Superparamagnetic FePt nanoparticles as excellent MRI contrast agents, Journal of Magnetism and Magnetic Materials, 320 (2008) L79-L83. https://doi.org/10.1016/j.jmmm.2008.01.026

[148] T.-H. Shin, Y. Choi, S. Kim, J. Cheon, Recent advances in magnetic nanoparticle-based multi-modal imaging, Chemical Society Reviews, 44 (2015) 4501-4516. https://doi.org/10.1039/C4CS00345D

[149] J.-H. Lee, Y.-M. Huh, Y.-w. Jun, J.-w. Seo, J.-t. Jang, H.-T. Song, S. Kim, E.-J. Cho, H.-G. Yoon, J.-S. Suh, Artificially engineered magnetic nanoparticles for ultra-sensitive molecular imaging, Nature medicine, 13 (2007) 95-99. https://doi.org/10.1038/nm1467

[150] J. Gao, X. Chen, Z. Zhao, Octapod iron oxide nanoparticles as high performance T2 contrast agents for magnetic resonance imaging, in, Google Patents, 2018.

[151] V.V. Mody, A. Cox, S. Shah, A. Singh, W. Bevins, H. Parihar, Magnetic nanoparticle drug delivery systems for targeting tumor, Applied Nanoscience, 4 (2014) 385-392. https://doi.org/10.1007/s13204-013-0216-y

[152] J. Estelrich, E. Escribano, J. Queralt, M.A. Busquets, Iron oxide nanoparticles for magnetically-guided and magnetically-responsive drug delivery, International journal of molecular sciences, 16 (2015) 8070-8101. https://doi.org/10.3390/ijms16048070

[153] A.K. Hauser, R.J. Wydra, N.A. Stocke, K.W. Anderson, J.Z. Hilt, Magnetic nanoparticles and nanocomposites for remote controlled therapies, Journal of Controlled Release, 219 (2015) 76-94. https://doi.org/10.1016/j.jconrel.2015.09.039

[154] H. Guo, W. Chen, X. Sun, Y.-N. Liu, J. Li, J. Wang, Theranostic magnetoliposomes coated by carboxymethyl dextran with controlled release by low-frequency alternating magnetic field, Carbohydrate polymers, 118 (2015) 209-217. https://doi.org/10.1016/j.carbpol.2014.10.076

[155] D.C. Chan, D.B. Kirpotin, P.A. Bunn Jr, Synthesis and evaluation of colloidal magnetic iron oxides for the site-specific radiofrequency-induced hyperthermia of cancer, Journal of Magnetism and Magnetic Materials, 122 (1993) 374-378. https://doi.org/10.1016/0304-8853(93)91113-L

[156] A. Jordan, P. Wust, H. Fählin, W. John, A. Hinz, R. Felix, Inductive heating of ferrimagnetic particles and magnetic fluids: physical evaluation of their potential for hyperthermia, International journal of hyperthermia, 9 (1993) 51-68. https://doi.org/10.3109/02656739309061478

[157] A. Jordan, R. Scholz, P. Wust, H. Fähling, R. Felix, Magnetic fluid hyperthermia (MFH): Cancer treatment with AC magnetic field induced excitation of biocompatible superparamagnetic nanoparticles, Journal of Magnetism and Magnetic materials, 201 (1999) 413-419. https://doi.org/10.1016/S0304-8853(99)00088-8

[158] A. Jordan, R. Scholz, K. Maier-Hauff, M. Johannsen, P. Wust, J. Nadobny, H. Schirra, H. Schmidt, S. Deger, S. Loening, Presentation of a new magnetic field therapy system for the treatment of human solid tumors with magnetic fluid

hyperthermia, Journal of magnetism and magnetic materials, 225 (2001) 118-126. https://doi.org/10.1016/S0304-8853(00)01239-7

[159] N. Kawai, A. Ito, Y. Nakahara, M. Futakuchi, T. Shirai, H. Honda, T. Kobayashi, K. Kohri, Anticancer effect of hyperthermia on prostate cancer mediated by magnetite cationic liposomes and immune-response induction in transplanted syngeneic rats, The prostate, 64 (2005) 373-381. https://doi.org/10.1002/pros.20253

[160] J.t. Jang, H. Nah, J.H. Lee, S.H. Moon, M.G. Kim, J. Cheon, Critical enhancements of MRI contrast and hyperthermic effects by dopant-controlled magnetic nanoparticles, Angewandte Chemie International Edition, 48 (2009) 1234-1238. https://doi.org/10.1002/anie.200805149

[161] K. Maier-Hauff, R. Rothe, R. Scholz, U. Gneveckow, P. Wust, B. Thiesen, A. Feussner, A. von Deimling, N. Waldoefner, R. Felix, Intracranial thermotherapy using magnetic nanoparticles combined with external beam radiotherapy: results of a feasibility study on patients with glioblastoma multiforme, Journal of neuro-oncology, 81 (2007) 53-60. https://doi.org/10.1007/s11060-006-9195-0

[162] P. Wust, U. Gneveckow, P. Wust, U. Gneveckow, M. Johannsen, D. Böhmer, T. Henkel, F. Kahmann, J. Sehouli, R. Felix, Magnetic nanoparticles for interstitial thermotherapy–feasibility, tolerance and achieved temperatures, International Journal of Hyperthermia, 22 (2006) 673-685. https://doi.org/10.1080/02656730601106037

[163] B. Thiesen, A. Jordan, Clinical applications of magnetic nanoparticles for hyperthermia, International journal of hyperthermia, 24 (2008) 467-474. https://doi.org/10.1080/02656730802104757

[164] J.-H. Lee, J.-t. Jang, J.-s. Choi, S.H. Moon, S.-h. Noh, J.-w. Kim, J.-G. Kim, I.-S. Kim, K.I. Park, J. Cheon, Exchange-coupled magnetic nanoparticles for efficient heat induction, Nature nanotechnology, 6 (2011) 418-422. https://doi.org/10.1038/nnano.2011.95

[165] T. Sadhukha, T.S. Wiedmann, J. Panyam, Inhalable magnetic nanoparticles for targeted hyperthermia in lung cancer therapy, Biomaterials, 34 (2013) 5163-5171. https://doi.org/10.1016/j.biomaterials.2013.03.061

[166] R. Di Corato, A. Espinosa, L. Lartigue, M. Tharaud, S. Chat, T. Pellegrino, C. Ménager, F. Gazeau, C. Wilhelm, Magnetic hyperthermia efficiency in the cellular environment for different nanoparticle designs, Biomaterials, 35 (2014) 6400-6411. https://doi.org/10.1016/j.biomaterials.2014.04.036

[167] O. Veiseh, J.W. Gunn, M. Zhang, Design and fabrication of magnetic nanoparticles for targeted drug delivery and imaging, Advanced drug delivery reviews, 62 (2010) 284-304. https://doi.org/10.1016/j.addr.2009.11.002

[168] D. Singh, J.M. McMillan, X.-M. Liu, H.M. Vishwasrao, A.V. Kabanov, M. Sokolsky-Papkov, H.E. Gendelman, Formulation design facilitates magnetic nanoparticle delivery to diseased cells and tissues, Nanomedicine, 9 (2014) 469-485. https://doi.org/10.2217/nnm.14.4

[169] N. Lee, D. Yoo, D. Ling, M.H. Cho, T. Hyeon, J. Cheon, Iron oxide based nanoparticles for multimodal imaging and magnetoresponsive therapy, Chemical reviews, 115 (2015) 10637-10689. https://doi.org/10.1021/acs.chemrev.5b00112

[170] S.J. Mattingly, M.G. O'Toole, K.T. James, G.J. Clark, M.H. Nantz, Magnetic nanoparticle-supported lipid bilayers for drug delivery, Langmuir, 31 (2015) 3326-3332. https://doi.org/10.1021/la504830z

[171] Y. Ding, S.Z. Shen, H. Sun, K. Sun, F. Liu, Y. Qi, J. Yan, Design and construction of polymerized-chitosan coated Fe_3O_4 magnetic nanoparticles and its application for hydrophobic drug delivery, Materials Science and Engineering: C, 48 (2015) 487-498. https://doi.org/10.1016/j.msec.2014.12.036

[172] Y. Chen, J. Nan, Y. Lu, C. Wang, F. Chu, Z. Gu, Hybrid Fe_3O_4-poly (acrylic acid) nanogels for theranostic cancer treatment, Journal of biomedical nanotechnology, 11 (2015) 771-779. https://doi.org/10.1166/jbn.2015.2001

[173] Y. Yang, L. Yang, Q.-y. Sun, Archaeal and bacterial communities in acid mine drainage from metal-rich abandoned tailing ponds, Tongling, China, Transactions of Nonferrous Metals Society of China, 24 (2014) 3332-3342. https://doi.org/10.1016/S1003-6326(14)63474-9

[174] B. Bahrami, M. Hojjat-Farsangi, H. Mohammadi, E. Anvari, G. Ghalamfarsa, M. Yousefi, F. Jadidi-Niaragh, Nanoparticles and targeted drug delivery in cancer therapy, Immunology letters, 190 (2017) 64-83. https://doi.org/10.1016/j.imlet.2017.07.015

[175] J.-M. Shen, X.-M. Guan, X.-Y. Liu, J.-F. Lan, T. Cheng, H.-X. Zhang, Luminescent/magnetic hybrid nanoparticles with folate-conjugated peptide composites for tumor-targeted drug delivery, Bioconjugate chemistry, 23 (2012) 1010-1021. https://doi.org/10.1021/bc300008k

[176] C.A. Monnier, D. Burnand, B. Rothen-Rutishauser, M. Lattuada, A. Petri-Fink, Magnetoliposomes: opportunities and challenges, European Journal of Nanomedicine, 6 (2014) 201-215. https://doi.org/10.1515/ejnm-2014-0042

[177] G. Wang, Y. Ma, L. Zhang, J. Mu, Z. Zhang, X. Zhang, H. Che, Y. Bai, J. Hou, Facile synthesis of manganese ferrite/graphene oxide nanocomposites for controlled targeted drug delivery, Journal of Magnetism and Magnetic Materials, 401 (2016) 647-650. https://doi.org/10.1016/j.jmmm.2015.10.096

[178] Y. Teng, P.W. Pong, One-Pot Synthesis and Surface Modification of Lauric-Acid-Capped $CoFe_2O_4$ Nanoparticles, IEEE Transactions on Magnetics, 54 (2018) 1-5. https://doi.org/10.1109/TMAG.2018.2834524

[179] T. Indira, P. Lakshmi, Magnetic nanoparticles–a review, International Journal of Pharmaceutical sciences and nanotechnology, 3 (2010) 1035-1042. https://doi.org/10.1016/j.jmmm.2014.01.016

[180] Z. Karimi, Y. Mohammadifar, H. Shokrollahi, S.K. Asl, G. Yousefi, L. Karimi, Magnetic and structural properties of nano sized Dy-doped cobalt ferrite synthesized by co-precipitation, Journal of Magnetism and Magnetic Materials, 361 (2014) 150-156. https://doi.org/10.1016/j.jmmm.2014.01.016

[181] K.M. Krishnan, Biomedical nanomagnetics: a spin through possibilities in imaging, diagnostics, and therapy, IEEE transactions on magnetics, 46 (2010) 2523-2558. https://doi.org/10.1109/TMAG.2010.2046907

[182] H. Dong, S.-R. Du, X.-Y. Zheng, G.-M. Lyu, L.-D. Sun, L.-D. Li, P.-Z. Zhang, C. Zhang, C.-H. Yan, Lanthanide nanoparticles: from design toward bioimaging and therapy, Chemical reviews, 115 (2015) 10725-10815. https://doi.org/10.1021/acs.chemrev.5b00091

[183] H. Markides, M. Rotherham, A. El Haj, Biocompatibility and toxicity of magnetic nanoparticles in regenerative medicine, Journal of Nanomaterials, 2012 (2012). https://doi.org/10.1155/2012/614094

[184] A. Koner, D. Krndija, Q. Hou, D. Sherratt, M. Howarth, Acs Nano 2013, 7, 1137–1144; b) AB Chinen, CM Guan, JR Ferrer, SN Barnaby, TJ Merkel, CA Mirkin, Chem. Rev, 115 (2015) 10530-10574.

[185] Z. Ma, H. Liu, Synthesis and surface modification of magnetic particles for application in biotechnology and biomedicine, China Particuology, 5 (2007) 1-10. https://doi.org/10.1016/j.cpart.2006.11.001

[186] A.B. Salunkhe, V.M. Khot, S. Pawar, Magnetic hyperthermia with magnetic nanoparticles: a status review, Current topics in medicinal chemistry, 14 (2014) 572-594. https://doi.org/10.2174/1568026614666140118203550

[187] J. Lagendijk, Hyperthermia treatment planning, Physics in Medicine & Biology, 45 (2000) R61. https://doi.org/10.1088/0031-9155/45/5/201

[188] A. Jordan, R. Scholz, K. Maier-Hauff, F.K. van Landeghem, N. Waldoefner, U. Teichgraeber, J. Pinkernelle, H. Bruhn, F. Neumann, B. Thiesen, The effect of thermotherapy using magnetic nanoparticles on rat malignant glioma, Journal of neuro-oncology, 78 (2006) 7-14. https://doi.org/10.1007/s11060-005-9059-z

[189] R. Gordon, J. Hines, D. Gordon, Intracellular hyperthermia a biophysical approach to cancer treatment via intracellular temperature and biophysical alterations, Medical hypotheses, 5 (1979) 83-102. https://doi.org/10.1016/0306-9877(79)90063-X

[190] N. Hamilton, The small-signal frequency response of ferrites, High frequency electronics, (2011) 36-52.

[191] T.-J. Liang, H.-H. Nien, J.-F. Chen, Investigating the characteristics of cobalt-substituted MnZn ferrites by equivalent electrical elements, IEEE transactions on magnetics, 43 (2007) 3816-3820. https://doi.org/10.1109/TMAG.2007.903383

[192] R. Sharma, P. Thakur, P. Sharma, V. Sharma, Ferrimagnetic Ni^{2+} doped Mg-Zn spinel ferrite nanoparticles for high density information storage, Journal of Alloys and Compounds, 704 (2017) 7-17. https://doi.org/10.1016/j.jallcom.2017.02.021

[193] M. Pardavi-Horvath, Microwave applications of soft ferrites, Journal of Magnetism and Magnetic Materials, 215 (2000) 171-183. https://doi.org/10.1016/S0304-8853(00)00106-2

[194] O. Mounkachi, R. Lamouri, B. Abraime, H. Ez-Zahraouy, A. El Kenz, M. Hamedoun, A. Benyoussef, Exploring the magnetic and structural properties of Nd-doped Cobalt nano-ferrite for permanent magnet applications, Ceramics International, 43 (2017) 14401-14404. https://doi.org/10.1016/j.ceramint.2017.07.209

[195] M.T. Yagub, T.K. Sen, S. Afroze, H.M. Ang, Dye and its removal from aqueous solution by adsorption: a review, Advances in colloid and interface science, 209 (2014) 172-184. https://doi.org/10.1016/j.cis.2014.04.002

[196] G. Sharma, A. Kumar, S. Sharma, M. Naushad, P. Dhiman, D.-V.N. Vo, F.J. Stadler, $Fe_3O_4/ZnO/Si_3N_4$ nanocomposite based photocatalyst for the degradation of dyes from aqueous solution, Materials Letters, 278 (2020) 128359. https://doi.org/10.1016/j.matlet.2020.128359

[197] A. Kumar, G. Sharma, M. Naushad, A.a.H. Al-Muhtaseb, A. García-Peñas, G.T. Mola, C. Si, F.J. Stadler, Bio-inspired and biomaterials-based hybrid photocatalysts for

environmental detoxification: A review, Chemical Engineering Journal, 382 (2020) 122937. https://doi.org/10.1016/j.cej.2019.122937

[198] A. Kumar, S.K. Sharma, G. Sharma, C. Guo, D.-V.N. Vo, J. Iqbal, M. Naushad, F.J. Stadler, Silicate glass matrix@$Cu_2O/Cu_2V_2O_7$ p-n heterojunction for enhanced visible light photo-degradation of sulfamethoxazole: High charge separation and interfacial transfer, Journal of Hazardous Materials, 402 (2021) 123790. https://doi.org/10.1016/j.jhazmat.2020.123790

[199] E. Casbeer, V.K. Sharma, X.-Z. Li, Synthesis and photocatalytic activity of ferrites under visible light: a review, Separation and Purification Technology, 87 (2012) 1-14. https://doi.org/10.1016/j.seppur.2011.11.034

[200] P. Dhiman, A. Kumar, M. Shekh, G. Sharma, G. Rana, D.-V.N. Vo, N. AlMasoud, M. Naushad, Z.A. Alothman, Robust magnetic ZnO-Fe_2O_3 Z-scheme hetereojunctions with in-built metal-redox for high performance photo-degradation of sulfamethoxazole and electrochemical dopamine detection, Environmental Research, 197 (2021) 111074. https://doi.org/10.1016/j.envres.2021.111074

[201] A. Kumar, G. Sharma, M. Naushad, T. Ahamad, R.C. Veses, F.J. Stadler, Highly visible active Ag_2CrO_4/Ag/$BiFeO_3$@RGO nano-junction for photoreduction of CO_2 and photocatalytic removal of ciprofloxacin and bromate ions: The triggering effect of Ag and RGO, Chemical Engineering Journal, 370 (2019) 148-165. https://doi.org/10.1016/j.cej.2019.03.196

[202] K. Harish, H.B. Naik, R. Viswanath, Synthesis, enhanced optical and photocatalytic study of Cd–Zn ferrites under sunlight, Catalysis Science & Technology, 2 (2012) 1033-1039. https://doi.org/10.1039/c2cy00503d

[203] A. Di Paola, E. García-López, G. Marcì, L. Palmisano, A survey of photocatalytic materials for environmental remediation, Journal of hazardous materials, 211 (2012) 3-29. https://doi.org/10.1016/j.jhazmat.2011.11.050

[204] N.M. Mahmoodi, Zinc ferrite nanoparticle as a magnetic catalyst: synthesis and dye degradation, Materials Research Bulletin, 48 (2013) 4255-4260. https://doi.org/10.1016/j.materresbull.2013.06.070

[205] X. Guo, H. Zhu, Q. Li, Visible-light-driven photocatalytic properties of ZnO/$ZnFe_2O_4$ core/shell nanocable arrays, Applied Catalysis B: Environmental, 160 (2014) 408-414. https://doi.org/10.1016/j.apcatb.2014.05.047

[206] P. Xiong, J. Zhu, X. Wang, Cadmium sulfide–ferrite nanocomposite as a magnetically recyclable photocatalyst with enhanced visible-light-driven

Materials Research Foundations **112** (2021) 62-120 https://doi.org/10.21741/9781644901595-2

photocatalytic activity and photostability, Industrial & Engineering Chemistry Research, 52 (2013) 17126-17133. https://doi.org/10.1021/ie402437k

[207] P. Dhiman, N. Dhiman, A. Kumar, G. Sharma, M. Naushad, A.A. Ghfar, Solar active nano-$Zn_{1-x}Mg_xFe_2O_4$ as a magnetically separable sustainable photocatalyst for degradation of sulfadiazine antibiotic, Journal of Molecular Liquids, 294 (2019) 111574. https://doi.org/10.1016/j.molliq.2019.111574

Ferrite - Nanostructures with Tunable Properties and Diverse Applications Materials Research Forum LLC
Materials Research Foundations **112** (2021) 121-161 https://doi.org/10.21741/9781644901595-3

Chapter 3

Recent Advancement on Ferrite Based Heterojunction for Photocatalytic Degradation of Organic Pollutants: A Review

Pooja Shandilya[1]*, Shabnam Sambyal[1], Rohit Sharma[1], Amit Kumar[2,3], Dai-Viet N. Vo[4]

[1]School of Advanced Chemical Sciences, Shoolini University, Solan, Himachal Pradesh, India-173229

[2]International Research Centre of Nanotechnology for Himalayan Sustainability (IRCNHS), Shoolini University, India

[3]College of Materials Science and Engineering, Shenzhen Key Laboratory of Polymer Science and Technology, Guangdong Research Center for Interfacial Engineering of Functional Materials, Nanshan District Key Lab for Biopolymers and Safety Evaluation, Shenzhen University, Shenzhen 518055, PR China

[4]Center of Excellence for Green Energy and Environmental Nanomaterials (CE@GrEEN), Nguyen Tat Thanh University, 300A Nguyen Tat Thanh, District 4, Ho Chi Minh City, 755414, Vietnam

*pooja@shooliniuniversity.com, poojashandil03@gmail.com

Abstract

The progress of ferrites and ferrites-based nanocomposites has become extensively popular in the field of photocatalytic wastewater treatment. This class of compound exhibit several fascinating properties related with their high stability, low cost, ease of functionalization, and biocompatibility. Ferrites carry outstanding magnetic behavior that helps in their easy recovery from the aqueous system thus reducing the cost. The morphology and various properties such as magnetic, absorption, optoelectronic of magnetic ferrites can be varied and optimize by applying different synthetic routes and reaction conditions. With this background we have briefly presented and reviewed the latest development in the field of photodegradation of aqueous pollutants using ferrites based heterojunction. Especially, the type-II, Z-scheme and S-scheme based heterojunction for enhanced pollutant degradation under the exposure of light are thoroughly describe. Ferrites have inherent potential in water remediation applications hence many examples were consider to impart valuable knowledge to the readers.

Nevertheless, the large-scale utilization of these magnetic nanoparticles still needs to be explored. Therefore, the gaps, challenges and future prospective of ferrites nanoparticles are also explained to unveil the un-scrutinized standard of ferrites nanoparticles.

Keywords

Ferrite Photocatalyst, Degradation, Heterojunction, Organic Pollutant, Z-Scheme, S-Scheme

Contents

Materials Research Foundations **112** (2021) 121-161 https://doi.org/10.21741/9781644901595-3

1. Introduction

Rapid population sprouting and industrialization have sparked serious impact on environment, human health and aquatic life. Various industries like cosmetic, pharmaceutical, rubber, paint, food, battery, textile mill and domestic waste dispensing harmful effluents directly into the ecosystem [1, 2]. These effluents carry innumerable toxic organic and inorganic substances, consuming accessible dissolved oxygen and deteriorate the quality of water. Therefore, it is an urgent need to develop sustainable and effective method to improve the water standard. Among different conventional (coagulation, precipitation, filtration, biodegradation), established (evaporation, oxidation, membrane separation, solvent extraction) and emerging (advanced oxidation, bio sorption, nanofiltration) technologies, heterogeneous photocatalysis based on advanced oxidation process is most successful due to its environment friendliness, cost effectiveness and less energy consumption [3-5]. The photocatalyst generally utilize the renewable solar energy to initiate and amplify the photochemical reactions. This process produces charge carrier which are responsible for redox reaction on photocatalytic surface. During the photocatalysis, the photogenerated h^+ react with water molecules to generate $\cdot OH$ radical possessing strong oxidative potential ($E^o = +2.34$ V vs NHE) sufficient for the degradation various kind of aqueous pollutants (Figure 1 a) [6-9].

As a photocatalyst, ferrite based material has drawn great interest and promptly became one of the hottest topics among scientist witnessed by the increase in number of publications per year from 2012 to 2021, Figure 1b. The ferrites are generally combination of most abundant element iron and oxygen. The most commonly used ferrites are Fe (II) oxides, Fe (III) oxides or Fe (II, III) oxides which present in different phases and crystal structures. For example, Fe (II, III) exhibits cubic structure whereas Fe (III) shows polymorphism mainly in four different phases which are α-Fe_2O_3, β-Fe_2O_3, γ-Fe_2O_3 and ε-Fe_2O_3. The ferrites can be fabricated through various synthetic method with distinct morphologies like nanoparticles, nanotubes, nano-spheres, nanowires, nanosheets etc. The typical ferrites generally show several advantageous features like superparamagnetic, high biocompatibility, ease of functionalization, optoelectronic tunability and high stability [10-11]. These outstanding properties of magnetic nano-ferrites paved special attention in environmental remediation via photodegradation. Liu et al. fabricated α-Fe_2O_3 via SDS-assisted grinding method showing Fenton like mechanism in the existence of H_2O_2 under visible irradiation and degraded 99.6% methylene blue (MB) in 40 minutes [12, 13]. Under visible irradiation reactive oxidation species (ROS) is generated responsible for photodegradation. The photoinduced electrons were captured by Fe^{3+} ions reducing it to Fe^{2+} ions to form Fenton reagent Fe^{2+}/H_2O_2 that participate in photodegradation process.

For instance, Dhiman and co-workers fabricated $ZnFe_2O_4$ nanostructures via hydrothermal routes and studied for photodegradation of both cationic (safranine-O) in addition anionic remazol intense yellow) dye [14]. They controlled reaction conditions to prepared different morphologies like porous nanorods, nanoparticles, nano-flower and hollow microsphere of $ZnFe_2O_4$ and studied morphology based photocatalytic efficiency. The photocatalytic efficiency followed the order hollow microsphere (115.23 m^2g^{-1}) < nano-flower (92.73 m^2g^{-1}) < nanoparticles (121.74 m^2g^{-1}) < porous nanorods (138.43 m^2g^{-1}). These findings suggested that porous nanorods revealed maximum photodegradation activity toward both cationic and anionic dyes due to better distribution of charged species, high surface area and porous nature [15, 16]. Despite of different morphologies, ferrites nanoparticles also exhibit superparamagnetic, ferromagnetic or ferromagnetic characteristics. Among which superparamagnetic behaviour of ferrites are best for water purification on the note magnetic nanoparticles should possess high magnetization saturation and low value of coercivity and remanence magnetization. This is because superparamagnetic nanoparticles do not aggregate and can be dispersed freely in liquid medium even in absence of magnetic field. The magnetic behaviour shut off the use of tedious and lengthy process of centrifugation and filtration.

In photocatalysis, ferrites offer the additional advantages of high visible light absorption due to their narrow band gap of ~2.o eV (Figure 1c) [17-30]. The narrow band-gap semiconductors suffer high recombination rate thus, heterojunction were constructed to facilitate the direction of charge migration. The p-n junction between $CaFe_2O_4$ and WO_3 nanocomposite for photocatalytic decomposition of acetaldehyde under visible irradiation were reported [31]. The low recombination rate of electron/hole pairs and high photocatalytic performance was detected due to the presence of internal electric field and bend bending induced at the interface of two different semiconductor materials. The presence of defect sites also helps in charge separation by trapping the electrons and allowing large number of potential charge species available for photo-oxidation and photoreduction of pollutants [32-35].

In this manuscript we have tried to present a comprehensive knowledge on the properties, synthetic techniques and photodegradation ability of ferrite base nanocomposite. We start with the introductory part elaborating the need of environment remediation technique, basic principle and role of photocatalysis in degradation and advantageous features of ferrites based nanomaterials. Then we have discussed the important properties of ferrites like magnetic, absorption, optoelectronic, targeted delivery and hyperthermia [36, 37]. Thereafter, various synthetic approach: hydrothermal, sol-gel citrate, chemical precipitation, conventional and other methods employed for the fabrication of nano-ferrites with various morphologies and dimensionality were considered. Finally, we have discussed the usefulness of magnetic separable nanomaterials in photocatalytic degradation of organic pollutants for wastewater treatment. We have briefly lit up the various heterojunction type-II, Z-scheme and S-scheme based photocatalyst validating high photo-efficiency. In conclusion the various gaps, challenges and future prospective of ferrites based heterojunction were explicate.

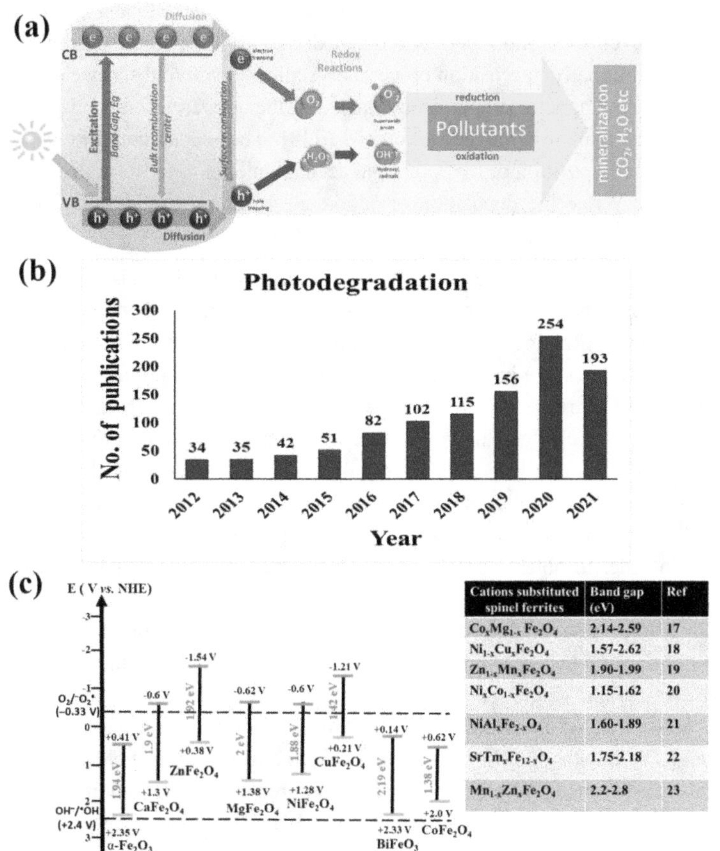

Cations substituted spinel ferrites	Band gap (eV)	Ref
$Co_xMg_{1-x}Fe_2O_4$	2.14-2.59	17
$Ni_{1-x}Cu_xFe_2O_4$	1.57-2.62	18
$Zn_{1-x}Mn_xFe_2O_4$	1.90-1.99	19
$Ni_xCo_{1-x}Fe_2O_4$	1.15-1.62	20
$NiAl_xFe_{2-x}O_4$	1.60-1.89	21
$SrTm_xFe_{12-x}O_4$	1.75-2.18	22
$Mn_{1-x}Zn_xFe_2O_4$	2.2-2.8	23

Figure 1: (a) General mechanism of photodegradation [6] (b) Reported publication of ferrites in last 10 years source of information https://www.sciencedirect.com by using keywords "Ferrites in photodegradation" (c) Band gap values of spinel ferrites and cations substituting ferrites [17-30].

2. Properties

2.1 Magnetic property

Strong magnetic behavior of ferrite is one of the most important functional characteristics responsible for easy, economical and selective separation. The magnetic separation is superior to other conventional techniques like centrifugation and filtration. The magnetism in ferrites is affected by several factors like oxidation state and distribution of cations, electronic configuration between octahedral and tetrahedral sites within the crystal lattice. Additionally, fabrication process, crystallinity, size and surface or interface effect also influence magnetic properties by changing the saturation magnetization (M_s), coercivity (H_c) and remanent magnetization (M_r) [38]. The use of complex fabrication method thermal decomposition or hydrothermal accompanied non-uniform distribution of cations in ferrites. While the deposition or solid-state reaction provides high inversion degree to nanoparticles in ferrites like their bulk counterpart [39].

In 2017, Li and co-workers gave the correlation between particle size and magnetic performance of Fe_3O_4 nanoparticles [40]. It was found with increase in particle size the M_s increase without affecting shape or crystallinity of particles. Whereas, the other parameters like M_r and H_c are influenced by particle size and reached maximum value of 13 emu/g and 190 Oe, respectively when particle size reached to 76 nm. The further increase in particle size changes the cubic structure from single domain to multi-domain and also reduces the M_r and H_c value. Eventually, H_c and M_r calculated for sphere like Fe_2O_3 nanoparticles was smaller than cube like nanoparticles. These findings clearly demonstrated that special attention needs to be given in crystalline properties and size of nanoparticles in order to obtain satisfactory magnetic performance. Although, ferrites in bulk-phase remains in multi-domains structure but when size decreased to 15-100 nm, ferrites nanoparticles possessed single domain structure with superparamagnetic behavior where the atomic spin aligned in same direction. In contrary to other conventional paramagnetic materials single domain Fe_2O_3 do not possess any residual magnetic moment in the absence of an external applied field. This unique feature guarantee ferrite nanoparticles to be outstanding in various biomedical and industrial fields. Besides this magnetic anisotropy related with chemical composition, cation distribution or symmetry of interstitial positions is also contributed to effectively enhance or tailored magnetic characteristics of ferrites.

2.2 Absorption property

Absorption is one of the highly potential physical treatment method to eliminate various organic and inorganic pollutant from waste water. This technique involves the attachment of organic or inorganic contaminant on the absorbent surface in order to carry out the mineralization process. However, absorption capacity of ferrites is much less as compare to other carbon-based adsorbent like graphene or activated carbon. However, by controlling the synthetic approach, morphology, size, chemical composition of ferrites, the sorption kinetics and absorption efficiency of ferrites can be modified. Generally, the absorption mechanism of ferrites nanoparticles depends upon several factors including electrostatic interaction, π-π interaction, hydrogen bonding, ion exchange, chemisorption, surface complexation and formation of new bonds. The surface of ferrites possesses large number of hydroxyl groups (Fe-OH) which are considered as crucial binding sites for various pollutants. Thus, by altering the density of these hydroxyl binding sites and their availability for adsorbate through various techniques, the absorption capacity found to be increase exceptionally [38, 41]. For example, Ahalya et al. doped manganese ferrite ($MnFe_2O_4$) with different concentration of cobalt ions ($Co_xMn_{1-x}Fe_2O_4$) [42]. The highest absorption capacity was observed for the absorption of Cr (VI) at x=0.6 due to increase in size and crystalline nature of nanoparticles. Further increase in dopant concentration, form more $CoFe_2O_4$ phases which have less absorption capacity than $MnFe_2O_4$.

Also, Ni-Zn ferrites fabricated by sol-gel method were used for efficient removal of Congo red dye at room temperature [43]. The absorption mechanism of doped nano-ferrites could be explained by ion exchange process. When the ferrites are doped with different metals ions, it affected the crystallinity, size, inter planner spacing and concentration of free electrons in crystal lattice due to difference in size of doped metal and parent metal cations. This modifies the absorption capacity of doped nano-ferrites. Interestingly, free electrons present in crystal lattice decrease the oxidation state of pollutant, and thereby successively increase the magnitude of absorption (Figure 2a). Apart from this polymer or surfactant coating are also well known strategies to enhance absorption ability of ferrite nanoparticles beneficial in pollutant degradation [44, 45]. During the surface coating single or double-layer formation occur on ferrite surface by long chain compound containing different functional groups. These reactive function group represents active sites for adhesion of various pollutant on surface of nano-ferrites. Beside, these functional groups in polymer matrix also act as potential binding sites for the adsorptional removal of pollutants (Figure 2b).

2.3 Optoelectronic property

Optoelectronic property is a basic characteristic of semiconducting materials based on quantum mechanical effects of light on the electronic system. Being a promising semiconductor, ferrites have achieved special attention in photocatalysis and sensor applications due to their optoelectronic tunability, high stability, high charge density, highly abundant and cost-effectiveness. The band-gap value of spinel ferrites nanoparticles approximately lies between 1.4 to 2.7 eV depending upon the type of cations present in the nanostructure. This feature of ferrites is highly promising in the field of photocatalysis. On irradiation, the photogenerated charge carrier directly participate in generation of superoxide and hydroxyl radical species. These reactive species owns high redox potential and therefore, easily degrade large number of pollutants. Despite of several advantageous optoelectronic feature, ferrites bear high recombination rate and low specific surface area. Therefore, different modification strategies like heterojunction formation, morphological modification, doping have been explored to increase density of active sites and reduce the diffusion length. These modifications increase the separation of charge carriers and increase their accessibility for surface reactions [46, 47].

Recently, Kang and groups constructed Fe_2O_3/C-g-C_3N_4 based Z-scheme photocatalyst for degradation of rhodamine B [48]. It was observed through various studies that electron-hole recombination was remarkably suppressed after heterojunction formation. In Fe_2O_3/C-g-C_3N_4 photocatalyst, the VB holes of Fe_2O_3 transport to VB of g-C_3N_4. Whereas, photogenerated electrons migrated from CB of Fe_2O_3 to CB of g-C_3N_4. This causes recombination of useless charge carrier while photoexcited electrons and holes with high redox potential are available for photodegradation process. Similarly, Huo et al. synthesized Z-scheme based α-Fe_2O_3/MIL-101 (Cr) hybrid for complete removal of carbamazepine in the presence of visible illumination [49]. The photogenerated carriers efficiently separated by Z-scheme pathway minimizing recombination rate and maximized photodegradation performance. These findings suggested that the optoelectronic properties of ferrites could be improved by various technique which enhance their usability in photocatalysis.

2.4 Targeted delivery properties

The target specific drug delivery is a novel and beneficial approach in the field of nano-medicine. These properties were established on the magnetic behavior of substances, therefore owing to their superparamagnetic characteristic's ferrites are considered as potential candidate for target specific drug delivery application. The direction and orientation of these substances could be controlled by applying an external magnetic field. The other specific feature of ferrites like high chemical stability, large number of active sites tunable size and shape and choices in functionalization make them highly suitable for targeted delivery [50]. For instance, Maiti et al. fabricated surface functionalized multifunctional $ZnFe_2O_4$ as a drug carrier for hydrophilic and hydrophobic anti-cancer drug molecules namely Daunorubicin and Curcumin, respectively [51]. It was

Materials Research Foundations **112** (2021) 121-161 https://doi.org/10.21741/9781644901595-3

observed that $ZnFe_2O_4$ proved excellent carrier for both kind of drugs and displayed high efficacy for targeting and damaging harmful cancer cells (HeLa).

In general, the release of drugs from their carries at specific sites depending upon various external stimulus (magnetic field, temperature, presence of light and ultrasound) and internal stimulus (pH or presence of enzyme). Thus, target specific drug delivery by magnetic nanoparticles is becoming beneficial in several ways as it improves pharmacokinetic and pharmacodynamics profile of drugs, intensify drug stability and aqueous solubility and reduces toxic effect and drug concentration in non-targeted tissues. Also, the biocompatibility, biodegradability and non-toxicity along with easy navigation via applied magnetic field, the magnetic nanoparticles have explored in various applications like magnetic resonance imaging, hyperthermia and magnetofection [52].

2.5 Hyperthermia properties

Generally, hyperthermia signifies the overheating conditions, when temperature of the system is raised above normal temperature. The ferrites nanoparticles possessed magnetic hyperthermia properties in which the electromagnetic energy is converted into heat energy via energy dissipation under the effect of alternating current magnetic field. The two types of relaxation process contribute in the energy dissipation of magnetic nanoparticles called Brown relaxation and Neel relaxation. Hyperthermia is a promising approach for the treatment of cancerous cells where specific tissue or organ for tumor therapy is heated to 41-46 °C with the help of laser wavelength, microwave or radio frequency. Since, the tumor cells show low thermal resistance due to poor development of blood vessels as compared to healthy tissue which mark hyperthermia a very successful approach for curing cancer. However, during the whole body and regional hyperthermia few clinical side effects were also observed by healthy tissues. Therefore, ferrites can be used to resolve this issue as it provides localized hyperthermia which only heat up the tumor zone. The magnetic ferrites nanoparticles not only demonstrate localized hyperthermia effect but also supplies control over the release kinetic of drugs and consequently proved fascinating compound in cancer treatment [53, 54]. For instance, Ebrahimisadr et al. examined the effect of concentration of Fe_2O_4 nanoparticles on the hyperthermia property and observe the maximum temperature reaches to 58 °C when concentration of nanoparticles is 12.5%. At the same time when concentration decreases to 1% in suspension, hyperthermia effect also reduces and temperature raised to 9 °C only (Figure. 2 c). These finding concludes that heat generation effect increased with increase in concentration. Whereas no significant effects of change in concentration were observed on specific absorption rate value. This result displayed that magnetic energy dissipation or specific absorption rate of ferrofluid are independent of concentration (Figure 2 d).

3. Synthesis

The ferrite nanocomposite fabricated by different synthetic methods regulate the size, shape, aspect ratio orientation, surface area, stability and purity. The diverse morphologies, surface area and porosity obtained through different synthetic routes: hydrothermal, precipitation, sol-gel, thermal decomposition and microwave assisted method are beneficial in various photocatalytic applications. Hence, proper synthetic techniques must be applied to deliver desired morphology with high yield. Also, cost effective, mild reaction conditions and most importantly green synthesis of ferrites should be the requisite.

3.1 Hydrothermal method

Hydrothermal process operated under high temperature and pressure is environmental-friendly method for large production of ferrites nanomaterials [55]. The various factors like precursor, pH, temperature, reactant concentration and additives are applied for dispersion and crystal morphology. Wu and groups synthesized g-C_3N_4/$CuFe_2O_3$ with Fe-Ni by hydrothermal method to remove arsenate from water [56]. The pH and temperature show great impact on shape and size of nanostructure. Surface morphology of nanocomposite examined by SEM in which Fe-Ni foam substance possess smooth surface (Figure 3a). The Cu present in the form of $Cu_2Fe_2O_4$, mass of $Cu_2Fe_2O_4$ on Fe-Ni depends upon stoichiometric relationship between $Cu_2Fe_2O_4$ and Cu. Therefore, nanocomposite substance shows higher activity which attributed to efficient charge separation. In another study $BiFe_2O_3$ were fabricated by using $Bi(NO_3)_3.5H_2O$ and $Fe(NO_3)_3.9H_2O$ precursor by adding KOH solution that assist in co-precipitation of Bi^{3+} and Fe^{3+} ions [57]. While, increase concentration of KOH converts the irregular structure (2.7μm) to brick like (2.6μm) of $BiFe_2O_3$. Also, particle size has a great impact on photocatalytic activity, smaller nanoparticles exhibit higher efficiency against methyl orange. $NiFe_2O_4$/$BiPO_4$ nanocomposite degraded 98% of tetracycline and 99% of rhodamine-B [58, 59]. The obtained heterostructure attain tiny particle size, pore volume and large specific surface area. The large surface area accountable by huge active sites determined by BET isotherm. From Figure 3b shows pore size distribution of $BiPO_4$ nanorods on $NiFe_2O_4$ nanoplates, the BET area increases with $BiPO_4$ substance and exhibiting high surface area $24.1085 m^2/g$ in $NiFe_2O_4$/$BiPO_4$. From the discussion, we can conclude that hydrothermal method is a suitable for optimizing the particle size of ferrites nanoparticle in such a way so as to accomplish higher photocatalytic efficiency.

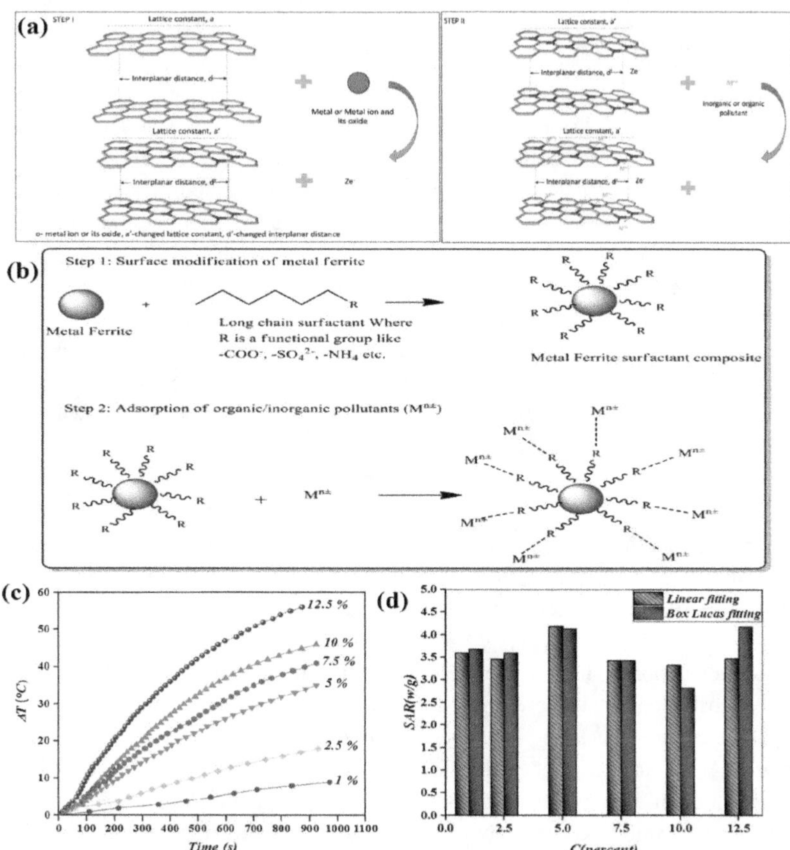

Figure 2: (a) General adsorption mechanism of metal and their oxide doped nano ferrite (b) General adsorption mechanism for surfactant coated Nano ferrites [41] (c) Temperature rise of Fe_3O_4 suspension with different concentrations in an AC magnetic field (d) SAR as a function of concentration [54].

3.2 Sol-gel method

Sol-gel process comprise with addition of metal precursor with citric acid to form gel. Precursor dissolved in ethanol and water at pH 9 until gel solution is formed afterward the gel dried at 450-800°C. Golshan et al., prepared $TiO_2@CuFe_2O_4$ by sol-gel method and determine the crystal phase structure of photocatalyst by XRD [60]. Different diffraction pattern of crystal structure with variety of crystal lattice were shown in Figure 3c. Different ferrite have dissimilar diffraction pattern due to metal cation of varied size giving different spectrum. The similar pattern of diffraction peak of TiO_2 and $CuFe_2O_4$ shows high purity whereas, strong and narrow diffraction peak of $TiO_2@CuFe_2O_4$ signifies excellent crystalline degree. Mg doped Co-Fe_2O_3 nanocomposite synthesized by auto-combustion process whose magnetic hysteresis curve analyzed at room temperature [61]. Magnetic parameter for instance coercivity, magnetization and remanence were examined by increasing Mg concentration in Co-Fe_2O_4. The saturation magnetization for Co and Mg ferrite were 77.94emu/g and 35.22emu/g. As mentioned, Neel model magnetization of tetrahedral sub-lattice was anti-parallel to octahedral. Although, Mg doped ferrites, non-magnetic Co^{2+} and Mg^{2+} ion to increase magnetic moment and Mg^{2+} ion having octahedral sites (Figure 3d). Many researchers are curious on knowing the influence of temperature and calcination time on photocatalytic mechanism.

3.3 Chemical-precipitation method

Chemical co-precipitation is used to fabricate ferrites by same mixing metal in aqueous solution with the addition of surfactant followed by thermal treatment. The pH is maintained at 7-10 so as to precipitate ferrite nanoparticle and then calcined at high temperature. Ternary $Bi_{12}O_{17}Cl_2/Ag/AgFeO_2$ nanocomposite fabricated by in-situ deposition-precipitation method in the presence of ethanol reducing Ag^+ to Ag^0 (Figure 3e). The heterostructure generate ROS to carry out the degradation of tetracycline [62]. X-ray photoelectron spectrum examined chemical state and surface elemental composition. Additionally, weight percentage of each species analyzed on the basis of atomic concentration of elements. Besides, heterostructure having Bi, Cl show negative shift and Ag, Fe exhibit positive movement in XPS. The strong interaction between $AgFeO_2$ and $Bi_{12}O_{17}Cl_2$ assist in electron migration from $AgFeO_2$ to $Bi_{12}O_{17}Cl_2$ surface. Further, Ni doped $CoFe_2O_3$ nanocomposite was fabricated by chemical co-precipitation method applied for the degradation of organic dye [63]. The scanning electron microscopy (SEM) studied morphology of surface and surface coated microstructure. SEM shows irregular morphology due to distorted crystalline structure and having porous nature (Figure 3f). The pH exhibit important role on crystalline structure with little effect on composition of substance. As nickel ions added in cobalt ferrite, it alters the lattice parameter.

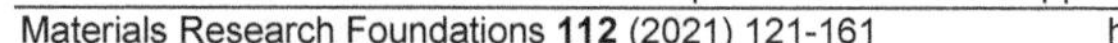
Materials Research Foundations **112** (2021) 121-161 https://doi.org/10.21741/9781644901595-3

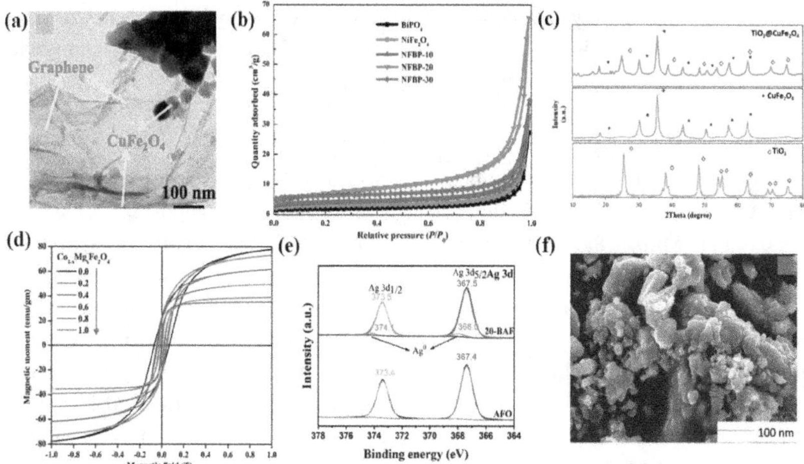

Figure 3: (a) TEM images of $CuFe_2O_3$/g-C_3N_4 [56](b) pore-size determination with various sample of $BiPO_4$ and $NiFe_2O_3$[58](c) XRD spectra of $CuFe_2O_3$, $NiFe_2O_3$@TiO_2 and TiO_2 [61](d) hysterias loop of $Co_{1-x}MgFe_2O_4$ [61](e) XPS spectra of $AgFeO_2$[62] (f) SEM images of Ni-$CoFe_2O_3$[63].

3.4 Conventional solid-state method

In solid state technique powders of metal salt and iron for instance Fe_2O_3, $CaFe_2O_4$, and $CaCO_3$ are heated up to 1100°C. Spinel ferrites MFe_2O_4 (M = Zn, Cd and Mg) along with Fe^{3+} in d^5 electron configuration can be synthesized through solid state mechanism [64]. For spinel compound crystallographic cationic sites are tetrahedral and octahedral in nature. The photodegradation of acetic acid liberate enormous amount of CO_2 under xenon lamp (300W). Before the reaction pH of aqueous solution was at 2.4 and remain constant throughout the reaction. Further, Cd doped Zn ferrite ($Co_{0.5}ZnO_{0.5}Cd_{0.2}Fe_{1.8}O_4$) synthesized by providing different calcination temperature [65]. Nanocomposite were fabricated by low-cost solid state reaction process and calcinated at high temperature (Figure 4a). While increasing temperature up to 1000°C lattice cell dimension increases by decreasing micro-strain. The cubic crystalline structure of nanoparticle identifies by Williamson and Hall size and strain broadening. Though, broadening of peak with additive component possess distinct θ value depending on the size and strain. From size strain the isotropic and micro contribution can be determined whereas, value of constant

Materials Research Foundations **112** (2021) 121-161 https://doi.org/10.21741/9781644901595-3

(K) depend upon shape of crystal. However, this method has many benefits such as good crystallinity, simple synthesis still bear certain limitation of large particle size along with the formation of additional undesirable phase.

3.5 Other method

Various other methods such as combustion, electro-spinning and poly-condensation method have also been describe for the synthesis of ferrites nanocomposite. The simple, fine homogenous Mn-Zn ferrites nanoparticle co-doped with Sm^{3+}- Gd^{3+} ion synthesize by combustion method possess effective magnetic, structural, and electrical properties [66]. Addition of co-doped ion increases crystalline structure, saturation magnetization and tunable electrical property. Based on the fact the production of porous substance suitable for catalyst-based or thermal insulation application. Electro-spinning method was used to prepare $CoFe_2O_4$, with metal nitrate precursor, polymeric carrier and PVP [67]. The magnetic property of magnetic fibers are related to substitution ion and thermal temperature. In addition, three-dimensional ferrite prepared by one step ultrasonic spray pyrolysis by using sulfonated polystyrene as sacrificial reagent [68]. Acetone plays important role by increasing selectivity and stability of structure with active sites. The N_2 adsorption-desorption isotherm exhibiting relative pressure (P/Po) (0.2-1.0) shows porous like structure. Mesoporous formation with pore size less than 20 nm ascribed to particle packing and outgassing due to decomposition of sphere templates (Figure 4b). It is known that refractory substance prepared by using self-propagating high temperature combustion process. Zhao et al., synthesized $NiFe_2O_4/Ag_3PO_4$ nanoparticle fabricated by ion-exchange technique by adding amount of Ag-ammonia and $NiFe_2O_4$ solution [69]. In the synthesis, nanoparticle was withdrawn at regular interval and filtered through membrane to remove the photocatalyst. It comprises of agglomerated grain with irregular, spherical structure with diameter 2.5 mm and $NiFe_2O_4$ diameter 2 mm (Figure 4c). Correspondingly, g-$C_3N_4/NiFe_2O_3$ fabricated through poly-condensation of urea at temperature up to 550 °C [70]. From Figure 4d EDX spectra indicate the presence of Ni, Fe C, N and O comprise of g-$C_3N_4/NiFe_2O_3$.

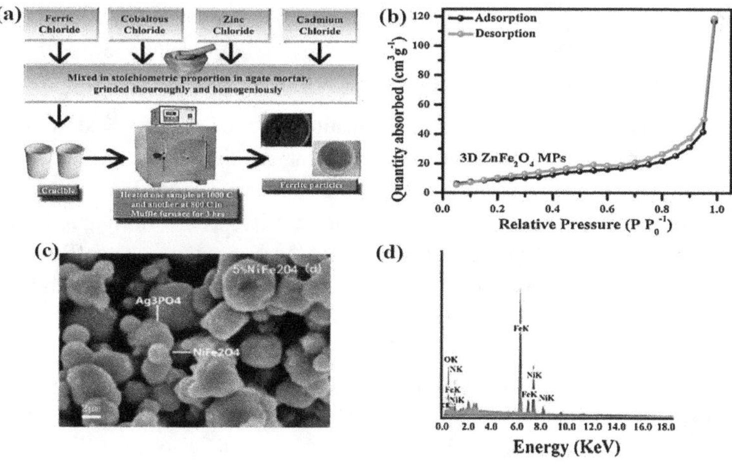

Figure 4: (a) Synthetic method of ferrite particles [65] (b) Pore size determination of 3D $ZnFe_2O_3$ MPs [68] (c) SEM images of $Ag_3PO_4/NiFe_2O_4$ [69] (d) EDX spectra of GCN/$NiFe_2O_4$ [70].

4. Application in photodegradation

4.1 Bare and doped ferrite

Spinel ferrite nanocomposite $NiFe_2O_4$, $ZnFe_2O_4$, $SnFe_2O_4$ and $CoFe_2O_4$ have shown strong visible light activity, magnetic recyclability, non-toxicity, low cost, and environment friendly [71,72]. Among metal oxide semiconductor, Fe_2O_3 considered as stable semiconductor for photocatalytic reaction due to low band-gap (2.3eV), recyclability, earth abundant high harvesting light and excellent stability. In photocatalytic mechanism of Zn ferrites-based heterojunction, photo-illuminated electron is migrated to CB reducing Fe^{3+} to Fe^{2+} also generating $\cdot SO_4^{2-}$ on CB and ·OH radicals at VB (Figure 5a) [71]. As pH is 6 total organic carbon and chemical oxygen demand is 50.5% and 78.6% in 300 min. Zn doped in cobalt ferrite material studied by diffraction peak shift towards lower range [73]. The absorption peak at 232 nm is ascribed to Fe^{3+} and O charge migration in octahedral coordination (Figure 5 b). Absorbance band at 300 nm arises due to the charge migration between oxygen and cobalt in tetrahedral sites and band at 477 and 627nm indicates Co^{2+} and Fe^{3+} ion in tetrahedral and octahedral geometry. In cobalt doped $ZnFe_2O_3$, valance band edge present at 0.38eV and CB edge at

2.43eV but in Co doped Fe_2O_3 CB edge was less negative than that of redox potential of O_2 (-0.33 V). Therefore, electron in CB do not generate $^{\cdot}O_2^-$ while, electron in CB of cobalt doped $ZnFe_2O_3$ react with H_2O_2 to generate $^{\cdot}OH$ radical (Figure 5c).

The doping in ferrites supress the recombination rate of charges and lowers the luminescence intensity. Despite of huge accomplishment ferrites as single component does not have acceptable performance. This is due to rapid recombination of charge carrier and small hole dispersion distance of around 2-4 nm. The various modification is required to enhance surface area, number of active sites, promote photoefficiency and reduce recombination. Different support material were utilized to magnify adsorptional ability by increasing surface area and to improve thermal and chemical stability. The various support material such as SiO_2, g-C_3N_4, carbon nanotubes, graphene, graphene aerogel, carbon quantum dots, metal organic framework were used. These support material may assist in reducing the agglomeration of nanoparticles and behave as electron scavenger thereby minimizing recombination process. It also provide additional adsorption site for pollutant and improves the absorption range of photocatalyst apart, good support material also prevent the formation of inappropriate heterojunction. Moreover, heterojunction construction for outstanding performance can be an effective approach to enhance photocatalytic efficiency and discussed in next segment.

4.2 Type-II based ferrite heterojunction

Heterojunction construction appears to be an effective strategy to simultaneously utilize high redox potential and broad absorption range which is merely not possible in either bare or doped photocatalysts. To date, numerous type-II based ferrite heterojunction were constructed to enhance photocatalytic activity. The ferromagnetic $NiFe_2O_4$ with antiparallel spin between Ni^{2+} and Fe^{3+} ions present at octahedral and tetrahedral sites, respectively possess an inverse spinel structure. $NiFe_2O_4$ a narrow band gap semiconductor exhibit broad absorption range beneficial for photocatalytic processes. The visible light driven $NiFe_2O_4/Bi_2O_3$ nanocomposite used for antibiotic tetracycline degradation fabricated by microwave-assisted hydrothermal method [74]. The suitable band potential value where, photogenerated electron in CB of $NiFe_2O_4$ were transfer to CB of Bi_2O_3 due to more negative CB of $NiFe_2O_4$ than Bi_2O_3 that supress the recombination rate of charges at the interface (Figure 5d).

From UV absorption $NiFe_2O_4$ display sharp absorption edge at 750 nm whereas, the heterojunction exhibit absorption range of 430-600 nm, corresponds to the band-gap of 1.70eV ($NiFe_2O_4$), 2.80eV (Bi_2O_3) and 1.67eV ($NiFe_2O_4/Bi_2O_3$) heterojunction. Bare $NiFe_2O_4$ degraded 47% of antibiotic whereas, degradation efficiency reaches to 90% in min in the presence of heterostructure with 50% of $NiFe_2O_4$. Increasing the concentration

Materials Research Foundations **112** (2021) 121-161 https://doi.org/10.21741/9781644901595-3

of $NiFe_2O_4$ above 50% may be due to the increased recombination rate. The degradation efficiency was further confirmed by determining total organic carbon (TOC) which is nearly 52% in 90 min. This confirms the mineralization of tetracycline into CO_2 and H_2O molecule. Further, the role of main active species is realize by performing scavenging experiment. The EDTA and isopropyl alcohol were added as holes, $\cdot OH$ radical scavenger. The percentage of degradation reaches to 65% and 89% with the addition of holes and $\cdot OH$ radical scavengers. The obtained results validates the role of holes as main active species in photodegradation.

CdS quantum dots decorated by $CaFe_2O_4$@$ZnFe_2O_4$ prepared by sol-gel auto combustion followed by deposition method and 83% of norfloxacin photodegraded in 90 min under solar irradiation [75]. The maximum degradation were noticed at pH=3, at this pH norfloxacin exist as positively charged species whereas, at this pH heterojunction surface is negatively charged about -12.07 mV as confirmed by zeta potential analysis. The heterojunction formation were establish by band bending (Fig 5e and f) where, the band edge potential and Fermi level was calculated by equation 1 so as to convert normal hydrogen electrode value to vacuum energy level.

$$E_{NHE} = -4.5 - E_{vac} \quad (1)$$

The work function is related to work function and vacuum energy level by equation 2.

$$\phi = E_{vac} - E_F \quad (2)$$

After heterojunction formation, the band edges of $ZnFe_2O_4$ bends by 1.63 eV while the band structure of p-type $CaFe_2O_4$ remains intact. This band bending direct the charge transfer in one direction called iso-energetic electron transfer. The iso-energetic electron transfer between CdS and $CaFe_2O_4$ where, the electrons and holes migrate from CB and VB CdS to the CB and VB of $CaFe_2O_4$, respectively. The obtained results were also supported by enhanced photocurrent density from 5.9 to 10.39 mA/cm^2. Generally, charge transfer mechanism is based upon columbic interaction at the interface of donor and acceptor semiconductor. However, charge transfer pathway in this ternary heterojunction is claimed to happen via iso-energetic electron transfer. The enhanced interfacial charge transfer significantly reduces recombination by increasing lifetime of charge carriers. Here, h+ and $\cdot O_2^-$ is the major active species however, outstanding degradation were observed with $\cdot SO_4^-$ radical. The development of heterojunction especially type-II attracts great interest of most of the researcher, nowadays it still having some drawbacks. Even through kinetics point of view electron transferred from SC2 to SC1 was blocked by electrostatic repulsions between CB of two heterojunction, holes migrated from SC1 to SC2 prohibited by same repulsion between their holes. Simultaneously, thermodynamic, formation of type-II heterojunction minimizes the redox potential of photocatalyst, reduce photocatalytic reaction.

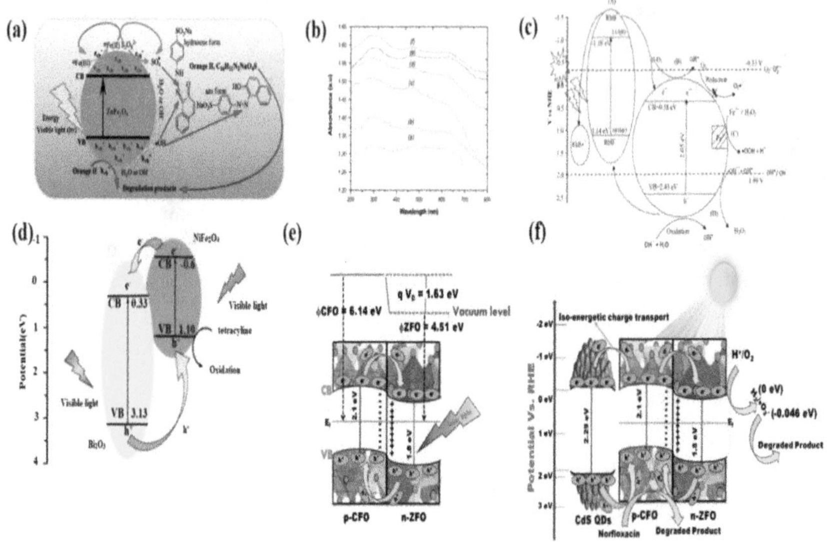

Figure 5: (a) Photo-degradation structure of $ZnFe_2O_3$ [71] (b) Schematic diagram of CZF4 [73] (c) UV-vis absorption spectra of different CZF5 samples [73](d) Charge migration mechanism of Bi_2O_3 and $NiFe_2O_4$ [74] (e) Band bending mechanism in $CaFe_2O_4$@$ZnFe_2O_4$-heterojunction (f) Iso-energetic charge transfer in CdS/$CaFe_2O_4$@$ZnFe_2O_4$ [75].

4.3 Z-scheme based ferrite heterojunction

Z-scheme based heterojunction attract great attention due to their enhance photoefficiency, high charge separation owing to directional migration of charge carrier, and high reduction and oxidation ability. For the comprehensive understanding of Z-scheme heterojunction, its historical development from 1st generation to recent 3rd generation must be explore in detail (Figure 6) [76]. Firstly, Bard et al. in 1979 proposed the concept of Z-scheme [77]. The three generation comprise of (i) Traditional Z-scheme, (ii) All-solid-state Z-scheme, and (iii) Direct Z-scheme. The traditional Z-scheme also called liquid-phase Z-scheme the two semiconductor are combine with reversible redox mediator (Fe^{3+}/Fe^{2+} and IO^{3-}/I^{-}). The redox ion pair facilitates the transfer of electrons from the CB of SC-II to the VB of SC-I thereby recombining the futile electron-hole pair. In this manner, electron and holes remain with the SC-I and SC-II with higher reduction and oxidation potential, respectively. The 1st generation of Z-scheme limits due to the reversible nature of redox ion pair hence both reduction and oxidation will compete with each other and reducing the photo-conversion efficiency. Moreover, it is only applicable in liquid phase that limits its implementation in other phases.

The 2nd generation, all-solid-state Z-scheme proposed in 2006 by Tada and group [78]. This generation have noble metal (Au and Ag) as electron mediator instead of reversible redox mediator therefore, it greatly inhibit the backward reaction. Though, employing rare noble metal is not a cost-effective technique consequently cannot be applied to large-scale. However, light shielding effect of redox and electron mediator in both 1st and 2nd generation greatly reduce the light absorption ability of photocatalyst. In 3rd generation of Z-scheme the two semiconductor are in direct contact with each other where, the charge migration is driven by the generation of internal electric field at the interface [79]. Direct Z-scheme completely inherit all the advantages of 1st and 2nd generation such as efficient charge separation, utilization of band edges with high redox potential. Owing to the various advantages direct Z-scheme heterojunction has been widely explored.

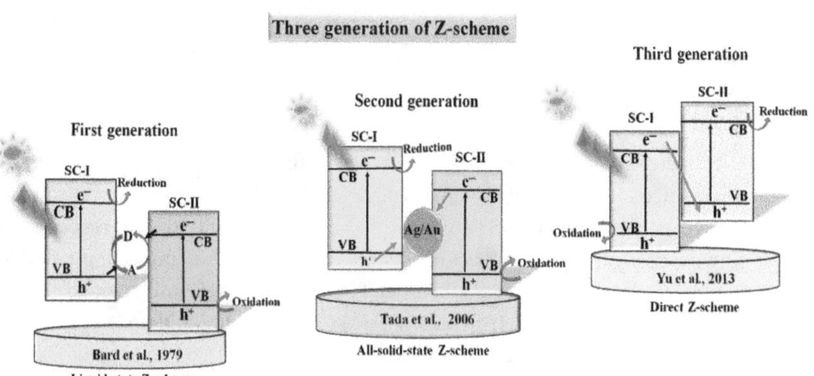

Figure 6: Hierarchy of the progression in Z-scheme heterojunction from 1st generation to 3rd generation.

The development of Z-scheme encouraged various researchers for the fabrication of ferrite based photocatalytic system. Liu and co-worker utilized simple hydrothermal method for the preparation of $ZnS/Fe_2O_3/GO$ [80]. It reveals that heterojunction photodegraded 96% and 90% of MB in 40 and 480 min under UV light and visible light, respectively. The obtained results ascribed to excellent light harvesting ability, large surface area, enhanced interfacial charge separation and low charge migration resistance. ZnS quantum dots exhibit quantum confinement which is responsible for its optical and electronic properties and hence beneficial for photocatalytic application. Here, graphene oxide act as supportive material for ZnS QDs and Fe_2O_3 nanoparticles so as to minimize the agglomeration thereby surface area remains intact. Apart, GO also act as electron mediator and making charge migration process more facile. The surface modification of ZnS and Fe_2O_3 were done by using thioglycolic acid (-COOH) and citric acid ((-OH) containing so as to covalently bind both ZnS and Fe_2O_3 together. The photocatalytic mechanism involves the recombination of photogenerated electron from CB of Fe_2O_3 to VB of ZnS. The electron in the CB of ZnS reduces O_2 to generate $^{\cdot}O_2^-$ radical meanwhile, VB holes on Fe_2O_3 oxidize H_2O molecule to generate $^{\cdot}OH$ radical (Figure 7a). The heterojunction is easily separable within 95s under magnetic field.

Wang and groups fabricated nanorods $g\text{-}C_3N_4/Fe_2O_3$ heterojunction with addition of Fe-melamine supra-molecular act as sacrificial template [81]. The structure of metal organic supramolecular network made by alternating ligand, metal and fabrication condition like solvent, ligands, temperature, and molar ratio of metal ions. $g\text{-}C_3N_4$ nanoparticle attract wide attention due to high photochemical stability and good electronic structure. The high specific surface area of $Fe_2O_3/g\text{-}C_3N_4$ ($11.96 m^2g^{-1}$) is reported than bare $g\text{-}C_3N_4$ ($5.2 m^2g^{-1}$) as examined by adsorption isotherm. It was observed that heterostructure completely degrade RhB in 20 min in comparison to $g\text{-}C_3N_4$ which degraded only 20% of dye.

$NiFe_2O_4/BiPO_4$ heterojunction were synthesized via hydrothermal method in which 1D $BiPO_4$ nanorods anchored on $NiFe_2O_4$ nanoplates [82]. The constructed 1D and 2D heterojunction degraded 99% and 98% of RhB and TC in 100 and 60 min, respectively. From trapping experiment, number of scavenger reagent like tert-butanol, isopropanol for $^{\cdot}OH^-$, ammonium oxalate (AO), triethanolamine (TEOA), ethylenediaminetetraacetic acid (EDTA) for h^+ and benzoquinone (BQ) for $^{\cdot}O_2^-$ can be used. On adding TEOA (90%) high degradation efficiency is achieved whereas, on addition of IPA and BQ the photodegradation efficiency was decreased up to 45% and 51% respectively (Figure 7b). Besides, radical trapping test $^{\cdot}O_2^-$ and $\cdot OH$ radical were active species and the direction of charge migration verified Z-scheme mechanism.

It was recently reported that, highly magnetic double Z-scheme Fe_2O_3@$NiFe_2O_3$/P-g-C_3N_4, fabricated by co-precipitation method followed by calcination at 400°C [83]. The low cost, metal free and eco-friendly heterojunction degraded 90% of CIP under visible light. In inverse spinel, the Ni^{2+} and Fe^{3+} ions in octahedral sites and tetrahedral sites with antiparallel spins produce ferrimagnetism. The Fe_3O_4@$NiFe_2O_3$/P-g-C_3N_4 heterojunction have dual Z-scheme mechanism, one Z-scheme exist between $NiFe_2O_3$ and P-g-C_3N_4 while another one operates between Fe_3O_4 and $NiFe_2O_3$ (Figure 7c). Further, the flat band potential and type of semiconductor can be confirmed by Moot-schottky plot where, the positive MS plot represent n-type semiconductor and negative MS plot represent p-type. As mentioned in this research article the value for $NiFe_2O_3$, P-g-C_3N_4 and Fe_3O_4 is +1.54, -1.47 and -0.6 V w.r.t. Ag/AgCl electrode. The shifting of holes form Ni^{3+} to Ni^{2+} responsible for p-type behaviour. Similarly, transfer of electron from Fe^{2+} to Fe^{3+} impart n-type nature to Fe_3O_4. Linear-sweep voltammetry (LSV) display charge separation efficiency analysed by photocurrent densities. On illuminating Fe_3O_4 surface with light, it generates anodic current of 1.377 $mAcm^{-2}$ confirming n-type behaviour Figure 7d. The heterojunction under positive bias potential of +1.5 V display sharp rise in photocurrent density to 1.87 $mAcm^{-2}$. The integrated synergistic effect of bare semiconductor enhance the photocurrent response and hence charge separation efficiency.

Ternary construction of Z-scheme g-C_3N_4/ZnO@Fe_2O_3 synthesized by direct pyrolysis followed by sol-gel method photodegraded of 99.3% tartrazine in 35 min under visible light [84]. The band gap value for g-C_3N_4, ZnO, ZnO@Fe_2O_3 and g-C_3N_4/ZnO@Fe_2O_3 were 2.7, 3.1, 2.5 and 2.6 eV, respectively (Figure 7e). The rate of recombination and charge carrier separation were studied by Pl spectroscopy. The lower PL emission peak represents lower recombination rate in heterojunction as compared to bare photocatalysts (Figure 7f). The Z-scheme based photocatalyst exhibit higher charge migration for degradation of organic pollutants than conventional type-II heterojunction photocatalyst.

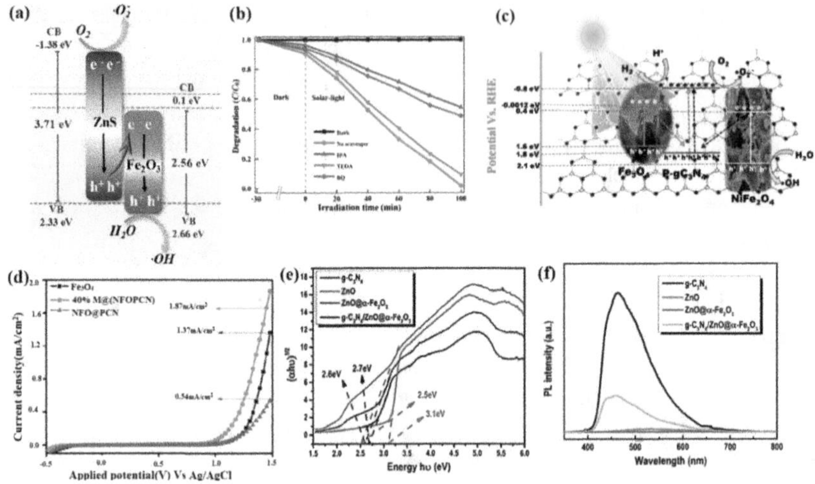

Figure 7: (a) Charge migration of ZnS/Fe_2O_3 [80] (b) charge trapping experiment [82](c) Double Z-scheme of Fe_2O_3@$NiFe_2O_3P$-g-C_3N_4[83](d) LSV plots of Fe_2O_3@$NiFe_2O_3P$-g-C_3N_4, Fe_3O_4 and $NiFe_2O_3P$-g-C_3N_4.[83](e) PL spectra of g-C_3N_4/ZnO@α-Fe_2O_3[84](f) optical band-gap of g-C_3N_4, ZnO, ZnO@α-Fe_2O_3 and g-C_3N_4/ZnO@αFe_2O_3 nanocomposite [84].

4.4 S-scheme based ferrite heterojunction

The S-scheme heterojunction were constructed by combining reduction and oxidation photocatalysts having staggered type band structure similar to type-II heterojunction. The reduction photocatalysts generally used for solar fuel generation where VB holes are futile and needs to be removed by sacrificial agent. Whereas, oxidation photocatalysts were greatly utilized in photodegradation in which CB electrons are useless. Though, the arrangement of two semiconductor the S-scheme is similar to type-II heterojunction but differ in charge migration pathway. Here, the electron diffuses at the interface from semiconductor with lower to higher work function and continue till the Fermi level of two semiconductor get equilibrated [85]. The electron diffusion creates internal electric field at the interface and alignment in Fermi level that exhibit upward and downward band bending at SC 1 and SC 2, respectively (Figure 8a). Band bending and internal electric field demonstrated driving force to proceed recombination of useless electron if SC 2 and holes of SC 1. However, electrons and holes with high redox potential were persisted in CB of SC 1 and VB of SC 2, respectively beneficial for spatial charge separation. S-scheme charge migration mechanisms resemble 'step' (low CB to high CB) as in macroscopic system.

Further, such type of charge migration can be established by various characterization technique such as XPS, ESR, AFM, M-S plot and PL. As after charge migration the semiconductor surface display increase or decrease in electron density after contact which alters the effective nuclear charge and hence the binding energy of element. In S-scheme heterojunction effective electron-hole movement and separation is possible due to effective interfacial contact between reduction and oxidation semiconductor [86]. The close contact between semiconductors can be analyse by AFM technique. Figure 8b, c microscale thickness of WO_3 (3.0nm) and $g\text{-}C_3N_4$ (2.5nm) showing strong electrostatic interfacial contact in $WO_3/g\text{-}C_3N_4$ assist in facile charge migration. Due thin layered structure of heterojunction formation the strong columbic force increases the internal electric field. The work function as calculated by DFT for WO_3 (6.23eV) is larger than $g\text{-}C_3N_4$ (4.18eV) semiconductor (Figure 8d and e). Noticeably, $g\text{-}C_3N_4$ shows upward band-banding with strong reduction potential by releasing $\cdot O_2^-$ and WO_3 exhibit downward banding with involving ·OH radical. This support the S-type charge migration pathway in $WO_3/g\text{-}C_3N_4$ heterojunction.

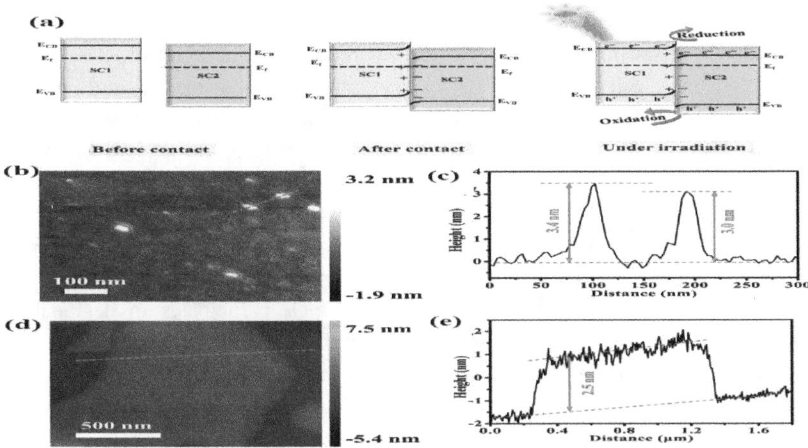

Figure 8: (a) S-scheme based mechanism [85] AFM images of (b, c) [86] WO_3 nanosheets and (d, e) $g\text{-}C_3N_4$ nanosheets [86].

In the other context, M-S plot characterized S-scheme in which TiO_2 (3.2eV) having larger band-gap than PDA (1.6 eV) having high LUMO of PDA then CBM of TiO_2 (Figure 9a) [87]. The electron migrated from PA to TiO_2 unless Fermi level equilibrates and Fermi level of TP_3 nanocomposite shift upward compare to TiO_2. It examined that coupling of more negative band potential in TP_3 with PDA can increase reduction potential of TiO_2. Therefore, high charge density of TP_3 was efficient for enhancing the charge separation. S-scheme also characterized by electron EPR spectra in which DMPO-·OH signal observed for PDA due to negligible oxidation potential of holes and strong signal in T and TP_3 because holes exist in VB of TiO_2 (Figure 9b). Further, in DMPO-$\cdot O_2^-$ signal of PDA due to photo-excited electron in LUMO were recombine with holes in HOMO owing to narrow band gap (Figure 9c). Thereafter, DMPO-$\cdot O_2^-$ peaks of T and TP3 exhibit reduction capability and reduces O_2 to $\cdot O_2^-$. According to this photo-excited holes exist in VB of TiO_2 so by the charge migration mechanism they explained S-scheme.

For instance, non-toxic S-scheme based $Sn_2Fe_2O_4/ZnFe_2O_4$ heterojunction prepared by optimizing $Sn_2Fe_2O_4$ content to 5% via solvothermal method [88]. The heterojunction photo-catalytically degraded 93% of tetracycline whereas, COD decrease to 77%. The direction of charge migration in S-scheme heterojunction is kinetically and thermodynamically favourable. Transient photocurrent and EIS measurement analysed separation and movement of charge carrier around heterojunction. As clearly observe in Figure 9d the large arc radius exhibits lower separation and migration efficiency thus excessive addition of $SnFe_2O_4$ enhance the recombination rate of electron-hole pair in $SnFe_2O_4/ZnFe_2O_4$. $SnFe_2O_4$ content increases 5% of composite exhibiting smallest arc radius with fast separation and migration of charges so it leads to increase in charge transfer resistance. The electron migration pathway in $SnFe_2O_4/ZnFe_2O_4$ heterojunction follow S-scheme. Here, due to formation internal electric field electron in the CB of $ZnFe_2O_4$ recombine with VB holes of $SnFe_2O_4$. As the VB edge potential of both $SnFe_2O_4$ and $ZnFe_2O_4$ is 1.54 and 1.55 eV, respectively which is not sufficient for ·OH generation thus remaining holes in the VB of $SnFe_2O_4$ participates in degradation process. While, electron in the CB of $SnFe_2O_4$ successfully reduces O_2 molecule to $\cdot O_2^-$ radical. The trapping experiment also support the $\cdot O_2^-$ radical and holes as active species in degradation.

Another step-scheme $ZnFe_2O_4/g\text{-}C_3N_4$ heterojunction was utilized as adsorbent and photocatalyst for the removal of U(V1) [89]. Nearly 94% and 1892mg/g of U(V1) is adsorbed and photo-catalytically reduced from pH=5-6. The property of adsorbent influenced by pore volume and large specific surface area that give more active sites for adsorption of pollutant. Also, pH of the solution greatly influence the ionic state of uranium, and surface charge on adsorbent. At low pH 3 to 4 uranium exist as UO_2^{2+} and the surface of $ZnFe_2O_4$/g-C_3N_4 is also positively charged. This greatly reduces the adsorption capacity due to the repulsive forces between the adsorbate and adsorbent as well as the competitive adsorption between UO_2^{2+} and H^+. As the pH increases positive charge of uranium decreases and the surface of heterojunction also become negatively charge. Thus, repulsive forces changes to attractive forces at higher pH which significantly enhance the removal efficiency at pH 5-6. XPS spectra of $ZnFe_2O_4$/g-C_3N_4 is obtained after adsorption and photoreduction of U(VI) (Figure 9e).

The two peak appeared at 392.26 and 381.75 eV ascribed for U $4f_{5/2}$ and U $4f_{7/2}$ indicative of U (VI) absorption on the heterostructure surface. Subsequently, two peaks at 390.66 and 379.95 eV were assigned for U $4f_{5/2}$ and U $4f_{7/2}$ of U(IV) after photoreduction of U(VI) into U (IV). Further, to reveals the adsorptional mechanism adsorption kinetics were studied. The adsorption mechanism obeys pseudo second order kinetics as higher correlation coefficient (R^2=0.996) were obtained in comparison to pseudo first order kinetics (R^2=0.969). Also, the calculated adsorption capacity (q_e= 219 mg/g) is near to the experimental value (q_e= 189 mg/g) when applied pseudo second order model. The kinetic study reveals that chemisorption step is the rate determining step. Further, Langmuir has prominent coefficient (R_2 = 0.993) than Freundlich (R_2 = 0.933) that demonstrating monolayer adsorption. The S-scheme type electron migration were also supported by DFT calculation, the work function of g-C_3N_4 (4.63eV) higher than that of $ZnFe_2O_4$ (3.95eV). Therefore, initially electron diffuses from $ZnFe_2O_4$ to g-C_3N_4 until their Fermi level equilibrated and generates built-in electric filed. The built-in electric filed assist in recombining the CB electron of g-C_3N_4 with the VB holes of $ZnFe_2O_4$. Such type of electron migration pathway renders the electron and holes with strong reducing and oxidizing ability.

The ternary rGO/g-C_3N_4/$CoFe_2O_4$ heterojunction fabricated by hydrothermal method obeying S-scheme type charge migration pathway degraded 97% of nitrophenol under visible light [90]. The increasing degradation efficiency shows high surface area, effective charge separation and mobility of nanocomposite. The 2D rGO nano-sheet contains sp^2 hybridized carbon atom that provide fast charge carrier mobility and large active surface area. The photoexcited CB electron of $CoFe_2O_4$ transferred to VB of g-C_3N_4 via rGO surface (Figure 9f). The CB of g-C_3N_4 has negative potential compared with standard reduction potential (-0.33V), that can reduces O_2 to $\cdot O_2^-$ radical. The magnetic nature of ternary heterojunction was examined by VSM hysteresis loop. The heterojunction exhibits ferromagnetic property at room temperature displaying magnetic saturation of 15.58emu/g for bare $CoFe_2O_4$ and 11.49 emu/g for ternary heterojunction. Combining non-magnetic rGO and g-C_3N_4 reduces the magnetic saturation of ternary

Materials Research Foundations **112** (2021) 121-161 https://doi.org/10.21741/9781644901595-3

heterojunction. Form the reported literature it can be concluded that S-scheme type charge migration is intrinsically more feasible over type -II and Z- Scheme. The effective charge carrier separation, recombination of useless electron-holes pair and utilization of high oxidation and reduction potential makes S-scheme more adaptable. Lastly, we have summarizes various ferrite based heterojunction obeying type-II, Z-scheme and S-scheme pathway for the pollutant photodegradation are listed in Table 1.

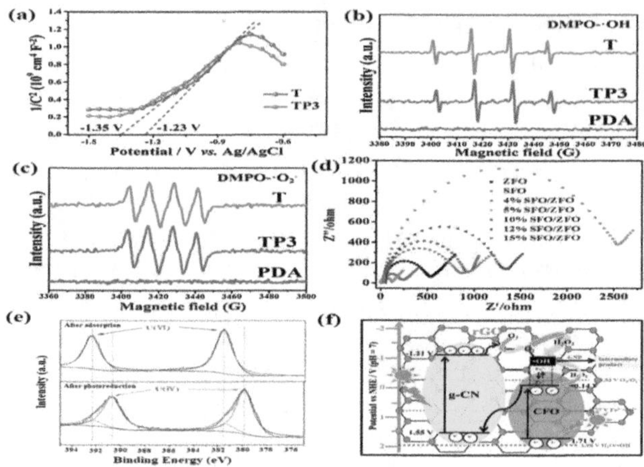

Figure 9: (a) Mott-Schottky plots of T and TP3 (b) EPR spectra of DMPO-·•OH (c) in aqueous and DMPO-•O_2^- in methanol dispersion in the presence of T, TP3 and PDA [87] (d) XPS of $ZnFe_2O_4$/g-C_3N_4 [89] (f) Charge migration mechanism of r-GO/g-CN/CF0 [90].

Table1: Summary of ferrites-based nanocomposite for the pollutant photodegradation.

Ferrite based photocatalyst	Synthetic method	Pollutant	Photocatalytic Condition	Photocatalytic Performance	Type of charge migration	Reactive oxygen species	Recyclability	Ref.
$ZnFe_2O_4$	Reduction -oxidation method	Orange II	Light flux zone of Xe lamp. Vis- ≥ 420 nm pH = 3.0~9.0	78.6% of Orange II degradation	NA	SO_4^{2-}, ·OH	5 cycle	[71]
Fe_2TiO_4	Laser ablation approach	Rhodamine	150 W visible light lamp 485nm	80% of RhB degradation	NA	e^-, h^+	2 cycle	[91]
Ni–$ZnFe_2O_4$	Sol–gel method	Rhodamine	Conc. of Ni–$ZnFe_2O_4$ =0.5g Conc. of RhB= 250 ml of 10 mg/l Light flux zone of Xe lamp 552 nm	98% of RhB degradation	NA	·OH, $·O_2^-$, and h^+	5 cycle	[92]
$Co_{1-x}Zn_xFe_2O_4$	Microwave combustion method	Rhodamine	Conc. of dye - 6mg/l 150 W Halide lamp UV-330 nm pH 2.0	99.9% of RhB degradation	NA	·OH	5 cycle	[73]
Au–$Zn_xFe_2O_4$	Thermal evaporation	Rhodamine	Conc. of dye = 5μm of 10ml 150-W xenon lamp Vis - 554 nm	31% of RhB degradation	NA	·OH, $·O_2^-$, and h^+	3 cycle	[93]
TiO_2–$ZnFe_2O_4$	Citrate gel auto combustion method	Methyl-red and thymol blue	450 W Xe arc lamp 350 to 550 nm	90 % Methyl-red and thymol blue of degradation	Type-II	e^- and h^+	4 cycle	[94]
$CoFe_2O_4$-PANI	Hydrothermal method	Methyl Orange and Methyl blue	> 420 nm 500 W Hg and Xe lamp,		Type-II	·OH	3 cycle	[95]
$NiFe_2O_4/Bi_2O_3$	Hydrothermal	Tetracycline degradation	Conc.=100 ml of 10 mg/L	90% of TC degradation	Type-II	h^+	3 cycle	[74]

	method		150W xenon lamp 420nm					
TiO_2@$CuFe_2O_4$	Sol-gel method	2,4-dichlorophenoxyacetic acid (2,4-DP)	Conc.= 10 to 60 mg/L, 6W low pressure mercury UV-C lamp (λ = 254 nm pH = 6.5 ± 0.2	97.2% of 2,4-DP degradation	Type-II	SO_4^{2-}, $\cdot OH$, $\cdot O_2^-$	5 cycle	[60]
$ZnFe_2O_4$–TiO_2	Solvothermal method	Bisphenol A degradation	Conc.= 10 mg L^{-1} 10 W LED lamp 465 nm pH 7	>99% of BPA degradation	Type-II	h^+, $\cdot OH$	10 cycle	[96]
POPD-$CoFe_2O_4$	Microwave-assisted	Tetracycline	Conc. 100 mL, 20 mg/L XE300, power 300 W, 380 nm -780 nm	45.98 % of TC degradation	Type-II	e^- and h^+	5 cycle	[97]
ZnF_2O_4/g-C_3N_4	Polymerization method	Ciprofloxacin degradation	$\lambda \geq 400$ nm	92% of CIP degradation	Type-II	e^-, h^+·OH and $\cdot O_2^-$	4 cycle	[98]
ZnS/Fe_2O_3/rGO	Hydrothermal method	Methylene blue	150 µl of TGA Xe lamp 300 W with a 420 nm	96.45% of MB degradation	Z-scheme	$\cdot OH$, $\cdot O_2^-$ and h^+	5 cycle	[80]
$Bi_2Fe_4O_9$/Bi_2WO_6	Hydrothermal method	Rhodamine	10 mg L-1 of RhB 300 W xenon lamp (>420 nm	100% of RhB degradation	Z- scheme	$\cdot OH$, $\cdot O_2^-$, and h^+	5 cycle	[99]
$Bi_{12}O_{17}Cl_2$/Ag/$AgFeO_2$	In-situ deposition - precipitation method	Tetracycline	0.5 g L^{-1} 3.3 < pH < 7.68)	87.4% of TC degradation	Z-scheme	$\cdot O_2^-$	6 cycle	[62]
ZnO/NZF	Precipitation method	Rhodamine	30 mL of RhB A 250W high-pressure Hg lamp 500W	94% of RhB degradation	Z- scheme	$\cdot O_2^-$	NA	[102]

			Xenon lamp >420 nm					
$NiFe_2O_4/BiPO_4$	Hydrothermal method	Tetracycline and Rhodamine B	50 mL of TC	TC (98%) and RhB (99%) of degradation	Z- scheme	$\cdot OH$, $\cdot O_2^-$, and h^+	4 cycle	[82]
$ZnFe_2O_4/NCDs/Ag_2CO_3$	Hydrothermal and calcined method	Levofloxacin	Conc. of LFLX =0.60 g/L Conc. of $ZnFe_2O_4/NCDs/Ag_2CO_3$ =10 mg/L 300 W Xe lamp 420-760nm	88.75% of LFLX degradation	Z-scheme	$\cdot OH$ $\cdot O_2^-$ and h^+,	5 cycle	[100]
Fe_3O_4@$NiFe_2O_4$/P-g-C_3N_4	Sol-gel method	Ciprofloxacin	Conc. of CIP =20 mg L^{-1} Xe lamp with an excitation wavelength of 330 nm.	90% CIP degradation	Z-scheme	$\cdot OH$, $\cdot O_2^-$, h^+, and e^-	4 cycle	[83]
PANI/$AgFeO_2$	In situ growth	Tetracycline	20 mg L^{-1} of TC 300 W xenon lamp below 420 nm (3.3 < pH < 7.68	91.8% of TC degradation	Z-scheme	h^+, $\cdot OH$ and $\cdot O_2^-$	4 cycle	[101]
g-$C_3N_4/CoFe_2O_4$	Hydrothermal method	4 - Nitrophenol	Conc. of 4-NP=20 mg/l pH=7	97% of 4-NP degradation	S-scheme	$OH\cdot$	4 cycle	[90]
$SnFe_2O_4/ZnFe_2O_4$	Solvothermal method	Tetracycline	Conc. of TC=100 mL of 10 mg/L 300W xenon lamp 400 nm pH = 11	93.2% of TC degradation	S-scheme	$h^+ \cdot OH$ and $\cdot O_2^-$	4 cycle	[88]
$ZnFe_2O_4$/g-C_3N_4	Hummers' method	Hexavalent uranium	Conc. of (U(VI)) =699.3 mg/g 8 W LED pH 5.0	90% of (U(VI)) degradation	S-scheme	e^-, $h^+ \cdot OH$ and $\cdot O_2^-$	5 cycle	[89]

Conclusion and Outlook

In this manuscript, efforts have been put to explore the recent development of ferrites based nanocomposites explored for photocatalytic wastewater treatment. The basic information on the structural and properties with various synthetic methods of ferrites with versatile morphology, phases, crystalline structure and dimensionality were discussed so as to provide depth insight on ferrites nanocomposites. The influence on the morphology, properties and activity of ferrites nanoparticles by altering the synthetic approaches were discussed. The ferrites exhibit excellent superparamagnetic properties beneficial for its utilization in various photocatalytic processes, facile separation and improved reusability. Furthermore, there are several studies reported the usage of bare ferrites photocatalyst and modified ferrites that dispenses several advantages in photodegradation processes. Still there are few practical issues in the utilization of ferrites nanoparticles in large scale which further needs to be explored. The ferrites-based nanocomposite should be prepared without liberating its magnetic properties. Under different reaction conditions ferrites can undergo phase transformation which changes their microstructure properties, magnetic behavior and may introduce defects. Investigation in the field of metal defects and generation of oxygen vacancies in ferrites surface are still meagre. These defects have ability to alter the band structure of materials and traps the electrons that can enhance photoactivity of materials.

The ferrites can be modified by constructing heterojunction working via either type-II, Z-scheme or S-scheme mechanism. Fabrication of heterojunction were considered more effective due to the combination of synergistic effect of individual photocatalysts rather than doping or introducing defects. During heterojunction formation the semiconductor with appropriate band edge potential must be carefully chosen in such as way so as to impart higher photoefficiency. The ideal heterojunction should have high surface area, large no of active site, highly chemically and thermally stable, cost-effective, environmental friendly, good uniform dispersion of NPs, broad absorption range, facile charge migration, low rate of recombination, generation of both hydroxyl and superoxide radical, multi-dimension electron migration pathway, high porosity etc. In the near future, we believe outstanding ferrite based photocatalyst with superior photoefficiency may be fabricated and explored for various applications.

References

[1] E. Casbeer, VK. Sharma, XZ. Li, Synthesis and photocatalytic activity of ferrites under visible light: a review, Separation and Purification Technology, 87 (2012) 1-4. https://doi.org/10.1016/j.seppur.2011.11.034

[2] M. Ismael, Ferrites as solar photocatalytic materials and their activities in solar energy conversion and environmental protection: A review, Solar Energy Materials and Solar Cells 219 (2021) 110786. https://doi.org/10.1016/j.solmat.2020.110786

[3] A. Kumar, G. Sharma, M. Naushad, H. Ala'a, A. Kumar, I. Hira, T. Ahamad, AA. Ghfar, FJ. Stadler, Visible photodegradation of ibuprofen and 2, 4-D in simulated waste water using sustainable metal free-hybrids based on carbon nitride and biochar, Journal of environmental management, 231 (2019) 1164-75. https://doi.org/10.1016/j.jenvman.2018.11.015

[4] P. Shandilya, A. Sudhaik, P. Raizada, A. Hosseini-Bandegharaei, P. Singh, A. Rahmani-Sani, V.Thakur, AK. Saini, Synthesis of eu^{3+}– doped zno/bi_2o_3 heterojunction photocatalyst on graphene oxide sheets for visible light-assisted degradation of 2, 4-dimethyl phenol and bacteria killing, Solid State Sciences, 102 (2020) 106164. https://doi.org/10.1016/j.solidstatesciences.2020.106164

[5] PA. Raizada, S. Singh. Hybrid metal oxide semiconductors for waste water treatment, Environ Sci Eng, 4 (2017) 187-206.

[6] V. Binas, D. Venieri, D. Kotzias, G. Kiriakidis, Modified TiO_2 based photocatalysts for improved air and health quality, Journal of Materiomics, 3(1) (2017) 3-16. https://doi.org/10.1016/j.jmat.2016.11.002

[7] A. Kumar, A. Kumar, G. Sharma, H. Ala'a, M. Naushad, AA. Ghfar, C. Guo, FJ. Stadler, Biochar-templated g-C_3N_4/$Bi_2O_2CO_3$/$CoFe_2O_4$ nano-assembly for visible and solar assisted photo-degradation of paraquat, nitrophenol reduction and CO_2 conversion, Chemical Engineering Journal, 339 (2018) 393-410. https://doi.org/10.1016/j.cej.2018.01.105

[8] P. Dhiman, S. Sharma, A. Kumar, M. Shekh, G. Sharma, M. Naushad, Rapid visible and solar photocatalytic Cr (VI) reduction and electrochemical sensing of dopamine using solution combustion synthesized ZnO–Fe_2O_3 nano heterojunctions: mechanism Elucidation. Ceramics International, 46(8) (2020) 12255-68. https://doi.org/10.1016/j.ceramint.2020.01.275

[9] P. Shandilya, D. Mittal, M. Soni, P. Raizada, A. Hosseini-Bandegharaei, AK. Saini, P. Singh, Fabrication of fluorine doped graphene and $SmVO_4$ based dispersed and adsorptive photocatalyst for abatement of phenolic compounds from water and bacterial disinfection, Journal of Cleaner Production, 203 (2018) 386-99. https://doi.org/10.1016/j.jclepro.2018.08.271

[10] P. Shandilya, D. Mittal, A. Sudhaik, M. Soni, P. Raizada, AK. Saini, P. Singh. $GdVO_4$ modified fluorine doped graphene nanosheets as dispersed photocatalyst for mitigation of phenolic compounds in aqueous environment and bacterial disinfection, Separation and Purification Technology, 210 (2019) 804-16. https://doi.org/10.1016/j.seppur.2018.08.077

[11] P. Shandilya, D. Mittal, M. Soni, P. Raizada, JH. Lim, DY. Jeong, RP. Dewedi, AK. Saini, P. Singh, Islanding of $EuVO_4$ on high-dispersed fluorine doped few layered graphene sheets for efficient photocatalytic mineralization of phenolic compounds and bacterial

disinfection, Journal of the Taiwan Institute of Chemical Engineers, 93 (2018) 528-42. https://doi.org/10.1016/j.jtice.2018.08.034

[12] R. Suresh, S. Rajendran, PS. Kumar, DV. Vo, L. Cornejo-Ponce, Recent advancements of spinel ferrite based binary nanocomposite photocatalysts in wastewater treatment, Chemosphere, 274 (2021) 129734. https://doi.org/10.1016/j.chemosphere.2021.129734

[13] A. Kumar, A. Kumar, G. Sharma, H. Ala'a, M. Naushad, AA. Ghfar, FJ. Stadler, Quaternary magnetic BiOCl/g-C_3N_4/Cu_2O/Fe_3O_4 nano-junction for visible light and solar powered degradation of sulfamethoxazole from aqueous environment, Chemical Engineering Journal, 334 (2018) 462-78. https://doi.org/10.1016/j.cej.2017.10.049

[14] J. Liu, B. Wang, Z. Li, Z. Wu, K. Zhu, J. Zhuang, Q. Xi, Y. Hou, J. Chen, M. Cong, J. Li, Photo-Fenton reaction and H_2O_2 enhanced photocatalytic activity of α-Fe_2O_3 nanoparticles obtained by a simple decomposition route, Journal of Alloys and Compounds, 771 (2019) 398-405. https://doi.org/10.1016/j.jallcom.2018.08.305

[15] P. Shandilya, P. Raizada, A. Sudhaik, A. Saini, R. Saini, P. Singh, Metal and Carbon Quantum Dot Photocatalysts for Water Purification, In Water Pollution and Remediation: Photocatalysis (2021) (pp. 81-118). Springer, Cham. https://doi.org/10.1007/978-3-030-54723-3_3

[16] M. Dhiman, R. Sharma, V. Kumar, S. Singhal, Morphology controlled hydrothermal synthesis and photocatalytic properties of $ZnFe_2O_4$ nanostructures, Ceramics International, 42(11) (2016) 12594-605. https://doi.org/10.1016/j.ceramint.2016.04.115

[17] AG. Abraham, A. Manikandan, E. Manikandan, S. Vadivel, SK. Jaganathan, A. Baykal, PS. Renganathan, Enhanced magneto-optical and photo-catalytic properties of transition metal cobalt (Co^{2+} ions) doped spinel $MgFe_2O_4$ ferrite nanocomposites Journal of Magnetism and Magnetic Materials, 452 (2018) 380-8. https://doi.org/10.1016/j.jmmm.2018.01.001

[18] AA. Al-Ghamdi, FS. Al-Hazmi, LS. Memesh, FS. Shokr, LM. Bronstein, Evolution of the structure, magnetic and optical properties of $Ni_{1-x}Cu_xFe_2O_4$ spinel ferrites prepared by soft mechanochemical method. Journal of Alloys and Compounds, 712 (2017) 82-9. https://doi.org/10.1016/j.jallcom.2017.04.052

[19] A. Ashok, LJ. Kennedy, JJ. Vijaya, Structural, optical and magnetic properties of $Zn_{1-x}Mn_xFe_2O_4$ ($0 \leq x \leq 0.5$) spinel nano particles for transesterification of used cooking oil, Journal of Alloys and Compounds, 780 (2019)816-28. https://doi.org/10.1016/j.jallcom.2018.11.390

[20] AB. Nawale, NS. Kanhe, SA. Raut, SV. Bhoraskar, AK. Das, VL. Mathe, Investigation of structural, optical and magnetic properties of thermal plasma synthesized Ni-Co spinel

ferrite nanoparticles, Ceramics International, 43(9) (2017) 6637-47. https://doi.org/10.1016/j.ceramint.2017.02.022

[21] A. Lassoued, MS. Lassoued, B. Dkhil, S. Ammar, A. Gadri, Substituted effect of Al^{3+} on structural, optical, magnetic and photocatalytic activity of Ni ferrites, Journal of Magnetism and Magnetic Materials, 476 (2019) 124-33. https://doi.org/10.1016/j.jmmm.2018.12.062

[22] MA. Almessiere, Y. Slimani, H. Gungunes, A. Manikandan, A. Baykal, Investigation of the effects of Tm^{3+} on the structural, microstructural, optical, and magnetic properties of Sr hexaferrites, Results in Physics, 13 (2019) 102166. https://doi.org/10.1016/j.rinp.2019.102166

[23] C. Murugesan, K. Ugendar, L. Okrasa, J. Shen, G. Chandrasekaran, Zinc substitution effect on the structural, spectroscopic and electrical properties of nanocrystalline $MnFe_2O_4$ spinel ferrite. Ceramics International, 47(2) (2021) 1672-85. https://doi.org/10.1016/j.ceramint.2020.08.284

[24] S. Ida, K. Yamada, T. Matsunaga, H. Hagiwara, Y. Matsumoto, T. Ishihara. Preparation of p-type $CaFe_2O_4$ photocathodes for producing hydrogen from water, Journal of the american chemical society, 132(49) (2010) 17343-5. https://doi.org/10.1021/ja106930f

[25] S. Boumaza, A. Boudjemaa, A. Bouguelia, R. Bouarab, M. Trari, Visible light induced hydrogen evolution on new hetero-system $ZnFe_2O_4/SrTiO_3$, Applied Energy, 87(7) (2010) 2230-6. https://doi.org/10.1016/j.apenergy.2009.12.016

[26] X. Yuan, H. Wang, Y. Wu, X. Chen, G. Zeng, L. Leng, C. Zhang, A novel SnS_2–$MgFe_2O_4$/reduced graphene oxide flower-like photocatalyst: Solvothermal synthesis, characterization and improved visible-light photocatalytic activity, Catalysis Communications, 61 (2015) 62-6. https://doi.org/10.1016/j.catcom.2014.12.003

[27] Y. Xia, Z. He, J. Su, B. Tang, K. Hu, Y. Lu, S. Sun, X. Li, Fabrication of magnetically separable $NiFe_2O_4$/BiOI nanocomposites with enhanced photocatalytic performance under visible-light irradiation, RSC advances, 8(8) (2018) 4284-94. https://doi.org/10.1039/C7RA12546A

[28] R. Cheng, X. Fan, M. Wang, M. Li, J. Tian, L. Zhang, Facile construction of $CuFe_2O_4/g-C_3N_4$ photocatalyst for enhanced visible-light hydrogen evolution, Rsc Advances, 6(23) (2016) 18990-5. https://doi.org/10.1039/C5RA27221A

[29] X. Wang, N. Zhang, G. Wang, Visible light $Bi_2S_3/BiFeO_3$ photocatalyst for effective removal of Rhodamine B, InMATEC Web of Conferences (2018) (Vol. 238, p. 03007). EDP Sciences. https://doi.org/10.1051/matecconf/201823803007

[30] HY. Hafeez, SK. Lakhera, N. Narayanan, S. Harish, Y. Hayakawa, BK. Lee, B. Neppolian, Environmentally sustainable synthesis of a $CoFe_2O_4$–TiO_2/rGO ternary

photocatalyst: a highly efficient and stable photocatalyst for high production of hydrogen (solar fuel), ACS omega, 4(1) (2019) 880-91. https://doi.org/10.1021/acsomega.8b03221

[31] Z. Liu, ZG. Zhao, M. Miyauchi, Efficient visible light active $CaFe_2O_4/WO_3$ based composite photocatalysts: effect of interfacial modification, The Journal of Physical Chemistry C, 113(39) (2009) 17132-7. https://doi.org/10.1021/jp906195f

[32] A. Kumari, A. Kumar, G. Sharma, J. Iqbal, M. Naushad, FJ. Stadler, Constructing Z-scheme $LaTiO_2N/g\text{-}C_3N_4$@ Fe_3O_4 magnetic nano heterojunctions with promoted charge separation for visible and solar removal of indomethacin, Journal of Water Process Engineering, 36 (2020) 101391. https://doi.org/10.1016/j.jwpe.2020.101391

[33] A. Kumar, G. Sharma, M. Naushad, T. Ahamad, RC. Veses, FJ. Stadler, Highly visible active $Ag_2CrO_4/Ag/BiFeO_3$@ RGO nano-junction for photoreduction of CO_2 and photocatalytic removal of ciprofloxacin and bromate ions: the triggering effect of Ag and RGO, Chemical Engineering Journal. 370 (2019) 148-65. https://doi.org/10.1016/j.cej.2019.03.196

[34] A. Kumar, G. Sharma, M. Naushad, A. Kumar, S. Kalia, C. Guo, GT. Mola, Facile hetero-assembly of superparamagnetic $Fe_3O_4/BiVO_4$ stacked on biochar for solar photo-degradation of methyl paraben and pesticide removal from soil. Journal of Photochemistry and Photobiology A: Chemistry. 337 (2017) 118-31. https://doi.org/10.1016/j.jphotochem.2017.01.010

[35] P. Shandilya, P. Raizada, P. Singh, Photocatalytic Degradation of Azo Dyes in Water, In Water Pollution and Remediation: Photocatalysis (2021) 119-146 Springer, Cham. https://doi.org/10.1007/978-3-030-54723-3_4

[36] A.Guleria, R. Sharma, P. Shandilya, Photocatalytic and Adsorptional Removal of Heavy Metals from Contaminated Water using Nanohybrids. Photocatalysis: Advanced Materials and Reaction Engineering. 100 (2021) 113-60. DOI: https://doi.org/10.21741/9781644901359-4

[37] Singh P, Gautam S, Shandilya P, Priya B, Singh VP, Raizada P. Graphene bentonite supported $ZnFe_2O_4$ as superparamagnetic photocatalyst for antibiotic degradation, Adv. Mater. Lett, 8(3) (2017) 229-38. DOI: 10.5185/amlett.2017.1467

[38] TN. Pham, TQ. Huy, AT. Le, Spinel ferrite (AFe_2O_4)-based heterostructured designs for lithium-ion battery, environmental monitoring, and biomedical applications, RSC Advance, 10(52) (2020) 31622-61. http://dx.doi.org/10.1039/d0ra05133k

[39] FG. Da Silva, J. Depeyrot, AF. Campos, R. Aquino, D. Fiorani, D. Peddis, Structural and Magnetic Properties of Spinel Ferrite Nanoparticles. Journal of nanoscience and nanotechnology, 19(8) (2019) 4888-902. https://doi.org/10.1088/1742-6596/1436/1/012144

[40] Q. Li, CW. Kartikowati, S. Horie, T. Ogi, T. Iwaki, K. Okuyama, Correlation between particle size/domain structure and magnetic properties of highly crystalline Fe_3O_4 nanoparticles, Scientific reports, 7(1) (2017) 1-7. https://doi.org/10.1038/s41598-017-09897-5

[41] M. Kumar, HS. Dosanjh, J. Singh, K. Monir, H. Singh, Review on magnetic nanoferrites and their composites as alternatives in waste water treatment: synthesis, modifications and applications, Environmental Science: Water Research & Technology, 6(3) (2020) 491-514. https://doi.org/10.1039/C9EW00858F

[42] K. Ahalya, N. Suriyanarayanan, V. Ranjithkumar, Effect of cobalt substitution on structural and magnetic properties and chromium adsorption of manganese ferrite nano particles. Journal of Magnetism and Magnetic Materials, 372 (2014) 208-13. https://doi.org/10.1016/j.jmmm.2014.07.030

[43] R. Liu, H. Fu, H. Yin, P. Wang, L. Lu, Y. Tao, A facile sol combustion and calcination process for the preparation of magnetic Ni0.5Zn0.5Fe_2O_4 nanopowders and their adsorption behaviors of Congo red, Powder technology, 274 (2015) 418-25. https://doi.org/10.1016/j.powtec.2015.01.045

[44] NM. Mahmoodi, Surface modification of magnetic nanoparticle and dye removal from ternary systems, Journal of Industrial and Engineering Chemistry, 27 (2015) 251-9. https://doi.org/10.1016/j.jiec.2014.12.042

[45] H. Wang, S. Jia, H. Wang, B. Li, W. Liu, N. Li, J. Qiao, CZ. Li, A novel-green adsorbent based on betaine-modified magnetic nanoparticles for removal of methyl blue. Science Bulletin, 62(5) (2017) 319-25. https://doi.org/10.1016/j.scib.2017.01.038

[46] N. Guijarro, P. Bornoz, M. Prévot, X. Yu, X. Zhu, M. Johnson, X. Jeanbourquin,,F. Le Formal, K. Sivula, Evaluating spinel ferrites MFe_2O_4 (M= Cu, Mg, Zn) as photoanodes for solar water oxidation: prospects and limitations, Sustainable Energy & Fuels, 2(1) (2018)103-17. https://doi.org/10.1149/MA2017-01/32/1523

[47] A. Kezzim, N. Nasrallah, A. Abdi, M. Trari, Visible light induced hydrogen on the novel hetero-system $CuFe_2O_4/TiO_2$, Energy Conversion and Management, 52(8-9) (2011) 2800-6. https://doi.org/10.1016/j.enconman.2011.02.014

[48] MJ. Kang, H. Yu, W. Lee, HG. Cha, Efficient Fe_2O_3/Cg-C_3N_4 Z-scheme heterojunction photocatalyst prepared by facile one-step carbonizing process, Journal of Physics and Chemistry of Solids, 130 (2019) 93-9. https://doi.org/10.1016/j.jpcs.2019.02.017

[49] Q. Huo, X. Qi, J. Li, G. Liu, Y. Ning, X. Zhang, B. Zhang, Y. Fu, S. Liu, Preparation of a direct Z-scheme α-Fe_2O_3/MIL-101 (Cr) hybrid for degradation of carbamazepine under visible light irradiation. Applied Catalysis B: Environmental. 255 (2019) 117751. https://doi.org/10.1016/j.apcatb.2019.117751

[50] M. Amiri, M. Salavati-Niasari, A. Akbari, Magnetic nanocarriers: evolution of spinel ferrites for medical applications, Advances in colloid and interface science, 265 (2019) 29-44. https://doi.org/10.1016/j.cis.2019.01.003

[51] D. Maiti, A. Saha, PS. Devi, Surface modified multifunctional $ZnFe_2O_4$ nanoparticles for hydrophobic and hydrophilic anti-cancer drug molecule loading, Physical Chemistry Chemical Physics, 18(3) (2016) 1439-50. https://doi.org/10.1039/C5CP05840F

[52] I. Sharifi, H. Shokrollahi, S. Amiri, Ferrite-based magnetic nanofluids used in hyperthermia applications, Journal of magnetism and magnetic materials, 324(6) (2012) 903-15. https://doi.org/10.1016/j.jmmm.2011.10.017

[53] CE. Demirci Dönmez, PK. Manna, R. Nickel, S. Aktürk, J. van Lierop, Comparative heating efficiency of cobalt-, manganese-, and nickel-ferrite nanoparticles for a hyperthermia agent in biomedicines, ACS applied materials & interfaces, 11(7) (2019) 6858-66. https://doi.org/10.1021/acsami.8b22600

[54] S. Ebrahimisadr, B. Aslibeiki, R. Asadi, Magnetic hyperthermia properties of iron oxide nanoparticles: The effect of concentration, Physica C: Superconductivity and its applications, 549 (2018) 119-21. https://doi.org/10.1016/j.physc.2018.02.014

[55] J. Lu, CV. Ngo, SC. Singh, J. Yang, W. Xin, Z. Yu, C. Guo. Bioinspired hierarchical surfaces fabricated by femtosecond laser and hydrothermal method for water harvesting. Langmuir, 35(9) (2019) 3562-7. https://doi.org/10.1021/acs.langmuir.8b04295

[56] LK. Wu, H. Wu, HB. Zhang, HZ. Cao, GY. Hou, YP. Tang, GQ, Zheng. Graphene oxide/$CuFe_2O_4$ foam as an efficient absorbent for arsenic removal from water, Chemical Engineering Journal, 334 (2018) 1808-19. https://doi.org/10.1016/j.cej.2017.11.096

[57] T. Gao, Z. Chen, F. Niu, D. Zhou, Q. Huang, Y. Zhu, L. Qin, X. Sun, Y. Huang, Shape-controlled preparation of bismuth ferrite by hydrothermal method and their visible-light degradation properties, Journal of Alloys and Compounds, 648 (2015) 564-70. https://doi.org/10.1016/j.jallcom.2015.07.059

[58] R. Koutavarapu, MR. Tamtam, CR. Myla, M. Cho, J. Shim, Enhanced solar-light-driven photocatalytic properties of novel Z-scheme binary $BiPO_4$ nanorods anchored onto $NiFe_2O_4$ nanoplates: Efficient removal of toxic organic pollutants. Journal of Environmental Sciences, 102 (2021) 326-40. https://doi.org/10.1016/j.jes.2020.09.021

[59] S. Gautam, P. Shandilya, VP. Singh, P. Raizada, P. Singh, Solar photocatalytic mineralization of antibiotics using magnetically separable $NiFe_2O_4$ supported onto graphene sand composite and bentonite. Journal of Water Process Engineering. 14(86) (2016) 100. https://doi.org/10.1016/j.jwpe.2016.10.008

[60] M. Golshan, B. Kakavandi, M. Ahmadi, M. Azizi, Photocatalytic activation of peroxymonosulfate by TiO_2 anchored on cupper ferrite (TiO_2@ $CuFe_2O_4$) into 2, 4-D degradation: Process feasibility, mechanism and pathway, Journal of hazardous materials, 359(2018) 325-37. https://doi.org/10.1016/j.jhazmat.2018.06.069

[61] HS. Mund, BL. Ahuja, Structural and magnetic properties of Mg doped cobalt ferrite nano particles prepared by sol-gel method, Materials Research Bulletin, 85 (2017) 228-33. https://doi.org/10.1016/j.materresbull.2016.09.027

[62] J. Guo, L. Jiang, J. Liang, W. Xu, H. Yu, J. Zhang, S. Ye, W. Xing, X. Yuan, Photocatalytic degradation of tetracycline antibiotics using delafossite silver ferrite-based Z-scheme photocatalyst: Pathways and mechanism insight, Chemosphere, 270 (2021) 128651. https://doi.org/10.1016/j.jece.2021.105524

[63] J. Revathi, MJ. Abel, V. Archana, T. Sumithra, R .Thiruneelakandan. Synthesis and characterization of $CoFe_2O_4$ and Ni-doped $CoFe_2O_4$ nanoparticles by chemical Co-precipitation technique for photo-degradation of organic dyestuffs under direct sunlight. Physica B: Condensed Matter, 587 (2020) 412136. https://doi.org/10.1016/j.physb.2020.412136

[64] K. Tezuka, M. Kogure, YJ. Shan, Photocatalytic degradation of acetic acid on spinel ferrites MFe_2O_4 (M= Mg, Zn, and Cd), Catalysis Communications, 48 (2014) 11-4. https://doi.org/10.1016/j.catcom.2014.01.016

[65] AB. Kulkarni, SN. Mathad, Effect of Sintering Temperature on Structural Properties of Cd doped Co-Zn Ferrite. Journal of Nano-and Electronic Physics, 10 (2018) (1). https://doi.org/10.21272/jnep.10(1).01001

[66] VJ. Angadi, B. Rudraswamy, K. Sadhana, SR. Murthy, K. Praveena, Effect of Sm^{3+}–Gd^{3+} on structural, electrical and magnetic properties of Mn–Zn ferrites synthesized via combustion route, Journal of Alloys and Compounds, 656 (2016) 5-12. https://doi.org/10.1016/j.jallcom.2015.09.222

[67] MS. Cortés, A. Martinez-Luevanos, LA. García-Cerda, OS. Rodriguez-Fernandez, AF. Fuentes, J. Romero-García, SM, Montemayor Nanostructured pure and substituted cobalt ferrites: fabrication by electrospinning and study of their magnetic properties, Journal of Alloys and Compounds, 653 (2015) 290-7. https://doi.org/10.1016/j.jallcom.2015.08.262

[68] L. Lv, Y. Wang, P. Cheng, B. Zhang, F. Dang, L. Xu, Ultrasonic spray pyrolysis synthesis of three-dimensional $ZnFe_2O_4$-based macroporous spheres for excellent sensitive acetone gas sensor, Sensors and Actuators B: Chemical, 297 (2019) 126755. https://doi.org/10.1016/j.snb.2019.126755

[69] GY. Zhao, LJ. Liu, JR. Li, Q. Liu, Efficient removal of dye MB: through the combined action of adsorption and photodegradation from $NiFe_2O_4/Ag_3PO_4$, Journal of Alloys and Compounds, 664 (2016) 169-74. https://doi.org/10.1016/j.jallcom.2016.01.004

[70] A.Sudhaik, P. Raizada, P. Shandilya, P. Singh, Magnetically recoverable graphitic carbon nitride and $NiFe_2O_4$ based magnetic photocatalyst for degradation of oxytetracycline antibiotic in simulated wastewater under solar light, Journal of environmental chemical engineering, 6(4) (2018) 3874-83. https://doi.org/10.1016/j.jece.2018.05.039

[71] C. Cai, J. Liu, Z. Zhang, Y. Zheng, H. Zhang, Visible light enhanced heterogeneous photo-degradation of Orange II by zinc ferrite ($ZnFe_2O_4$) catalyst with the assistance of persulfate, Separation and Purification Technology, 165 (2016) 42-52. https://doi.org/10.1016/j.seppur.2016.03.026

[72] E. Ferdosi, H. Bahiraei, D. Ghanbari. Investigation the photocatalytic activity of $CoFe_2O_4$/ZnO and $CoFe_2O_4$/ZnO/Ag nanocomposites for purification of dye pollutants. Separation and Purification Technology, 211 (2019) 35-9. https://doi.org/10.1016/j.seppur.2018.09.054

[73] M. Sundararajan, V. Sailaja, LJ. Kennedy, JJ. Vijaya, Photocatalytic degradation of rhodamine B under visible light using nanostructured zinc doped cobalt ferrite: kinetics and mechanism. Ceramics International, 43(1) (2017) 540-8. https://doi.org/10.1016/j.ceramint.2016.09.191

[74] A. Ren, C. Liu, Y. Hong, W. Shi, S. Lin, P. Li, Enhanced visible-light-driven photocatalytic activity for antibiotic degradation using magnetic $NiFe_2O_4/Bi_2O_3$ heterostructures, Chemical engineering journal, 258 (2014) 301-8. https://doi.org/10.1016/j.cej.2014.07.071

[75] A. Behera, D. Kandi, S. Sahoo, K Parida, Construction of isoenergetic band alignment between CdS QDs and $CaFe_2O_4$@$ZnFe_2O_4$ heterojunction: A promising ternary hybrid toward norfloxacin degradation and H_2 energy production. The Journal of Physical Chemistry C, 123 (2019) 17112-126. https://doi.org/10.1021/acs.jpcc.9b03296

[76] J. Low, C. Jiang, B. Cheng, S. Wageh, AA. Al-Ghamdi, J. Yu, A review of direct Z-scheme photocatalysts, Small Methods, 1(5) (2017) 1700080. https://doi.org/10.1002/smtd.201700080

[77] A.J. Bard, Photoelectrochemistry and heterogeneous photocatalysis at semiconductor, Journal of photochemistry, 10(1) (1979) 59-75. https://doi.org/10.1016/0047-2670(79)80037-4

[78] H. Tada, T. Mitsui, T. Kiyonaga, T. Akita, K. Tanaka, All-solid-state Z-scheme in CdS-Au-TiO_2 three component nanojunction system, Nature Material, 5(10) (2006) 782-786. 10.1038/nmat1734.

[79] J. Yu, S. Wang, J. Low, W. Xiao, Enahnced photocatalytic performance of direct Z-scheme g-C_3N_4-TiO_2 photocatalysts for the decomposition of formaldehyde in air, Physical Chemistry Chemical Physics, 15 (2013) 16883-16889. https://doi.org/10.1039/C3CP53131G

[80] Q. Liu, J. Cao, Y. Ji, X. Li, W. Li, Y. Zhu, X. Liu, J. Li, J. Yang, Y. Yang, Construction of a direct Z-scheme ZnS quantum dot (QD)-Fe_2O_3 QD heterojunction/reduced graphene oxide nanocomposite with enhanced photocatalytic activity, Applied Surface Science, 506 (2020) 144922. https://doi.org/10.1016/j.apsusc.2019.144922

[81] J. Wang, C. Li, J. Cong, Z. Liu, H. Zhang, M. Liang, J. Gao, S. Wang, J. Yao, Facile synthesis of nanorod-type graphitic carbon nitride/Fe_2O_3 composite with enhanced photocatalytic performance, Journal of Solid-State Chemistry, 238 (2016) 246-51. https://doi.org/10.1016/j.jssc.2016.03.042

[82] R. Koutavarapu, MR. Tamtam, CR. Myla, M. Cho, J. Shim, Enhanced solar-light-driven photocatalytic properties of novel Z-scheme binary $BiPO_4$ nanorods anchored onto $NiFe_2O_4$ nanoplates: Efficient removal of toxic organic pollutants, Journal of Environmental Sciences, 102 (2021) 326-40. https://doi.org/10.1016/j.jes.2020.09.021

[83] P. Mishra, A. Behera, D. Kandi, S. Ratha, K. Parida, Novel magnetic retrievable visible-light-driven ternary Fe_3O_4@$NiFe_2O_4$/phosphorus-doped g-C_3N_4 nanocomposite photocatalyst with significantly enhanced activity through a double-Z-scheme system, Inorganic chemistry, 59(7) (2020) 4255-72. https://doi.org/10.1021/acs.inorgchem.9b02996

[84] S. Balu, S. Velmurugan, S. Palanisamy, SW. Chen, V. Velusamy, TC. Yang, ES. El-Shafey, Synthesis of α-Fe_2O_3 decorated g-C_3N_4/ZnO ternary Z-scheme photocatalyst for degradation of tartrazine dye in aqueous media, Journal of the Taiwan Institute of Chemical Engineers, 99 (2019) 258-67. https://doi.org/10.1016/j.jtice.2019.03.011

[85] Q. Xu, L. Zhang, B. Cheng, J. Fan, J. Yu. S-scheme heterojunction photocatalyst, Chem, (2020). https://doi.org/10.1016/j.chempr.2020.06.010

[86] Fu J, Xu Q, Low J, Jiang C, Yu J, Ultrathin 2D/2D WO_3/g-C_3N_4 step-scheme H_2-production photocatalyst. Applied Catalysis B: Environmental, 243 (2019) 556-65. https://doi.org/10.1016/j.apcatb.2018.11.011

[87] A. Meng, B. Cheng, H. Tan, J. Fan, C. Su, J. Yu, TiO_2/polydopamine S-scheme heterojunction photocatalyst with enhanced CO_2-reduction selectivity, Applied Catalysis B: Environmental, 289 (2021) 120039. https://doi.org/10.1016/j.apcatb.2021.120039

[88] J. Wang, Q. Zhang, F. Deng, X. Luo, DD, Dionysiou, Rapid toxicity elimination of organic pollutants by the photocatalysis of environment-friendly and magnetically recoverable step-scheme $SnFe_2O_4$/$ZnFe_2O_4$ nano-heterojunctions, Chemical Engineering Journal, 379 (2020) 122264. https://doi.org/10.1016/j.cej.2019.122264

[89] Z. Dai, Y. Zhen, Y. Sun, L. Li, D. Ding, $ZnFe_2O_4$/g-C_3N_4 S-scheme photocatalyst with enhanced adsorption and photocatalytic activity for uranium (VI) removal, Chemical Engineering Journal, 415 (2021) 129002. https://doi.org/10.1016/j.cej.2021.129002

[90] B. Palanivel, M. Lallimathi, B. Arjunkumar, M. Shkir, T. Alshahrani, KS. Al-Namshah, MS. Hamdy, S. Shanavas, M. Venkatachalam, G. Ramalingam, rGO supported g-C_3N_4/$CoFe_2O_4$ heterojunction: Visible-light-active photocatalyst for effective utilization of H_2O_2 to organic pollutant degradation and OH radicals production, Journal of Environmental Chemical Engineering, 9(1) (2021) 104698 https://doi.org/10.1016/j.jece.2020.104698

[91] A. Shukla, AK. Bhardwaj, BK. Pandey, SC. Singh, KN. Uttam, J. Shah, RK. Kotnala, R. Gopal, Laser synthesized magnetically recyclable titanium ferrite nanoparticles for photodegradation of dyes, Journal of Materials Science: Materials in Electronics, 28(20) (2017) 15380-6. https://doi.org/10.1007/s10854-017-7423-3

[92] SA. Jadhav, SB. Somvanshi, MV. Khedkar, SR. Patade, KM. Jadhav, Magneto-structural and photocatalytic behavior of mixed Ni–Zn nano-spinel ferrites: visible light-enabled active photodegradation of rhodamine B, Journal of Materials Science: Materials in Electronics, 31 (2020) 11352-65. https://doi.org/10.1016/j.jallcom.2020.157996

[93] H. Liu, H. Hao, J. Xing, J. Dong, Z. Zhang, Z. Zheng, K. Zhao, Enhanced photocatalytic capability of zinc ferrite nanotube arrays decorated with gold nanoparticles for visible light-driven photodegradation of rhodamine B, Journal of materials science, 51(12) (2016) 5872-9. https://doi.org/10.1007/s10853-016-9888-5

[94] PP. Hankare, RP. Patil, AV. Jadhav, KM. Garadkar, R. Sasikala, Enhanced photocatalytic degradation of methyl red and thymol blue using titania–alumina–zinc ferrite nanocomposite, Applied Catalysis B: Environmental, 107(3-4) (2011) 333-9. https://doi.org/10.1016/j.apcatb.2011.07.033

[95] P. Xiong, Q. Chen, M. He, X. Sun, X. Wang, Cobalt ferrite–polyaniline heteroarchitecture: a magnetically recyclable photocatalyst with highly enhanced performances, Journal of Materials Chemistry, 22(34) (2012) 17485-93. https://doi.org/10.1039/C2JM31522J

[96] TB. Nguyen, RA. Doong, Heterostructured $ZnFe_2O_4/TiO_2$ nanocomposites with a highly recyclable visible-light-response for bisphenol A degradation, RSC Advances, 7(79) (2017) 50006-16. https://doi.org/10.1039/C7RA08271A

[97] F. He, Z. Lu, M. Song, X. Liu, H. Tang, P. Huo, W. Fan, H.Dong, X. Wu, S.Han, Selective reduction of Cu^{2+} with simultaneous degradation of tetracycline by the dual channels ion imprinted POPD-$CoFe_2O_4$ heterojunction photocatalyst, Chemical Engineering Journal, 360 (2019) 750-61. https://doi.org/10.1016/j.cej.2018.12.034

[98] KK. Das, S.Patnaik, S. Mansingh, A. Behera, A. Mohanty, C. Acharya, KM. Parida, Enhanced photocatalytic activities of polypyrrole sensitized zinc ferrite/graphitic carbon nitride nn heterojunction towards ciprofloxacin degradation, hydrogen evolution and antibacterial studies, Journal of colloid and interface science, 561 (2020) 551-67. https://doi.org/10.1016/j.jcis.2019.11.030

[99] B. Li, C. Lai, G. Zeng, L. Qin, H. Yi, D. Huang, C. Zhou, X. Liu, M. Cheng, P. Xu, C. Zhang, Facile hydrothermal synthesis of Z-scheme $Bi_2Fe_4O_9/Bi_2WO_6$ heterojunction photocatalyst with enhanced visible light photocatalytic activity, ACS applied materials & interfaces, 10(22) (2018) 8824-36. https://doi.org/10.1021/acsami.8b06128

[100] L. Li, CG. Niu, H. Guo, J. Wang, M. Ruan, L. Zhang, C. Liang, HY. Liu, YY. Yang, Efficient degradation of Levofloxacin with magnetically separable $ZnFe_2O_4$/NCDs/Ag_2CO_3 Z-scheme heterojunction photocatalyst: Vis-NIR light response ability and mechanism insight. Chemical Engineering Journal, 383 (2020) 123192. https://doi.org/10.1016/j.cej.2019.123192

[101] S. Chen, D. Huang, G. Zeng, X. Gong, W. Xue, J. Li, Y. Yang, C. Zhou, Z. Li, X. Yan, T. Li, Modifying delafossite silver ferrite with polyaniline: Visible-light-response Z-scheme heterojunction with charge transfer driven by internal electric field, Chemical Engineering Journal, 370 (2019) 1087-100. https://doi.org/10.1016/j.cej.2019.03.282

[102] W. Wang, N. Li, K. Hong, H. Guo, R. Ding, Z. Xia, Z-scheme recyclable photocatalysts based on flower-like nickel zinc ferrite nanoparticles/ZnO nanorods: enhanced activity under UV and visible irradiation, Journal of Alloys and Compounds, 777(2019) 1108-14. https://doi.org/10.1016/j.jallcom.2018.11.075

Materials Research Foundations **112** (2021) 162-188 https://doi.org/10.21741/9781644901595-4

Chapter 4

Suitability of Ferrites for Biomedical Applications

Meenakshi Dhiman[1, *] and Shikha Rana[2]

[1]Chitkara University Institute of Engineering and Technology, Chitkara University, Punjab, India

[2]Department of Physics, Himachal Pradesh University, Shimla, India

meenakshi.dhiman@chitkara.edu.in

Abstract

This chapter aims at the latest amelioration in the biomedical application of spinel ferrites. Magnetic hyperthermia treatment of tumour cells and contrast resonance imaging of human body parts, are a few applications of spinel ferrites that have gained substantial interest nowadays. The thermo-therapeutic and radiology based diagnosis treatment has many benefits over traditional methods used for disease diagnosis and imaging treatment of chronic carcinogenic origin diseases. The various types of doped and undoped spinel ferrite nanoparticles are investigated and optimised for different biomedical applications. Through this chapter, an attempt is made to provide readers the knowledge about various required characteristics and latest used strategies for the successful implementations of spinel ferrite in biomedical applications at commercial levels.

Keywords

Spinel Ferrite, Magnetic, Functionalization, Hyperthermia, Drug Delivery, Magnetic Resonance Imaging

Contents

1. Introduction

Spinel ferrite nano ranged magnetic particles (SFNMs) are ferrimagnets with exemplary magnetic characteristics, that are highly dependent on particle size and shape anisotropy [1]. SFNMs are generally divalent ion doped spinel metal oxides structures which have plenty of applications in storage [2] , electromagnetic devices [3], catalysis [4] sensing [5], targeted drug delivery [6], diagnosis and imaging [7], thermotherapic cancer treatment [8] and waste water contamination removal [9]. The doped cations are randomly arranged on two crystallographic sites in spinel structure and their arrangement determines the optical, magnetic and electrical characteristics of the SFNMs [10,11,12,13]. The doped cations have particular site preference according to sizes of dopant ion and of interstitial site, preparation method and type of reaction conditions or environment [14,15,16].The positioning of dopant ion within the crystallographic structure i.e. cation distribution has great impact on the magnetic behaviour of the SFNMs. The electron spin depends on the lattice sites and it determines the overall magnetic moment of the lattice [17,18,19]. On the bases of cation distribution, the SFNMs are divided into three main types; spinel, inverse spinel and mixed spinel configurations [20]. Extensive studies of each type of SFNMs are still undergoing and much has been investigated in various in-vitro, ex-vitro and in-vivo biomedical applications [21,22,23]. The main required characteristics of SFNMs in biomedical area are super paramagnetic nature at 300 °K. They also have small magnetic and eddy losses, large resistivity and permeability, high chemical, mechanical and thermal stability and compatibility [24,25,26]. These features of SFNMs entice the researchers' attention for their diversified biomedical uses.

The biomedical field areas include techniques which are engineered and designed with biological entities. The resulting hybrid systems are employed in conventional methods to form multifunctional and targeted diagnosis, imagining, and delivery treatments [27]. Despite a lot of experimentation in biomedical areas, fast diagnosis, easy contrast imaging, site specific drug delivery and many other like needs of the medical science is still unfilled. Metal/Metal oxides, polymers, hydrogels, Lipids, liposomes, ceramics like many material groups are employed in medical areas either by natural way or chemical way [28,29]. From last few decades, the advancements in bioengineering techniques are focused on nanodevices, nanodrug carrying systems and nanomagnetic imaging techniques, formed by incorporating properties of nanomaterials with hybrid systems [30]. These nano hybrid systems have shown efficacious delivery with enhanced precision and sensitivity. The interest in biomedical applications of SFNMs increases with the fact that they have retained good magnetic properties even after use of various functionalization techniques on the bare SFNMs. This chapter includes discussions on the various possible functionalization, synthesis techniques and properties of SFNMs which are utilized in biomedical fields.

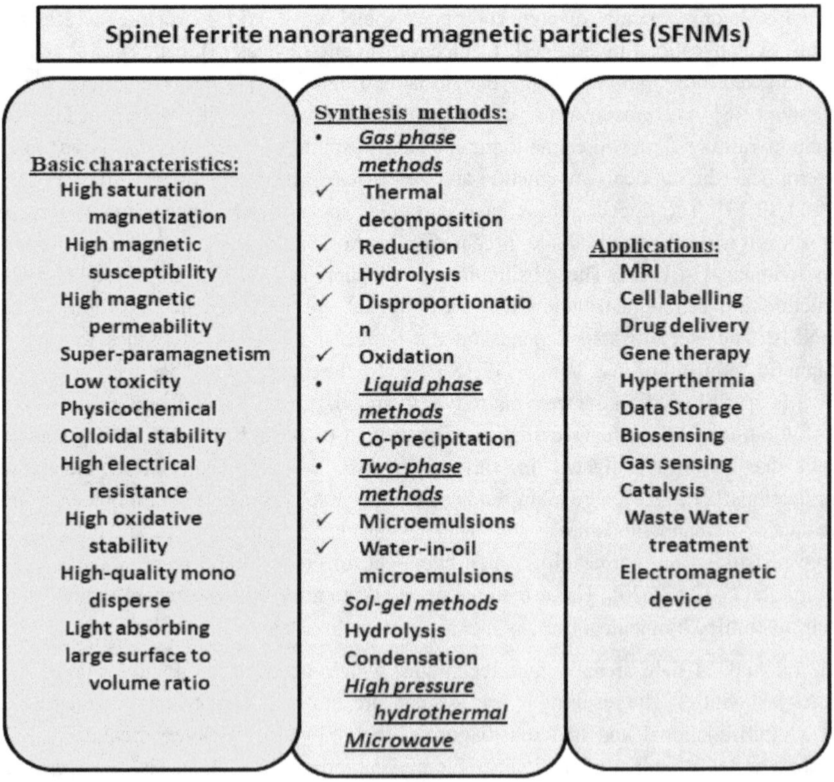

Spinel ferrite nanoranged magnetic particles (SFNMs)

Basic characteristics:
High saturation magnetization
High magnetic susceptibility
High magnetic permeability
Super-paramagnetism
Low toxicity
Physicochemical
Colloidal stability
High electrical resistance
High oxidative stability
High-quality mono disperse
Light absorbing
large surface to volume ratio

Synthesis methods:
- *Gas phase methods*
  - ✓ Thermal decomposition
  - ✓ Reduction
  - ✓ Hydrolysis
  - ✓ Disproportionation
  - ✓ Oxidation
- *Liquid phase methods*
  - ✓ Co-precipitation
- *Two-phase methods*
  - ✓ Microemulsions
  - ✓ Water-in-oil microemulsions

Sol-gel methods
Hydrolysis
Condensation
High pressure hydrothermal
Microwave

Applications:
MRI
Cell labelling
Drug delivery
Gene therapy
Hyperthermia
Data Storage
Biosensing
Gas sensing
Catalysis
Waste Water treatment
Electromagnetic device

Figure 1: Figure including basic characteristics, synthesis methods and applications of SFNMs.

2. Ferrites

The doped cubic SFNMs are widely studied ferromagnetic materials with settable magnetic properties. The various types of synthesis methods as mentioned in figure 1 are employed for preparation of SFNMs, which provides possibility to get engineered desired range SFNMs [31]. The SFNMs are generally of nano size and have large magnetic interactions which cause them to agglomerate easily. Since 1949, when Louis Néel explained magnetic interactions first time many theories and optimization techniques are used to control and explain intrinsic and extrinsic magnetic behaviour of the magnetic materials [32]. Krishnan et al. [33] have reviewed the magnetic characteristics dependence on the particle size of SFNMs. The synthesis, annealing and calcination temperature determines the particle size, phase formation and morphology of the prepared samples. SFNMs have been used in various Diagnostics and Therapeutic application as shown in figure 2. This common problem of agglomeration and bio stability of SFNMs is avoided by combining these magnetic entities with biopolymers, carbon based materials, silica, bioactive ingredients etc. [31]. Polyethylene glycol, Folic acid, Dextran and liposomes like ingredients are commonly employed nowadays for hybrid formations, which are applied in various biomedical applications of SFNMs [34,35]. Silica functionalized magnetic nanoparticles are assimilated with alkoxysilane surface active groups which expedite their multiple applications [35,4]. In any biomedical use, the desired features like controlled drug delivery, enhanced diagnostic behaviour and the interaction between the cell and hybrid system depends on the nature, size, morphology of the hybrid system, type of surface coating and charge. The toxic behaviour of SFNMs in clinical trials and medical diagnosis is highly determined by their surface chemistry, solution stability, size, dose and procedure used, internal structure, chemical composition, adsorption and opsonization rates [36].

Materials Research Foundations **112** (2021) 162-188 https://doi.org/10.21741/9781644901595-4

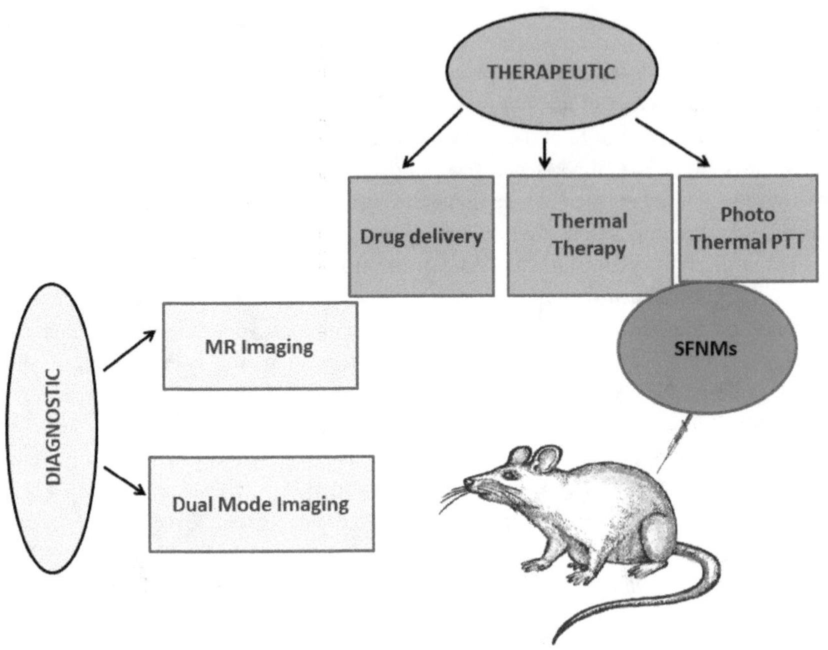

Figure 2: Figure showing various therapeutic and diagnostic applications of SFNMs.

3. Biomedical application

3.1 Hyperthermia

Magnetic hyperthermia refers to the elimination of cancer cells without significant damage to healthy cells, particularly overheating. The use of direct high temperature exposure for treatment of various body problems is known as 'Hyperthermia'. Hyperthermia means atypical rise in body temperature from external source, not similar to general body temperature rise situations like fever or heatstroke. Hyperthermia is a thermotherapic treatment which could be employed for removal of cyst, for control over infection and pain in the body. Also, the local or whole body hyperthermia treatment is chosen according to the level of cancer spreading in the patient body too, as cancer cells are less active at high temperature (>315 K) [37]. The various advancements in this thermotherapic treatment are differentiated by their power deposition method and the gross targeted body area. The Cancer treatment by a non-invasive thermotherapic method helps to regulate temperature of the tumour site with external governance through magnetic forces as shown in figure 3.Hyperthermia processes are generally done by using mobile heating carriers either externally or internally and water bath like heating source. Every used method has some advantages and disadvantages and used according to the level of cancer spread in the human body. Source heating is more effective and less invasive, whereas water bath heating is more pertinent for metastasis cancers. The chief limitations of hyperthermia treatment are local inhomogeneous thermal heating of affected region and abysmal heating of deeply located effected regions [8]. Hence, to deal with these limitations nanotechnology based treatment practices are widely studied nowadays. Nano-ranged materials are influencing every aspect of human life cycle by their exemplary characteristics. Every scientific field has some requirements in terms of desired shape, size, characteristics etc. and fortunately nowadays, excessive research on the properties and characteristics of nano-ranged materials through optimization leads to designing of desired materials. The various advancements in this thermotherapic treatment are differentiated by their power deposition method and the gross targeted body area [38].

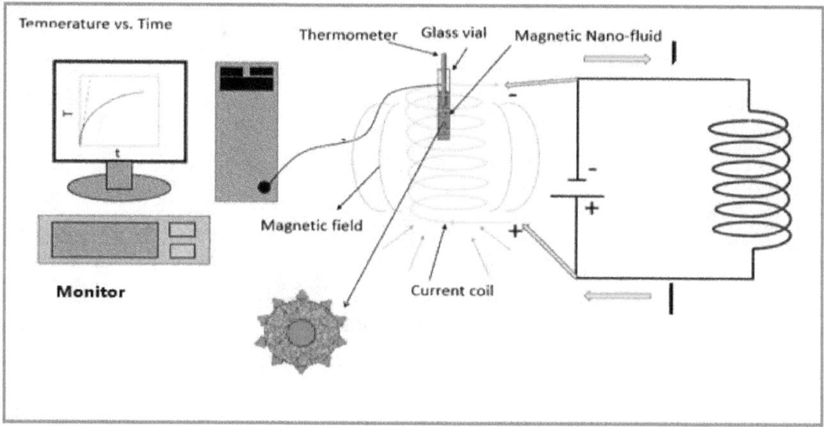

Figure 3 Illustration of Practical set up required for study of therapeutic nature of magnetic nanoparticles [41].

SFNMs are becoming most widely studied and employed class among all the nanomaterials due to unique physiochemical, magnetic and optical characteristics with easy functionalization possibilities at subatomic levels. SFNMs have shown multiple applications due to their small dimensional metal based inorganic structures which could

Materials Research Foundations **112** (2021) 162-188 https://doi.org/10.21741/9781644901595-4

be well governed by external magnetic forces [39]. Prudently, they have shown extraordinary reaction stability with desired physiochemical and magnetic nature for pharmaceutical and environmental applications. SFNMs are in medical use since last two centuries but with increase in possible research opportunity their use has become more approachable. SFNMs having high magnetic saturation values are found to have more heating capacity and hence provide required heat on the targeted cells [40]. The earliest used magnetic nanoparticles in 18thand 19thcenturies for hyperthermia treatment were super paramagnetic particles of magnetite [37]. W. Busch and W. B. Coley have used heat inducing phenomena for treatment of cancer in the late 19th century [41,42]. Later on R. K. Gilchrist and his co-researchers had used inductive heating for the treatment colon and rectum carcinoma in [43,44]. From last twenty years, many studies are going on doped magnetite structure known as SFNMs to study various synthesis and functionalization processes for enhancement of their magnetic behaviour which could be used for their in vitro and in vivo biomedical applications. In this part of the chapter we have provided an informative explanation about their intrinsic magnetic behaviour and their influence on thermo therapic uses.

Magnetic hyperthermia is generally accomplished by four distinct treatment methods as mentioned in table 1. According to table 1, intracellular and extracellular thermotherapies are the most efficient treatments for cancer treatment. In the first method, SFNMs like magnetic heat sources are placed deep within the tumor site with the help of external magnetic field whereas in the second one required heat is provided from outside with the help of sources like water bath or microwave etc. But in the second one extra heat is required as direct heating of the cancer cell is not possible. The intracellular hyperthermia is found to have more potent heat distribution but need sufficient clinical studies before their implementation in the human body. For all the strategies mentioned in the table 1 prescribed therapeutic agent are classified in three ranges, as in millimeter, micrometer and nanometer.

Human body has been observed to exhibit dissimilar reaction to carcinoma cells in comparison to healthy cells [45]. The multiplication rate of these cells is very less at particular thermal conditions and as a result, hyperthermia is given specifically at the affected human body areas in contrary to entire body. The SFNMs are known to have various types of B-H curves and hence their magnetic behaviour is also different. This magnetic loss with respect to variation in external frequency is widely investigated characteristic of SFNMs for thermotherapic use [46]. Magnetic nanometer therapeutic agents in a colloidal form are capable to reach tumor site and perform the required function without disturbing normal cells. Also, these nanocarriers have low Curie temperature which provides controlled heating with easy possible autogenous control

[47,48]. Magnetic nanocarrier use eddy current loss and hysteresis loss like phenomena to generate heat which are observed due to their multidomain structure and small size. To raise blood compatibility with decrease in agglomeration and smooth bonding with solution affinity groups, facile functionalization of SFNMs is generally done with various polymers, silica, fatty acid and protein like biocompatible groups [49,50].

Table 1: Commonly employed four basic hyperthermia methods [44].

Strategy	**Arterial Embolization**	**Direct Injection**	**Intracellular**	**Interstitial implantation**
Heat generation	Intravascular	Extracellular	Intracellular	Interstitial
Treatment method	Directly injected through Arterially embolizing	Injected directly in to the tumor site.	Directly injected through Arterial embolization or intravenous injection	Implantation of Heating seeds
Size of agent	10^{-6} to 10^{-9} m	10^{-6} to 10^{-9} m	10^{-9} m	10^{-3} nm
Merits	Effective temperature distribution and controlled temperature dose.	Low risk of arterial catheteriza-tion.	Generally used in treatment of metastases cases.	Less cell burning.

Fatty acids like Oleic acid has shown marvellous anti-cancer activity and hence nowadays it is widely studied in combined hyperthermia based targeted drug delivery in cancer patients [51]. Prashant B. Kharat et al. [52] has used coprecipitation method to synthesize Oleic Acid functionalized Cobalt ferrite samples with mean particle size of 10 nm. The functionalized sample has high saturation magnetization (76.86emu/g) with small coercive field and remanence values. Both the samples were investigated for induction heating behaviour which is found to rise with the concentration of the samples. The SAR value is calculated for bare sample as 248 W/g and for functionalized one it is found to be 216 W/g with very small 1 mg/mL concentration, which makes the prepared sample a potential candidate for hyperthermia treatment. Also, Sandeep B. Somvanshi and his group [53] have synthesized a composite oleic acid and magnesium ferrite. They have also investigated physiochemical behaviour of the formed composite and found the average crystallite size of bare and coated Magnesium ferrite particle as 18 nm and 20 nm respectively. The room temperature magnetic saturation values are 41emu/gm and 38emu/gm approximately, with very small remanence ratios for bare and coated ones respectively. With the coating of Oleic acid colloidal stability is found to increase with increase in the Specific Absorption rate (SAR) as 98 W/g in comparison to 91 W/g of bare magnesium ferrite. The maximum SAR comes with least dose of 8 mg/mL of the composite of Oleic acid with Magnesium ferrite. Cigdem E. Demirci and his research group [54] have prepared various ferrite coated oleic acid composites with probable thermotherapic efficacy. The prepared composites are further coated with N-(trimethoxysilylpropyl)-ethylene diamine-tri-acetate to increases their convective diffusion capacity and their heating capacity was observed in two carrier solution with different concentration. The maximum SLP was obtained as 315 W/g and 295 W/g for CoFe2O4 and $MnFe_2O_4$ or $NiFe_2O_4$ respectively at frequency 195 kHz and applied magnetic field 50 kA/m, which confirm the potential use of the formed composites as hopeful hyperthermia agent. Carbon nanotubes provides high surface area and could easily enter in cellular membrane [55]. In the same context, Papori Seal et al. [56] has incorporated manganese ferrite with multiwalled carbon nano tube by co-precipitation method followed by stirring. The crystallite size of the unfunctionalized and functionalized sample is found to be 4.5 and 6.9 nm. The SAR values are found to increase from 42.22 W/g to 53.14 W/g after functionalization. The heating effect of the colloidal suspension of the bare and functionalized sample is observed up to ~ 45 °C and ~ 46 °C in 30 minutes at 231 kHz working frequency, which also confirms their potential hyperthermic behaviour.

Table 2: Recent articles of SFNMs in hyperthermia studies.

SFNMs Particle	Preparation technique	Coating	Morphology (size)	Magnetic behaviour	Hyperthermic results	Ref
Mn-Zn Fe2O4	Co-Precipitation	Lauric Acid	Spherical (11nm)	SPM, Ms = 4kA/m, Magnetic	SAR = 456.4 kW/kg for 0.25 mg/mL with excellent cytotoxicity for HaLe cells	[57]
Mn-Ga Fe2O4	Sol-gel auto combustion	Oleic Acid-Pluronic F127	Near spherical (15-34 nm)	Ms = 33.5-30.2 emu/gm Mr = 0.59-0.61 emu/gm Hc = 3.41-3.06 Oe	Max SAR = 160.9 W/g for 3 mg/mL of the coated sample at frequency 354 kHz and magnetic field = 10.2 kA/m and non-Hemolytic nature with low toxicity	[58]

Fe3O4@CoFe2O4 and Fe3O4@ZnCoFe2O4	Two step Co-Precipitation	-	Core shell structure(9-12 nm)	Ms = 33.7 – 59.3 emu/gm Mr = 2.7 – 10.75 emu/gm Hc = 40.5-225 Oe	Best SLP value = 379.2 W/g at frequency 97 kHz and magnetic field = 50 kA/m	[59]
$Co_xMn_{1-x}Fe_2O_4$	One pot oxidative hydrolysis method	Europium(III) with phen and TTA	Near spherical 25.4 nm	Ms = 58-73 emu/gm	At frequency 335 kHz and magnetic field = 12kA/m and SAR values 201 W/gm for 0.5 mg/mL concentration.	[60]
Hematite and $NiFe_2O_4$	Hydrothermal	-------------	Spherical and Hexagonal shapes (50 and 7 nm)	Super-paramagnetism ,Ms = 2.36 and 18.8 emu/gm Hc = 18 and 5 Oe Mr = 0.08 and 0.09 emu/gm	At Frequency 332.8 Hz and magnetic field = 170Oe and SAR values 3.5 and 4.5 W/gm respectively.	[61]
Li-Zn Fe_2O_4	Sol gel	RGO	Spherical (15 nm)	Ms = 62 emu/gm (bare) and 17 emu/gm (composite)	At frequency 290 kHz and magnetic field 335Oe and SAR = 502 W/gm (bare) and 334 W/gm (composite) for 1 mg/mL concentration.	[62]
Zn Fe_2O_4	coprecipitation	Chitosan	Core shell structure (12 nm)	Ms = 87 emu/gm	At frequency 360 kHz and applied magnetic field = 33.3 kA/m and max SAR value = 264 W/g for 0.18mg/mL	[63]
Mn-Zn Fe_2O_4	Sol gel auto combustion	PEG	Spherical (15.7 nm)	Ms = 30.1 - 44.4 emu/gm Mr = 15.6 – 5.7 emu/gm Hc = 497-124 Oe	At applied frequency = 425 kHz and Magnetic field = 180 Oe and SAR = 8.6 – 3.5W/gm	[64]

Materials Research Foundations **112** (2021) 162-188 https://doi.org/10.21741/9781644901595-4

3.2 Target specific drug delivery

From last half a century, carcinoma cells are becoming a foremost reason of human casualty and hence become utmost investigated research area in medical science. Till date, mainly chemotherapy treatment is employed for cancer treatment but extended selectivity, less solution solubility, developed drug resistance and biocompatibility and inimical effects on healthy cells like limitations of old chemotherapy are need to be addressed [65]. With capability of nanoparticles to control and change pharmacokinetics, nanoparticles have become first choice for fabrication of nanodrug systems. Nanodrugs have upgraded solution solubility, enhanced half-life and improved bio distribution results in comparison to tradition chemo drugs [66]. They also have more approaches towards cancerous tissues by entering endothelial layers in the tumor surroundings which have diameter ranging from 30 to 470 Angstroms. They show possible amassing in the interstitial tumor liquid due to enhanced permeability and retention aftermath showing low toxic behaviour. This effect was first observed, reported and investigated by Maeda et al. [67,68]. This effect offers cancer site targeted drug delivery through nanodrug carriers. The standardized drug delivery through magnetic nanoranged particles on the felicitous tumor site without any noticeable activity loss, after entering the effected region the drug reaches to the malignant cells. The magnetic nano ranged particles are transported to the tumor site mainly by two ways i.e active and passive targeting ways [69]. Active drug delivery way includes delivery of drug on specific site by functionalization of SFNMs with ligand or antibody groups for easy molecular recognition. Active drug delivery has increased the curative rate of chemo drug by enhancing the drug effectiveness. The various types of specific groups like folic acid, proteins and amide groups, RNA, gene, etc. are recently investigated for tumor specific hybrid system [70]. Passive drug delivery depends on the fact that nano ranged particles below 200 nm can comfortably leak through the micro vessels of malignant sites. Target specific drug delivery involves either ligand or antibody arbitrated methods [71,72,73].

Magnetic nanoparticles with different size and shape possess different physiochemical properties and also behave differently when used in the same systems. The various shapes and characteristics of the magnetic nanomaterials are governed by reaction media parameters like temperature, pH, precursor concentration, preparation time etc. [74]. Surface engineered and functionalized SFNMs with desired dimension and formulations are recently employed in clinical studies of drug delivery. Various multifunctional, specifically assembled core/shell hybrid magnetic ceramic groups are investigated as nanocarriers for in vivo target specific drug delivery in last three decades [75]. The type and concentration of the functional groups present on the nanocarrier surface is decided by the coating type and procedure employed [76]. This type of bonding is generally used for small size molecules. The augmentation of these magnetic nanocarriers is needed utmost to fulfil various requirements of medical science. They provide feasible formulations for regulated drug transmittance and stowing profiles [77]. Several drug loading methods are a result of either chemical bonding or physical interplay between functional group on the surface of the nano-carriers and conjoined ligand [78]. The chemical and solvent stability of SFNMs is increased by coating with various surfactants, polymers, micelles, peptides etc. The bonding through covalent bond is more firmed and generally present between conjoined ligand and thiol, amine and carboxylic acid like groups present on the surface of nanocarriers [79]. They decrease the side effects of the used cyto-toxic drug by controlling their dosage and systematic delivery in the human body [80]. For last few decades, nanoengineered structures which are stable and could release drug on the correct location with controlled release mechanism are being efficiently investigated. The physical interaction involves comparatively fast bond formation due to electrostatic, hydrophobic or hydrophilic interactions [81]. The magnetic nanocarrier based targeted drug delivery was initially investigated in 1970's and further with the advancement in surface engineering dimension of drug carrier, reduced dimensions from micro to nano levels [82]. For sustained release of drug, various external (magnetic field, temperature, light) or internal (pH medium, enzyme) stimuli have been used as magnetic nanocarriers also provide opportunity for drug release in real time imagining [83,84,85].

Among different types of organic and inorganic nanohybrid systems, magneto-liposome is a promising hybrid magnetic entity which could be easily guided externally by magnetic field and could easily perform targeted isolation of tumor cells and effective drug delivery like functions [85]. Figure 4 shows illustration of target specific drug delivery governed externally with magneto-Liposome injection. The magnetic nature of this system is determined by particle size of magnetic nanoparticles. Recently, Beatriz D. Cardoso et al. [86] had synthesized a hybrid magnesium ferrite-liposome nano system. The synthesized nanoparticles have SPM nature and crystallite size around 25 nm. The curcumin loading efficiency on the composite was investigated by studying quenching effect in fluorescence emission method. The encapsulation of curcumin is found to be high (>89%). The formed nano-system has shown excellent characteristics required for drug delivery.

Materials Research Foundations **112** (2021) 162-188 https://doi.org/10.21741/9781644901595-4

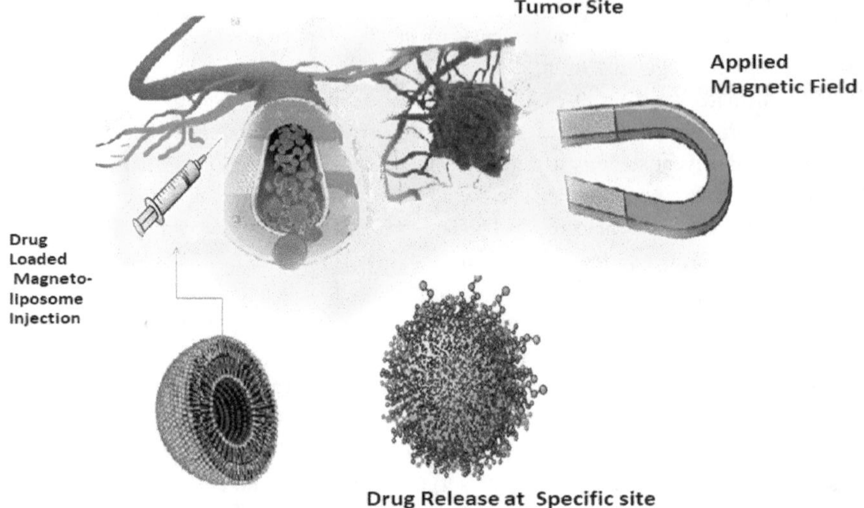

Figure 4: Illustration of target specific delivery of drug with magneto-Liposome injection governed with external magnetic field.

K. Sriram et al. [87] have prepared and investigated nanodrug efficiency of a hybrid system consisted of vanillin- Chitosan functionalized calcium ferrite. Taguchi method was employed to find drug encapsulation efficiency of the prepared composite. The various ratios of curcumin and vanillin-chitosan were investigated and the highest encapsulation rate was observed as 98.3%. The formed composites were also studied for in-vitro cytotoxic behaviour against MCF-7 breast cells. Their biocompatible nature was confirmed by enhanced cell viability of 1,929 fibroblast cells. Recently, J. Panda and his co-workers [88] has synthesized nano drug carrier system consisted of Cobalt ferrite functionalized with Oleic acid and Poly lactic acid and glycolic acid. The efficient loading of Docetaxel drug was found to be 8.4% with efficiency rate 81.8% on the prepared ferromagnetic sample of size 21 nm. The synthesized nanodrug carrier has good internalization efficiency and antiproliferative behaviour with very less toxicity against human RBCs.

Camp thothecin is a plant derived chemo drug usually used for the treatment of various types of cancers [89]. Isarel V.M.V. Enoch et al. [90] studied drug loading and delivery efficiency of SPM manganese ferrite sample having size approximately 93 nm. The crystalline nature of as- prepared and modified samples was confirmed by XRD results and the saturation magnetization values were found to be 38 emu/g for as-prepared and 33 emu/g for modified sample. The drug loading efficiency on coated sample was 86% of β – cyclodextrin-PEG-Manganese ferrite and 60% of PEG-Manganese ferrite samples. The increase in drug loading confirms encapsulation of Camptothecin in β – cyclodextrin. The in- vitro drug release profile shows 54.5 %and 69.4% drug release at pH 7.4 and 6.0 respectively. The drug release profile shows slow release of drug which can be attributed to firm binding between polymer and drug. The cytotoxic effect of prepared and coated samples is studied on HCT-15 and HEK293 by using MTT assay. All the observed results confirmed the potential application of the prepared sample for efficient target specific drug delivery. The same group of researcher [91] has also studied anticancer efficacy of the composite of β – cyclodextrin with Nickel ferrite loaded with Campthothecin drug. The cytotoxic nature of the prepared composite was investigated on various cell lines. The drug loading efficacy was found to be 80% and drug release rate increase with decrease in pH value, which confirms promising use of functionalized ferrite in efficient nano drug delivery systems.

3.3 Magnetic resonance imaging

It is an imaging technique which produces image of body part by utilizing the nuclear magnetic resonance phenomena shown by water molecules or fat present in human body and help in differentiating between normal and pathological tissues. The various types of atoms like Hydrogen, Helium, Carbon, Oxygen etc. can be utilized for the MR imaging but the commonly used is Hydrogen [92]. The nuclear spin of hydrogen atom gets aligned according to applied magnetic field either in parallel or anti-parallel way. The intensity difference between both the possible alignments is calculated by the given formula:

$$\Delta E = \frac{\gamma h B}{2\pi} \quad (1)$$

Where, γ is gyromagnetic ratio, B is applied magnetic field and h is Planck's constant [93].This difference is very small and leads to less sensitivity due to feeble detection of hydrogen molecules. On the application of external resonant alternating radio frequency, the applied electromagnetic energy is absorbed by nuclei and this leads to alignment of nuclei in anti-parallel direction. Molecules regain their original spin alignment and come to relaxation state after removal of radio frequency [94]. The recovery of longitudinal magnetization (parallel alignment) and decay of transverse magnetization (antiparallel

Materials Research Foundations **112** (2021) 162-188 https://doi.org/10.21741/9781644901595-4

alignment) are two possible NMR relaxation phenomenon and time taken for these two processes T1 and T2 determine the signal intensity [95]. There is a variety of contrast agents that are used in MRI to enhance the imaging quality by improving relaxation time. The use of magnetic nanomaterials in MRI provides better control over longitudinal and transverse relaxation times due to magnetic interactions between hydrogen atom spin and electronic spin of the used magnetic nanomaterials [96]. The efficiency of the contrast agent is calculated as the slope of the graph between the equivalent ion concentration and any of the relaxation time (r_1 and r_2), which is also known as 'relaxivity' [97]. The efficiency of the contrast agent is determined by type, shape, size, hydrophobic nature and clustering effect like factors. The relaxivity depends on the bonding between paramagnetic ion and water molecules and dephasing of parallel aligned nuclear spins [98]. In case of bulk magnetic contrast agents the dephasing of aligned nuclear spin phenomena dominates. As we move towards nanosized contrast agents, their magnetic behaviour highly depends on the size as their multidomain structure that becomes single domain below 10 nm. The critical size of single domain can be calculated by

$$D_c = \frac{36\sqrt{AK}}{\mu_o {M_s}^2} \qquad (2)$$

Where A and K are exchange and anisotropy constant, μ_o is permeability and M_s is magnetic saturation value[99]. The size of magnetic nanomaterials is found to control the r_2 values. The latest surface engineering of magnetic nanoparticles provides better control on r1 and r2 values by changing the surface morphology and core-shell composition [100,101]. The type of coating determines the relaxivity and protons behaviour and generally decreases with decrease in dimension of the contrast agent [102].The SFNMs based Contrast agents have core shell structures in which core is magnetic to provide enhancement to signal and outer shell has water loving groups and bioactive targeting materials to enhance compatibility [103]. Georgia Basina et. al. [104] have prepared hydrophilic functionalized magnetite and manganese ferrite samples and confirmed their crystallinity and surface morphology by XRD and HRTEM results. The prepared samples have particle size between 10-12 nm and they are modified with organic coating of tri-block copolymer Phronic F-127. The values of transverse relaxation times T_2 were investigated and relaxivity values are observed as 324 and 453 m/M/sec for manganese ferrite and magnetite samples. The obtained results have confirmed the potential application of the prepared sample as MRI contrast agent. Yuxiang Sun and his co-workers [105] have synthesized oleic acid functionalized Manganese-zinc ferrite nanoparticles of different shapes (Sphere shaped, worm shaped and agglomerated cluster type). They have also loaded the widely used Paclitaxel drug on these samples for dual

Materials Research Foundations **112** (2021) 162-188 https://doi.org/10.21741/9781644901595-4

mode in vivo MRI and drug release studies. The drug loading rate and release rate is found to vary with the shape and observed to be good for agglomerated cluster and worm shaped samples. After investigated in vivo tumor the ranostics studies, the researchers have confirmed the potential in vivo use of the prepared samples for MRI studies. L. Ansari et al. [106] has prepared a hybrid of Zinc- Cobalt ferrite functionalized with DMSA for in vivo MRI studies. The prepared samples have size between 5 - 10 nm and MR imaging study was done on rats. The T2 values of the bare and modified samples were found to increase with decrease in ferrite sample and relaxivity value increase with increase in ferrite sample concentration. The r2 values are observed as 32.85 and 168.96 mmol/L/sec for bare and modified samples. The investigated results confirmed the potential use of prepared samples as MRI contrast agent.

Conclusion

The interest of researchers in investigating magnetic nanostructures in arenas of biomedical and healthcare sector application is majorly due to their nontoxic behaviour, biocompatible nature and above all, the ease at which these can be maneuvered externally with the application of electrical and magnetic fields, makes them potential players. SFNMs and their composites with desired size and engineered structures and shapes are recently employed for various biomedical applications. Although Fe_3O_4 has conventionally been employed for customary clinical studies, but with easy functionalization methods and various tailored characteristics SFNMs are widely studied. The efficiency of SFNMs has clearly found to be depended on the surface chemistry and by applying biopolymers as stabilizer for core-shell their functionality can still be increased. A significant breakthrough has been made by superparamagnetic particles in achieving high efficacy in cancer treatment as heating agents in Magnetic field assisted Hyperthermia Treatment. During secured procedures, superparamagnetic Nano particles are used to enhance effectiveness through homogeneous heating of the target tissues but adequate control is required as overheating or inhomogeneous heating among healthy cells might occur. Generally the small size of SFNMs brings about many benefits, such as uniform dispersion and distribution of nanoparticles resulting in uniform heat generation, prevention of agglomeration, fine tuning of shapes and sizes, biophysical properties like hydrophilicity, hydrophobicity etc. result in increased functionality further provides several advantages like lowering dosage, effective pharmaceutical activity, improved drug stability, minimization of side effects and efficient drug accumulation, which are much desired in biomedical application systems.

Although these SFNMs are much studied and investigated for their applications in MRI, hyperthermia, targeted drug delivery etc. biomedical applications. However, there is still a vast scope identified and suggested by the researchers for improvisation in the efficacy of nanomaterials as drug delivery systems, in advanced nano formulations for bio-applications and extension in cancer treatment agents.

References

[1] M. Dhiman, S. Rana, M. Singh, and J. K. Sharma, "Magnetic studies of mixed Mg–Mn ferrite suitable for biomedical applications," *Integr. Ferroelectr.*, vol. 202, no. 1, pp. 29–38, 2019, https://doi.org/10.1080/10584587.2019.1674821

[2] Y. Fu *et al.*, "Copper ferrite-graphene hybrid: A multifunctional heteroarchitecture for photocatalysis and energy storage," *Ind. Eng. Chem. Res.*, vol. 51, no. 36, pp. 11700–11709, 2012, https://doi.org/10.1021/ie301347j

[3] P. Bhattacharya, S. Dhibar, G. Hatui, A. Mandal, T. Das, and C. K. Das, "Graphene decorated with hexagonal shaped M-type ferrite and polyaniline wrapper: A potential candidate for electromagnetic wave absorbing and energy storage device applications," *RSC Adv.*, vol. 4, no. 33, pp. 17039–17053, 2014, https://doi.org/10.1039/c4ra00448e

[4] S. Rana, A. Sharma, A. Kumar, S. S. Kanwar, and M. Singh, "Utility of Silane-Modified Magnesium-Based Magnetic Nanoparticles for Efficient Immobilization of Bacillus thermoamylovorans Lipase," *Appl. Biochem. Biotechnol.*, vol. 192, no. 3, pp. 1029–1043, 2020, https://doi.org/10.1007/s12010-020-03379-7

[5] N. S. Chen, X. J. Yang, E. S. Liu, and J. L. Huang, "Reducing gas-sensing properties of ferrite compounds MFe2O4 (M = Cu, Zn, Cd and Mg)," *Sensors Actuators, B Chem.*, vol. 66, no. 1, pp. 178–180, 2000, https://doi.org/10.1016/S0925-4005(00)00368-3

[6] M. S. Dahiya, V. K. Tomer, and S. Duhan, *Metal-ferrite nanocomposites for targeted drug delivery*. Elsevier Inc., 2018

[7] N. Alghamdi *et al.*, "Structural, magnetic and toxicity studies of ferrite particles employed as contrast agents for magnetic resonance imaging thermometry," *J. Magn. Magn. Mater.*, no. October, 2019, https://doi.org/10.1016/j.jmmm.2019.165981

[8] A. Meidanchi and H. Ansari, "Copper Spinel Ferrite Superparamagnetic Nanoparticles as a Novel Radiotherapy Enhancer Effect in Cancer Treatment," *J. Clust. Sci.*, vol. 32, no. 3, pp. 657–663, 2021, https://doi.org/10.1007/s10876-020-01832-5

[9] Y. J. Tu, C. K. Chang, C. F. You, and S. L. Wang, "Treatment of complex heavy metal wastewater using a multi-staged ferrite process," *J. Hazard. Mater.*, vol. 209–210, pp. 379–384, 2012, https://doi.org/10.1016/j.jhazmat.2012.01.050

[10] G. Kumar, R. Rani, V. Singh, S. Sharma, K. M. Batoo, and M. Singh, "Magnetic study of nano-crystalline cobalt substituted mg-mn ferrites processed via solution combustion technique," *Adv. Mater. Lett.*, vol. 4, no. 9, pp. 682–687, 2013, https://doi.org/10.5185/amlett.2013.1409

[11] N. Rezlescu and E. Rezlescu, "Dielectric properties of copper containing ferrites," *Phys. Status Solidi*, vol. 23, no. 2, pp. 575–582, 1974, https://doi.org/10.1002/pssa.2210230229

[12] M. Dhiman, S. Rana, Sanansha, N. Kumar, M. Singh, and J. K. Sharma, "Influence of Ho3+ substitution on structural and magnetic properties of Mg–Mn ferrites," *J. Mater. Sci. Mater. Electron.*, vol. 32, no. 7, pp. 8756–8766, 2021, https://doi.org/10.1007/s10854-021-05547-9

[13] H. Wang, Y. Song, X. Ye, H. Wang, W. Liu, and L. Yan, Asymmetric Supercapacitors Assembled by Dual Spinel Ferrites@Graphene Nanocomposites as Electrodes, vol. 1, no. 7. 2018

[14] M. Dhiman, S. Rana, K. Batoo, J. K. Sharma, and M. Singh, "Synthesis and characterization of Y and Sm doped Mg nanoferrites," *Integr. Ferroelectr.*, vol. 184, no. 1, pp. 151–157, 2017, https://doi.org/10.1080/10584587.2017.1368634

[15] H. Shokrollahi and K. Janghorban, "Influence of additives on the magnetic properties, microstructure and densification of Mn-Zn soft ferrites," *Mater. Sci. Eng. B Solid-State Mater. Adv. Technol.*, vol. 141, no. 3, pp. 91–107, 2007, https://doi.org/10.1016/j.mseb.2007.06.005

[16] M. Tadic, S. Kralj, M. Jagodic, D. Hanzel, and D. Makovec, "Magnetic properties of novel superparamagnetic iron oxide nanoclusters and their peculiarity under annealing treatment," *Appl. Surf. Sci.*, vol. 322, pp. 255–264, 2014, https://doi.org/10.1016/j.apsusc.2014.09.181

[17] Y. Matsuoka McClain, *Handbook of Modern.* 1981

[18] A. Globus, H. Pascard, and V. Cagan, "and Fundamental Properties in Ferrites," no. 4, pp. 163–168, 1977

[19] R. M. Bozorth, E. F. Tilden, and A. J. Williams, "Anisotropy and magnetostriction of some ferrites," *Phys. Rev.*, vol. 99, no. 6, pp. 1788–1798, 1955, https://doi.org/10.1103/PhysRev.99.1788

[20] S. Singhal and K. Chandra, "Cation distribution and magnetic properties in chromium-substituted nickel ferrites prepared using aerosol route," *J. Solid State Chem.*, vol. 180, no. 1, pp. 296–300, 2007, https://doi.org/10.1016/j.jssc.2006.10.010

[21] M. Gorgizadeh, N. Azarpira, M. Lotfi, F. Daneshvar, F. Salehi, and N. Sattarahmady, "Sonodynamic cancer therapy by a nickel ferrite/carbon nanocomposite on melanoma tumor: In vitro and in vivo studies," *Photodiagnosis Photodyn. Ther.*, vol. 27, no. March, pp. 27–33, 2019, https://doi.org/10.1016/j.pdpdt.2019.05.023

[22] P. Sharma *et al.*, "Nanomaterial Fungicides: In Vitro and In Vivo Antimycotic Activity of Cobalt and Nickel Nanoferrites on Phytopathogenic Fungi," *Glob. Challenges*, vol. 1, no. 9, p. 1700041, 2017, https://doi.org/10.1002/gch2.201700041

[23] P. L. Venugopalan, R. Sai, Y. Chandorkar, B. Basu, S. Shivashankar, and A. Ghosh, "Conformal cytocompatible ferrite coatings facilitate the realization of a nanovoyager in human blood," *Nano Lett.*, vol. 14, no. 4, pp. 1968–1975, 2014, https://doi.org/10.1021/nl404815q

[24] M. S. Al-Qubaisi *et al.*, "Cytotoxicity of nickel zinc ferrite nanoparticles on cancer cells of epithelial origin," *Int. J. Nanomedicine*, vol. 8, pp. 2497–2508, 2013, https://doi.org/10.2147/IJN.S42367

[25] M. Maaß, A. Griessner, V. Steixner, and C. Zierhofer, "Reduction of eddy current losses in inductive transmission systems with ferrite sheets," *Biomed. Eng. Online*, vol. 16, no. 1, pp. 1–18, 2017, https://doi.org/10.1186/s12938-016-0297-4

[26] M. Ansari, A. Bigham, and H. A. Ahangar, "Super-paramagnetic nanostructured CuZnMg mixed spinel ferrite for bone tissue regeneration," *Mater. Sci. Eng. C*, vol. 105, no. August, p. 110084, 2019, https://doi.org/10.1016/j.msec.2019.110084

[27] G. C. Lavorato, R. Das, J. Alonso Masa, M. H. Phan, and H. Srikanth, "Hybrid magnetic nanoparticles as efficient nanoheaters in biomedical applications," *Nanoscale Adv.*, vol. 3, no. 4, pp. 867–888, 2021, https://doi.org/10.1039/d0na00828a

[28] A. Vashist, A. Vashist, Y. K. Gupta, and S. Ahmad, "Recent advances in hydrogel based drug delivery systems for the human body," *J. Mater. Chem. B*, vol. 2, no. 2, pp. 147–166, 2014, https://doi.org/10.1039/c3tb21016b

[29] K. M. L. Taylor-Pashow, J. Della Rocca, R. C. Huxford, and W. Lin, "Hybrid nanomaterials for biomedical applications," *Chem. Commun.*, vol. 46, no. 32, pp. 5832–5849, 2010, https://doi.org/10.1039/c002073g

[30] T. N. Pham, T. Q. Huy, and A. T. Le, "Spinel ferrite (AFe2O4)-based heterostructured designs for lithium-ion battery, environmental monitoring, and

biomedical applications," *RSC Adv.*, vol. 10, no. 52, pp. 31622–31661, 2020, https://doi.org/10.1039/d0ra05133k

[31] L. H. Reddy, J. L. Arias, J. Nicolas, and P. Couvreur, "Magnetic nanoparticles: Design and characterization, toxicity and biocompatibility, pharmaceutical and biomedical applications," *Chem. Rev.*, vol. 112, no. 11, pp. 5818–5878, 2012, https://doi.org/10.1021/cr300068p

[32] J. S. Smart, "The Néel Theory of Ferrimagnetism," *Am. J. Phys.*, vol. 23, no. 6, pp. 356–370, 1955, https://doi.org/10.1119/1.1934006

[33] K. M. Krishnan, "Biomedical nanomagnetics: A spin through possibilities in imaging, diagnostics, and therapy," *IEEE Trans. Magn.*, vol. 46, no. 7, pp. 2523–2558, 2010, https://doi.org/10.1109/TMAG.2010.2046907

[34] L. Zhou *et al.*, "Development of carbon nanotubes/CoFe2O4 magnetic hybrid material for removal of tetrabromobisphenol A and Pb(II)," *J. Hazard. Mater.*, vol. 265, pp. 104–114, 2014, https://doi.org/10.1016/j.jhazmat.2013.11.058

[35] A. A. Abd Elrahman and F. R. Mansour, "Targeted magnetic iron oxide nanoparticles: Preparation, functionalization and biomedical application," *J. Drug Deliv. Sci. Technol.*, vol. 52, no. May, pp. 702–712, 2019, https://doi.org/10.1016/j.jddst.2019.05.030

[36] C. Su, "Environmental implications and applications of engineered nanoscale magnetite and its hybrid nanocomposites: A review of recent literature," *J. Hazard. Mater.*, vol. 322, pp. 48–84, 2017, https://doi.org/10.1016/j.jhazmat.2016.06.060

[37] I. Sharifi, H. Shokrollahi, and S. Amiri, "Ferrite-based magnetic nanofluids used in hyperthermia applications," *J. Magn. Magn. Mater.*, vol. 324, no. 6, pp. 903–915, 2012, https://doi.org/10.1016/j.jmmm.2011.10.017

[38] C. S. S. R. Kumar and F. Mohammad, "Magnetic nanomaterials for hyperthermia-based therapy and controlled drug delivery," *Adv. Drug Deliv. Rev.*, vol. 63, no. 9, pp. 789–808, 2011, https://doi.org/10.1016/j.addr.2011.03.008

[39] A. Sohail, Z. Ahmad, O. A. Bég, S. Arshad, and L. Sherin, "Revue sur le traitement par hyperthermie médiée par nanoparticules," *Bull. Cancer*, vol. 104, no. 5, pp. 452–461, 2017, https://doi.org/10.1016/j.bulcan.2017.02.003

[40] Suriyanto, E. Y. K. Ng, and S. D. Kumar, "Physical mechanism and modeling of heat generation and transfer in magnetic fluid hyperthermia through Néelian and Brownian relaxation: a review," *Biomed. Eng. Online*, vol. 16, no. 1, p. 36, 2017, https://doi.org/10.1186/s12938-017-0327-x

[41] Z. Hedayatnasab, F. Abnisa, and W. M. A. Wan Daud, "Investigation properties of superparamagnetic nanoparticles and magnetic field-dependent hyperthermia therapy," *IOP Conf. Ser. Mater. Sci. Eng.*, vol. 334, no. 1, 2018, https://doi.org/10.1088/1757-899X/334/1/012042

[42] "coley1891[42].pdf."

[43] R. K. GILCHRIST, R. MEDAL, W. D. SHOREY, R. C. HANSELMAN, J. C. PARROTT, and C. B. TAYLOR, "Selective inductive heating of lymph nodes," *Ann. Surg.*, vol. 146, no. 4, pp. 596–606, 1957, https://doi.org/10.1097/00000658-195710000-00007

[44] Z. Hedayatnasab, F. Abnisa, and W. M. A. W. Daud, "Review on magnetic nanoparticles for magnetic nanofluid hyperthermia application," *Mater. Des.*, vol. 123, pp. 174–196, 2017, https://doi.org/10.1016/j.matdes.2017.03.036

[45] D. C.F. Chan, D. B. Kirpotin, and P. A. Bunn Jr., "Synthesis and evaluation of colloidal magnetic iron oxides for the site-specific radiofrequency-induced hyperthermia of cancer," *J. Magn. Magn. Mater.*, vol. 122, no. 1–3, pp. 374–378, 1993, https://doi.org/10.1016/0304-8853(93)91113-L

[46] "shinkai1996[46].pdf."

[47] A. Jordan *et al.*, "Effects of Magnetic Fluid Hyperthermia (MFH) on C3H mammary carcinoma in vivo," *Int. J. Hyperth.*, vol. 13, no. 6, pp. 587–605, 1997, https://doi.org/10.3109/02656739709023559

[48] "J.1349-7006.1996.Tb03129.X.Pdf."

[49] R. Hergt *et al.*, "Physical limits of hyperthermia using magnetite fine particles," *IEEE Trans. Magn.*, vol. 34, no. 5 PART 2, p. 37453754, 1998

[50] H. Y. Son, A. Nishikawa, T. Ikeda, T. Imazawa, S. Kimura, and M. Hirose, "Effect of functional magnetic particles on radiofrequency capacitive heating: An in vivo study," *Japanese J. Cancer Res.*, vol. 93, no. 1, pp. 103–108, 2002, https://doi.org/10.1111/j.1349-7006.2001.tb01071.x

[51] S. Lee and K. Na, "Oleic acid conjugated polymeric photosensitizer for metastatic cancer targeting in photodynamic therapy," *Biomater. Res.*, vol. 24, no. 1, pp. 1–8, 2020, https://doi.org/10.1186/s40824-019-0177-7

[52] P. B. Kharat, S. B. Somvanshi, P. P. Khirade, and K. M. Jadhav, "Induction heating analysis of surface-functionalized nanoscale CoFe2O4for magnetic fluid hyperthermia toward noninvasive cancer treatment," *ACS Omega*, vol. 5, no. 36, pp. 23378–23384, 2020, https://doi.org/10.1021/acsomega.0c03332

[53] S. B. Somvanshi *et al.*, "Hyperthermic evaluation of oleic acid coated nano-spinel magnesium ferrite: Enhancement via hydrophobic-to-hydrophilic surface transformation," *J. Alloys Compd.*, vol. 835, p. 155422, 2020, https://doi.org/10.1016/j.jallcom.2020.155422

[54] C. E. Demirci Dönmez, P. K. Manna, R. Nickel, S. Aktürk, and J. Van Lierop, "Comparative Heating Efficiency of Cobalt-, Manganese-, and Nickel-Ferrite Nanoparticles for a Hyperthermia Agent in Biomedicines," *ACS Appl. Mater. Interfaces*, vol. 11, no. 7, pp. 6858–6866, 2019, https://doi.org/10.1021/acsami.8b22600

[55] C. Fisher, A. E. Rider, Z. Jun Han, S. Kumar, I. Levchenko, and K. Ostrikov, "Applications and nanotoxicity of carbon nanotubes and graphene in biomedicine," *J. Nanomater.*, vol. 2012, 2012, https://doi.org/10.1155/2012/315185

[56] P. Seal, M. Hazarika, N. Paul, and J. P. Borah, "MWCNT-$MnFe_2O_4$ nanocomposite for efficient hyperthermia applications," *AIP Conf. Proc.*, vol. 1942, 2018, https://doi.org/10.1063/1.5028714

[57] A. Bhardwaj, K. Parekh, and N. Jain, "In vitro hyperthermic effect of magnetic fluid on cervical and breast cancer cells," *Sci. Rep.*, vol. 10, no. 1, pp. 1–13, 2020, https://doi.org/10.1038/s41598-020-71552-3

[58] J. Sánchez, M. Rodríguez-Reyes, D. A. Cortés-Hernández, C. A. Ávila-Orta, and P. Y. Reyes-Rodríguez, "Heating capacity and biocompatibility of Pluronic-coated manganese gallium ferrites for magnetic hyperthermia treatment," *Colloids Surfaces A Physicochem. Eng. Asp.*, vol. 612, no. July 2020, 2021, https://doi.org/10.1016/j.colsurfa.2020.125986

[59] M. S. A. Darwish, H. Kim, H. Lee, C. Ryu, J. Y. Lee, and J. Yoon, "Engineering core-shell structures of magnetic ferrite nanoparticles for high hyperthermia performance," *Nanomaterials*, vol. 10, no. 5, pp. 1–16, 2020, https://doi.org/10.3390/nano10050991

[60] A. Das, S. Mohanty, R. Kumar, and B. K. Kuanr, "Tailoring the Design of a Lanthanide Complex/Magnetic Ferrite Nanocomposite for Efficient Photoluminescence and Magnetic Hyperthermia Performance," *ACS Appl. Mater. Interfaces*, vol. 12, no. 37, pp. 42016–42029, 2020, https://doi.org/10.1021/acsami.0c13690

[61] O. M. Lemine, N. Madkhali, M. Hjiri, N. A. All, and M. S. Aida, "Comparative heating efficiency of hematite (α-Fe_2O_3) and nickel ferrite nanoparticles for magnetic

hyperthermia application," *Ceram. Int.*, vol. 46, no. 18, pp. 28821–28827, 2020, https://doi.org/10.1016/j.ceramint.2020.08.047

[62] A. Mallick, A. S. Mahapatra, A. Mitra, J. M. Greneche, R. S. Ningthoujam, and P. K. Chakrabarti, "Magnetic properties and bio-medical applications in hyperthermia of lithium zinc ferrite nanoparticles integrated with reduced graphene oxide," *J. Appl. Phys.*, vol. 123, no. 5, 2018, https://doi.org/10.1063/1.5009823

[63] D. Lachowicz *et al.*, "Enhanced hyperthermic properties of biocompatible zinc ferrite nanoparticles with a charged polysaccharide coating," *J. Mater. Chem. B*, vol. 7, no. 18, pp. 2962–2973, 2019, https://doi.org/10.1039/c9tb00029a

[64] S. O. Aisida *et al.*, "The role of polyethylene glycol on the microstructural, magnetic and specific absorption rate in thermoablation properties of Mn-Zn ferrite nanoparticles by sol–gel protocol," *Eur. Polym. J.*, vol. 132, no. May, p. 109739, 2020, https://doi.org/10.1016/j.eurpolymj.2020.109739

[65] H. Nehoff, N. N. Parayath, L. Domanovitch, S. Taurin, and K. Greish, "Nanomedicine for drug targeting: Strategies beyond the enhanced permeability and retention effect," *Int. J. Nanomedicine*, vol. 9, no. 1, pp. 2539–2555, 2014, https://doi.org/10.2147/IJN.S47129

[66] S. Biswas and V. P. Torchilin, "Nanopreparations for organelle-specific delivery in cancer," *Adv. Drug Deliv. Rev.*, vol. 66, pp. 26–41, 2014, https://doi.org/10.1016/j.addr.2013.11.004

[67] H. Maeda, "The enhanced permeability and retention (EPR) effect in tumor vasculature: The key role of tumor-selective macromolecular drug targeting," *Adv. Enzyme Regul.*, vol. 41, no. 1, pp. 189–207, 2001, https://doi.org/10.1016/S0065-2571(00)00013-3

[68] H. Maeda, G. Y. Bharate, and J. Daruwalla, "Polymeric drugs for efficient tumor-targeted drug delivery based on EPR-effect," *Eur. J. Pharm. Biopharm.*, vol. 71, no. 3, pp. 409–419, 2009, https://doi.org/10.1016/j.ejpb.2008.11.010

[69] P. Kumari, B. Ghosh, and S. Biswas, "Nanocarriers for cancer-targeted drug delivery," *J. Drug Target.*, vol. 24, no. 3, pp. 179–191, 2016, https://doi.org/10.3109/1061186X.2015.1051049

[70] J. Su *et al.*, "Long Circulation Red-Blood-Cell-Mimetic Nanoparticles with Peptide-Enhanced Tumor Penetration for Simultaneously Inhibiting Growth and Lung Metastasis of Breast Cancer," *Adv. Funct. Mater.*, vol. 26, no. 8, pp. 1243–1252, 2016, https://doi.org/10.1002/adfm.201504780

[71] X. Yang *et al.*, "Tumor Microenvironment-Responsive Dual Drug Dimer-Loaded PEGylated Bilirubin Nanoparticles for Improved Drug Delivery and Enhanced Immune-Chemotherapy of Breast Cancer," *Adv. Funct. Mater.*, vol. 29, no. 32, pp. 1–12, 2019, https://doi.org/10.1002/adfm.201901896

[72] O. C. Olson and J. A. Joyce, "Cysteine cathepsin proteases: Regulators of cancer progression and therapeutic response," *Nat. Rev. Cancer*, vol. 15, no. 12, pp. 712–729, 2015, https://doi.org/10.1038/nrc4027

[73] Z. Karimi, L. Karimi, and H. Shokrollahi, "Nano-magnetic particles used in biomedicine: Core and coating materials," *Mater. Sci. Eng. C*, vol. 33, no. 5, pp. 2465–2475, 2013, https://doi.org/10.1016/j.msec.2013.01.045

[74] M. Angelakeris, "Magnetic nanoparticles: A multifunctional vehicle for modern theranostics," *Biochim. Biophys. Acta - Gen. Subj.*, vol. 1861, no. 6, pp. 1642–1651, 2017, https://doi.org/10.1016/j.bbagen.2017.02.022

[75] M. Arruebo, R. Fernández-Pacheco, M. R. Ibarra, and J. Santamaría, "Magnetic nanoparticles for drug delivery The potential of magnetic NPs stems from the intrinsic properties of their magnetic cores combined with their drug loading capability and the biochemical properties that can be bestowed on them by means of a suitab," vol. 2, no. 3, pp. 22–32, 2007, [Online]. Available: https://pdfs.semanticscholar.org/1844/8eb43dc235f82cb591983bc8df5ed799984c.pdf

[76] F. Xiong, S. Huang, and N. Gu, "Magnetic nanoparticles: recent developments in drug delivery system," *Drug Dev. Ind. Pharm.*, vol. 44, no. 5, pp. 697–706, 2018, https://doi.org/10.1080/03639045.2017.1421961

[77] J. Kudr *et al.*, "Magnetic nanoparticles: From design and synthesis to real world applications," *Nanomaterials*, vol. 7, no. 9, 2017, https://doi.org/10.3390/nano7090243

[78] S. Ahadian *et al.*, "Micro and nanoscale technologies in oral drug delivery," *Adv. Drug Deliv. Rev.*, vol. 157, pp. 37–62, 2020, https://doi.org/10.1016/j.addr.2020.07.012

[79] J. Cao, D. Huang, and N. A. Peppas, "Advanced engineered nanoparticulate platforms to address key biological barriers for delivering chemotherapeutic agents to target sites," *Adv. Drug Deliv. Rev.*, vol. 167, pp. 170–188, 2020, https://doi.org/10.1016/j.addr.2020.06.030

[80] F. Laffleur and V. Keckeis, "Advances in drug delivery systems: Work in progress still needed?," *Int. J. Pharm. X*, vol. 2, no. March, 2020, https://doi.org/10.1016/j.ijpx.2020.100050

[81] K. Park, "Controlled drug delivery systems: Past forward and future back," *J. Control. Release*, vol. 190, pp. 3–8, 2014, https://doi.org/10.1016/j.jconrel.2014.03.054

[82] P. Fomby *et al.*, "Stem cells and cell therapies in lung biology and diseases: Conference report," *Ann. Am. Thorac. Soc.*, vol. 12, no. 3, pp. 181–204, 2010, https://doi.org/10.1002/term

[83] B. T. Mai, S. Fernandes, P. B. Balakrishnan, and T. Pellegrino, "Nanosystems Based on Magnetic Nanoparticles and Thermo- or pH-Responsive Polymers: An Update and Future Perspectives," *Acc. Chem. Res.*, vol. 51, no. 5, pp. 999–1013, 2018, https://doi.org/10.1021/acs.accounts.7b00549

[84] Z. Wang, X. Deng, J. Ding, W. Zhou, X. Zheng, and G. Tang, "Mechanisms of drug release in pH-sensitive micelles for tumour targeted drug delivery system: A review," *Int. J. Pharm.*, vol. 535, no. 1–2, pp. 253–260, 2018, https://doi.org/10.1016/j.ijpharm.2017.11.003

[85] W. Il Choi, A. Sahu, F. R. Wurm, and S. M. Jo, "Magnetoliposomes with size controllable insertion of magnetic nanoparticles for efficient targeting of cancer cells," *RSC Adv.*, vol. 9, no. 26, pp. 15053–15060, 2019, https://doi.org/10.1039/c9ra02529d

[86] B. D. Cardoso *et al.*, "Magnetoliposomes containing magnesium ferrite nanoparticles as nanocarriers for the model drug curcumin," *R. Soc. Open Sci.*, vol. 5, no. 10, 2018, https://doi.org/10.1098/rsos.181017

[87] S. Kamaraj, U. M. Palanisamy, M. S. B. Kadhar Mohamed, A. Gangasalam, G. A. Maria, and R. Kandasamy, "Curcumin drug delivery by vanillin-chitosan coated with calcium ferrite hybrid nanoparticles as carrier," *Eur. J. Pharm. Sci.*, vol. 116, no. January, pp. 48–60, 2018, https://doi.org/10.1016/j.ejps.2018.01.023

[88] J. Panda *et al.*, "Anticancer potential of docetaxel-loaded cobalt ferrite nanocarrier: an in vitro study on MCF-7 and MDA-MB-231 cell lines," *J. Microencapsul.*, vol. 38, no. 1, pp. 36–46, 2021, https://doi.org/10.1080/02652048.2020.1842529

[89] X. H. Wang, M. Huang, C. K. Zhao, C. Li, and L. Xu, "Design, synthesis, and biological activity evaluation of campthothecin-HAA-Norcantharidin conjugates as antitumor agents in vitro," *Chem. Biol. Drug Des.*, vol. 93, no. 6, pp. 986–992, 2019, https://doi.org/10.1111/cbdd.13397

[90] I. V. M. V. Enoch, S. Ramasamy, S. Mohiyuddin, P. Gopinath, and R. Manoharan, "Cyclodextrin-PEG conjugate-wrapped magnetic ferrite nanoparticles for enhanced

drug loading and release," *Appl. Nanosci.*, vol. 8, no. 3, pp. 273–284, 2018, https://doi.org/10.1007/s13204-018-0798-5

[91] S. Ramasamy, I. V. M. V. Enoch, and S. Rex Jeya Rajkumar, "Polymeric cyclodextrin-dextran spooled nickel ferrite nanoparticles: Expanded anticancer efficacy of loaded camptothecin," *Mater. Lett.*, vol. 261, p. 127114, 2020, https://doi.org/10.1016/j.matlet.2019.127114

[92] O. Veiseh, J. W. Gunn, and M. Zhang, "Design and fabrication of magnetic nanoparticles for targeted drug delivery and imaging," *Adv. Drug Deliv. Rev.*, vol. 62, no. 3, pp. 284–304, 2010, https://doi.org/10.1016/j.addr.2009.11.002

[93] L. Wu, A. Mendoza-Garcia, Q. Li, and S. Sun, "Organic Phase Syntheses of Magnetic Nanoparticles and Their Applications," *Chem. Rev.*, vol. 116, no. 18, pp. 10473–10512, 2016, https://doi.org/10.1021/acs.chemrev.5b00687

[94] H. Zhang *et al.*, "Ultrasmall Ferrite Nanoparticles Synthesized via Dynamic Simultaneous Thermal Decomposition for High-Performance and Multifunctional T1 Magnetic Resonance Imaging Contrast Agent," *ACS Nano*, vol. 11, no. 4, pp. 3614–3631, 2017, https://doi.org/10.1021/acsnano.6b07684

[95] D. Ni, W. Bu, E. B. Ehlerding, W. Cai, and J. Shi, "Engineering of inorganic nanoparticles as magnetic resonance imaging contrast agents," *Chem. Soc. Rev.*, vol. 46, no. 23, pp. 7438–7468, 2017, https://doi.org/10.1039/c7cs00316a

[96] J. Wallyn and N. Anton, "Biomedical Imaging : Principles , Technologies , Clinical Aspects , C o n t r a s t A g e n t s , L i m i t a t i o n s a n d F u t u r e Tr e n d s in Nanomedicines," pp. 1–31, 2019

[97] Y. Miao *et al.*, "Composition-tunable ultrasmall manganese ferrite nanoparticles: Insights into their in vivo T1 contrast efficacy," *Theranostics*, vol. 9, no. 6, pp. 1764–1776, 2019, https://doi.org/10.7150/thno.31233

[98] Z. Zhou, L. Yang, J. Gao, and X. Chen, "Structure–Relaxivity Relationships of Magnetic Nanoparticles for Magnetic Resonance Imaging," *Adv. Mater.*, vol. 31, no. 8, pp. 1–32, 2019, https://doi.org/10.1002/adma.201804567

[99] D. Dheer, J. Nicolas, and R. Shankar, "Cathepsin-sensitive nanoscale drug delivery systems for cancer therapy and other diseases," *Adv. Drug Deliv. Rev.*, vol. 151–152, pp. 130–151, 2019, https://doi.org/10.1016/j.addr.2019.01.010

[100] Y. Wang *et al.*, "Engineering ferrite nanoparticles with enhanced magnetic response for advanced biomedical applications," *Mater. Today Adv.*, vol. 8, 2020, https://doi.org/10.1016/j.mtadv.2020.100119

[101] D. H. Kim, H. Zeng, T. C. Ng, and C. S. Brazel, "T1 and T2 relaxivities of succimer-coated MFe23+O4 (M=Mn2+, Fe2+ and Co2+) inverse spinel ferrites for potential use as phase-contrast agents in medical MRI," *J. Magn. Magn. Mater.*, vol. 321, no. 23, pp. 3899–3904, 2009, https://doi.org/10.1016/j.jmmm.2009.07.057

[102] F. Hyder and S. Manjura Hoque, "Brain tumor diagnostics and therapeutics with superparamagnetic ferrite nanoparticles," *Contrast Media Mol. Imaging*, vol. 2017, 2017, https://doi.org/10.1155/2017/6387217

[103] H. Shokrollahi, A. Khorramdin, and G. Isapour, "Magnetic resonance imaging by using nano-magnetic particles," *J. Magn. Magn. Mater.*, vol. 369, pp. 176–183, 2014, https://doi.org/10.1016/j.jmmm.2014.06.023

[104] G. Basina *et al.*, "Water-soluble spinel ferrites by a modified polyol process as contrast agents in MRI," *AIP Conf. Proc.*, vol. 1311, no. 2010, pp. 441–446, 2010, https://doi.org/10.1063/1.3530053

[105] Y. Sun *et al.*, "High-Performance Worm-like Mn-Zn Ferrite Theranostic Nanoagents and the Application on Tumor Theranostics," *ACS Appl. Mater. Interfaces*, vol. 11, no. 33, pp. 29536–29548, 2019, https://doi.org/10.1021/acsami.9b08948

[106] L. Ansari *et al.*, "Synthesis, Characterization and MRI Application of Cobalt-Zinc Ferrite Nanoparticles Coated with DMSA: An In-vivo Study," *Appl. Magn. Reson.*, vol. 52, no. 1, pp. 33–45, 2021, https://doi.org/10.1007/s00723-020-01220-2.

Materials Research Foundations **112** (2021) 189-217 https://doi.org/10.21741/9781644901595-5

Chapter 5

A Review on Synthesis and Characterizations of Mixed Nickel-Zinc Ferrites

Arminder Kaur[1], Pankaj Sharma[2], Sumit Bhardwaj[1], Munish Kumar[3], Indu Sharma[3], Khalid Mujasam Batoo[4], Gagan Kumar[1*]

[1]Department of Physics, Chandigarh University, Gharuan, Mohali, Punjab, India

[2]Department of Applied Sciences, National Institute of Technical Teachers Training and Research, Sector 26, Chandigarh, 160019, India

[3]Department of Physics, Career Point University, Hamirpur, HP, India

[4]King Abdullah Institute for Nanotechnology, King Saud University, P.O. Box 2455, Riyadh 11451, Saudi Arabia

*physics.bhargava@gmail.com

Abstract

Nanotechnology, when this word comes in mind, it gives deep thought of new development in communication, medical science, intelligent transport system and many more. Ferrites nanoparticles have great significance owing to their amazing chemical and physical properties. In modern era we are developing materials for microwave applications and communication devices. Before the discovery of semiconductor memory chips, ferrites were the major form for electronic memory used in computers. Scientist have been studying and working with nanoparticles in magnetically guided drug delivery. The reactivity of material increases by the use of nanoparticles of that material. The dielectric characteristics of ferrites lean on diverse factors for instance methods of preparations and chemical composition. In various studies it has been found that their conductivity has dependence on temperature, composition and frequency. Among the various kinds of ferrites, Ni–Zn ferrites are viewed as the most adaptable ferrites as a result of their novel characteristics for applications at high frequency. The Ni-Zn ferrites are exploited as core materials in a variety of EM devices as well as have broad range of industrial applications e.g. inductors, microwave devices, power supplies, high and low frequency transformer cores, electromagnetic interference (EMI) suppressions and antenna rods. These broad ranges of applications are owing to their high resistivity, low eddy currents, high saturation magnetization, chemical stability and high Curie temperature. In view of this, the present chapter deals with the research progress on nickel-zinc ferrites in the bulk as well as nano size.

Keywords

Ni-Zn Ferrites, Dielectric Constant, Saturation Magnetization

Contents

1. Introduction

Around 800 B.C. loadstone was discovered in Greek, which is a non-metallic solid. Such substance was utilized as navigation of North. It was found that loadstone can show magnetic properties. De- Magnate published paper related to study of magnetism in 1600 [1]. This is the origin of the era of ferrites. This word originated from the Lattin word "ferrum" meaning iron. Ferrites are ceramic substance and they are hard and grey or black in color. Ferrite materials consist of ferric ions with oxides. Due to high magnetic moment of ferric ions, ferrites show ferromagnetic properties. Owing to the fact of their moderate resistivity and minimum eddy current loss, they are highly versatile in nature [2]. They have remarkable potential to showcase in electronics, computer, catalysis, telecommunication, radar, pigments, radio, television, video tape.

Ferrites are categorised into four types:

a)Spinel ferrites

b) Garnet

Materials Research Foundations **112** (2021) 189-217 https://doi.org/10.21741/9781644901595-5

c)Ortho-ferrites

d) Hexagonal ferrites

Spinel Ferrites: Spinel ferrites are explained by MFe_2O_4 formula, in which M is described as divalent metal ions.

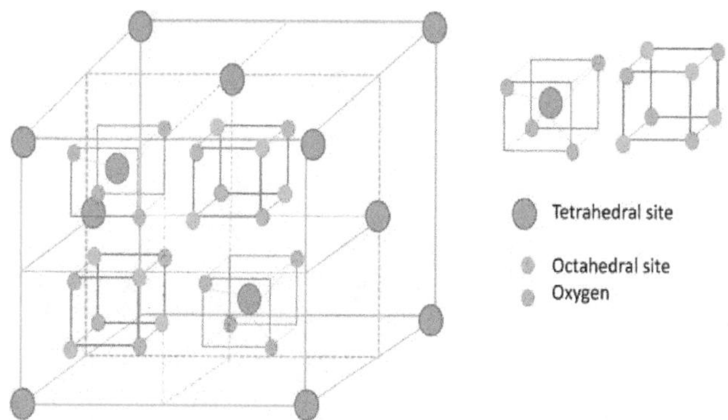

Figure 1: Spinel ferrites crystal structure having tetrahedral and octahedral sites (Reprinted from reference [3] with kind permission from Elsevier, Copyright 2017, Elsevier.

Spinel ferrites possess two interstitial sites name as tetrahedral (A) interstitial sites and octahedral interstitial sites (B). These interstitial sites have different preferences for different materials [4]. Hence, distribution of metal ions on these interstitial sites changes their properties enormously. Figure 1 illustrates the spinel ferrites crystal structure.
Garnet: Garnet ferrites are explained by $Me_3Fe_5O_{12}$ formula, in which Me is a trivalent ion. The crystal structure is cubic and it contains 8 molecules of $Me_3Fe_5O_{12}$. The ions are distributed over dodecahedral sites, tetrahedral sites and octahedral sites. Fe ions distributed over tetrahedral as well as octahedral sites in 3:2 ratio. Me is occupied by dodecahedral sites and surrounded by oxygen ions. Figure 2 illustrates the spinel ferrites crystal structure.

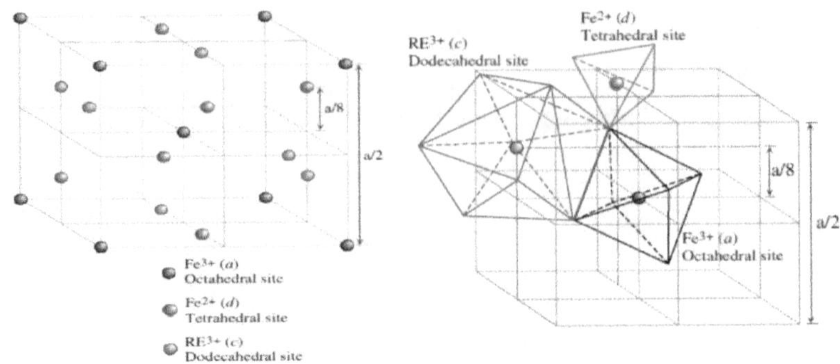

Figure 2: Garnet ferrites crystal structure (Reprinted from reference [5]with kind permission from Springer, Copyright 2009, Springer).

Ortho-Ferrites: Ortho-ferrites are represented by $RFeO_3$, in which R addresses the transition-metal ions or otherwise the rare-earth ions. These ferrites consist of the structure of perovskite (ABO), where A addresses the 12-coordinated O^{2-} site where generally the ions of lanthanide series resides while B addresses the site which is generally occupied by the transition-metal ion i.e. Fe^{3+} ions. The crystal structure of orthoferrite is shown in Figure 3. An ideal structure of perovskite structure is purely cubic. At room temperature, the mineral perovskite ($CaTiO_3$) is known to be orthorhombic and when the temperature is increased above than 900°C then becomes cubic. Other ceramics with the perovskite structure include $BaTiO_3$, $SrTiO_3$, and $KNbO_3$. The magneto-optical characteristics of ortho-ferrites are an immediate result of their structure. Generally, the unit cell of ortho-ferrites is orthorhombic to some extent rather than cubic similar to $BaTiO_3$. Although the structural unit cell of ortho-ferrite is nearly cubic, however the angle among the a-b plane and c-axis is larger than 90° to some extent. The cell of orthorhombic has a minor canting (~0.6°) in ortho-ferrite unit cell. An illustration of a typical ortho-ferrite is $YFeO_3$.

Materials Research Foundations **112** (2021) 189-217 https://doi.org/10.21741/9781644901595-5

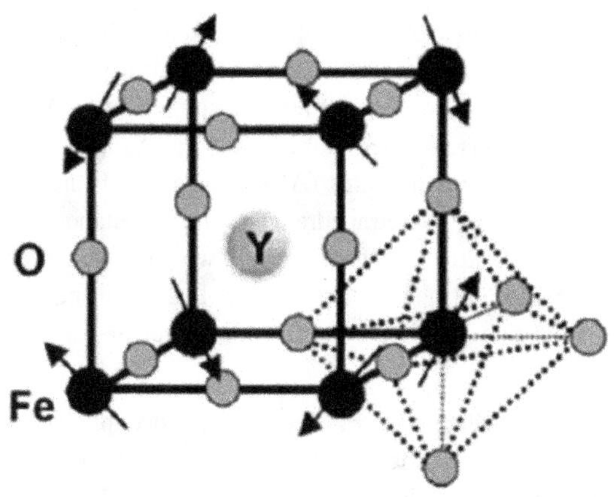

Figure 3: Representation of $YFeO_3$ unit cell.

It crystallizes orthorhombic unit cell with distorted perovskite structure. The location of Y^{3+} ions plays the major role to distort the structure of perovskite while Fe^{3+} ions reside in the octahedral surroundings. This structure can be imagined as a three-dimensional network of strings of FeO_6 octahedral. One of the anions (O^{2-}) forms the common apex of the two adjacent octahedral and display for the super-exchange bond (Fe-O-Fe) between two iron ions. Figure 3 illustrates the $YFeO_3$ unit cell.

Hexagonal ferrites: hexagonal ferrites are further divided into M, Y, W, Z, X and U compounds. All these ferrites have different structure with some similarity with each other.

Table 1: Comparison of various descriptions of the molecular units of hexagonal ferrites (Reprinted from reference[6] with kind permission from Elsevier, Copyright 2012, Elsevier).

Ferrite	Molecular Units	Molecular formula
S	S	CoFe2O4
M	M	$BaFe_{12}O_{19}$
Y	Y	$Ba_2Co_2Fe_{12}O_{22}$
W	M+2S	$BaCo_2Fe_{16}O_{27}$
Z	M+Y	$Ba_3Co_2Fe_{24}O_{41}$
X	2M+2S	$Ba_2Co_2Fe_{28}O_{46}$
U	2M+Y	$Ba_4Co_2Fe_{36}O_{60}$

The comparison of various descriptions of the molecular units of hexagonal ferrites is given in Table 1. Ferrites are largely popular in industry because of their huge applications and versatile nature. Among all ferrites, spinel ferrites MFe_2O_4 are quite well known in research field because of their stable structure. As a mentioned earlier, spinel ferrites contain tetrahedral interstitial sites (A) and octahedral interstitial sites (B) with MFe_2O_4 formula. There are total 64 tetrahedral sites and 32 octahedral sites in a unit cell. Cations are distributed differently in different spinel, which is the major cause of variable properties in spinel ferrites and hence their vast application in numerous industries. In consideration of distribution of divalent ions, these ferrites are divided into three categories.

a)Normal spinel ferrites: when all tetravalent sites are occupied by divalent metal ions and all octahedral sites are occupied by trivalent iron ions [Fe^{3+}]. Such kind of ferrites is known as normal spinel ferrites. For e.g. Ni-Zn ferrites, Mn-Zn ferrites, Zn ferrites, Co-Zn ferrites.

b) Inverse spinel ferrites: when tetrahedral sites are occupied by trivalent iron ions [Fe^{3+}] and octahedral sites are occupied by trivalent iron ions as well as divalent ions. Such kinds of ferrites are known as inverse spinel ferrites. For e.g., Ni-Co ferrites, Co ferrites.

c)Random spinel ferrites: when divalent ions and trivalent iron ions can occupy any octahedral and tetrahedral sites with no fixed proportion. Such ferrites are known as random spinel ferrites. For e.g. $(Mn^{2+}{}_{0.8}\,Fe_{0.2}{}^{3+})_A\,[Fe_{1.8}{}^{3+}Mn^{2+}{}_{0.2}]_BO_4$.

Therefore, chemical formula for spinel ferrites can be written as $(M_{1\text{-}a}\,M'_a)_A\,(M_a\,M'_{1\text{-}a})_B\,O_4$. Where M is divalent ions M' is trivalent ions, A is representing tetrahedral sites, B is representing octahedral sites, If a=0 then it is normal spinel ferrite, if a= 1then it is inverse spinel ferrite and if $0 < a < 1$ then it is random spinel ferrite

2. Nickel-zinc ferrites

F.A. Hezam et al. [6] prepared $Ni_{0.5}Mg_xZn_{0.5\text{-}x}Fe_2O_4$ $(0.0 < x < 0.5)$ by co-precipitation method. The size of nano ferrites was found to be 30 nm and structure was cubic spinel. Lattice constant and crystal size was calculated which was observed to be affected with the addition of Mg^{2+}. Magnetization of ferrites was also affected and it was found to be increased with increase in Mg^{2+} ions. Figure 4 illustrates the variation of lattice constant, crystallite size and saturation magnetization with Mg^{2+} ions content.

Materials Research Foundations **112** (2021) 189-217 https://doi.org/10.21741/9781644901595-5

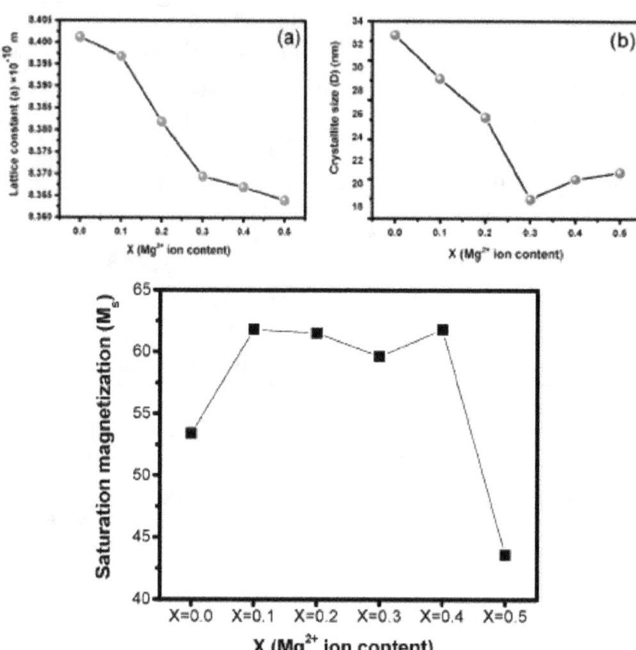

Figure 4: Variation of lattice constant, crystallite size and saturation magnetization with Mg^{2+} ions content (Reprinted from reference [7] with kind permission from Elsevier, Copyright 2021, Elsevier).

Superparamagnetic properties were showed by all ferrites sample conforming through sample magnetometer techniques. Mg^{2+} and Zn^{2+} have taken tetrahedral site resulting into super exchange interaction between Ni^{2+} and Fe^{3+}. J.N. Pavan Kumar Chintala et al. [8] synthesized $Ni_{0.65}Zn_{0.35}Fe_2O_4$ ferrite by using sol-gel technique. Sample structure was found to be cubic spinel. They found high value of saturation magnetization 176 emu/g at temperature 10 K. R. Verma et al. [9] fabricated $Zn_{1-y}Ni_yFe_2O_4$ (y= 0.0 to1.0) nano powder. With the help of XRD, sample magnetometer and U-V the structural, magnetic and optical properties were investigated. These studied confirmed that lattice parameter and coercivity decreased, saturation magnetization increased due to more addition nickel ions. Band gap was linearly varied with the concentration of nickel. S. Kumar M V et al. [10] prepared $Ni_{1-x}Zn_xGd_{0.1}Fe_{1.9}O_4$ ($0 \leq x \leq 1$) with the help of citrate auto-combustion synthesis process at low temperature. In this study, lattice constant and crystalline size reduces with addition of Zn^{2+} ions. They further reported that Zn ions occupied tetrahedral sites, which raise the tetrahedral site's radius and reduce octahedral site's radius while saturation magnetized has increased and coercivity force has reduced with increase in the concentration of Zn. D. Venkatesh et al. [10] fabricated $Ni_{0.5}Zn_{0.5-x}Cu_xFe_2O_4$ (where x = 0.05 to 0.25 in steps of 0.05) with citrate gel auto-combustion method. Cubic spinel structure was observed for all samples.

Table 2: Variation of magnetic and microstructure parameters (Reprinted from reference [10] with kind permission from Springer, Copyright 2019, Springer).

Composition (x)	Remanence M_r (emu/g)	Coercivity H_C (Oe)	Remanence ratio (M_r/M_s)	Grain diameter (nm)	Cation distribution
0.00	4.21	73.93	0.0690	130	$(Zn_{0.5}Fe_{0.4}Ni_{0.1})$ [Ni0.4Fe1.6] O4
0.05	13.51	98.64	0.2304	155	$(Zn_{0.45}Fe_{0.45}Ni_{0.1})$ $[Ni_{0.4}Fe_{1.55}Cu_{0.05}]O_4$
0.10	10.44	89.72	0.1902	195	$(Zn_{0.4}Fe_{0.5}Ni_{0.1})$ $[Ni_{0.4}Fe_{1.5}Cu_{0.1}]O_4$
0.15	12.21	101.67	0.2332	246	$(Zn_{0.35}Fe_{0.55}Ni_{0.1})$ $[Ni_{0.4}Fe_{1.45}Cu_{0.15}]O_4$
0.20	9.84	110.54	0.1976	200	$(Zn_{0.3}Fe_{0.7})$ $[Ni_{0.5}Fe_{1.3}Cu_{0.2}]O_4$
0.25	5.98	83.74	0.1345	183	$(Zn_{0.25}Fe_{0.75})$ $[Ni_{0.5}Fe_{1.25}Cu_{0.25}]O_4$

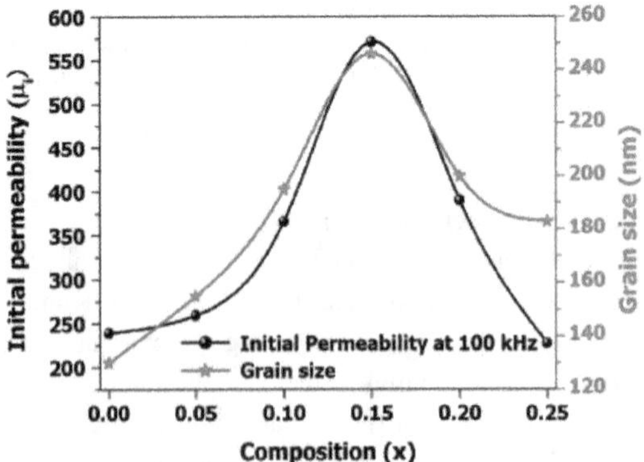

Figure 5: Variation of initial permeability and grain size with Cu^{2+} ions content (Reprinted from reference [11] with kind permission from Springer, Copyright 2019, Springer).

Materials Research Foundations **112** (2021) 189-217 https://doi.org/10.21741/9781644901595-5

The lattice constant and magnetization were observed to be decreased with the addition of Cu^{2+} ions and the maximum magnitude of saturation magnetization (61 emu/g) was found for $Ni_{0.5}Zn_{0.5}Fe_2O_4$ sample. Remanence and grain diameter were investigated to be increased with rise in copper concentration up to x=0.15. S. A. Jadhav et al. [12] studied $Ni_xZn_{1-x}Fe_2O_4$ ferrites (x=0.0 to 1), which was prepared by sol-gel method. The calculated lattice constant was varied from 8.39 Å to 8.32 Å while the crystalline size was changed from 23 nm to 15 nm. In this study photovoltaic potential of all sample also studied. Resistivity was observed to be decrease with rise in temperature, which showed semiconductor behaviour of all samples.

The increase in saturation magnetization was found with rise in Ni ions. Xiaobing Zhou et al. [13] observed that layer of $Ti_3C_2T_x$ on Ni-Zn ferrite can increase its electrical conductivity 6 times as compared to non-coated Ni-Zn ferrite. Such behaviour shows its huge application in beam-line higher-order mode load in advanced particle accelerators. P. Anantha Rao et al. [14] synthesized $Ni_{0.5}Zn_{0.5-x}Cd_xFe_2O_4$ (x = 0.0 to 0.4) to investigate its structural, magnetic, morphology and electrical properties. All samples were found to be cubic spinel. With addition of Cd ions, lattice constant was observed to be increased as ionic radius of cadmium is higher as compared to zinc.

However, FESEM study indicated an increase in grain size from 40 nm to 73 nm. FTIR study of samples revealed two major peaks at 600 cm^{-1} and 400 cm^{-1}, which told us that Cd^{2+} ions preferred tetrahedral site. From VSM study, it was found that magnetic saturation was decreased from 89.51 to 71.32 emu/g. In addition to it, remanent magnetization and coercive force were increased from 25.36 to 36.48 emu/g and 32.12 to 41.25 emu/g respectively. In DC resistivity study, semiconductor behaviour was observed in all samples as their resistivity declined with raise in temperature.

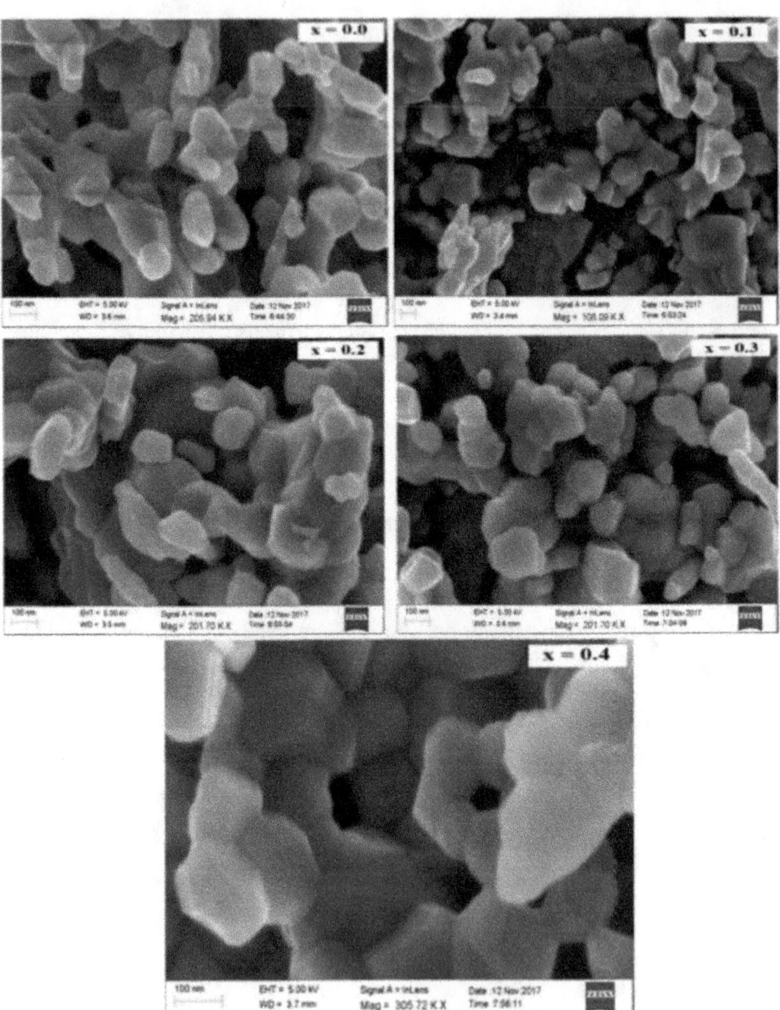

Figure 6: FESEM images of $Ni_{0.5}Zn_{0.5-x}Cd_xFe_2O_4$ nanoferrites (Reprinted from reference [13] with kind permission from Elsevier, Copyright 2020, Elsevier).

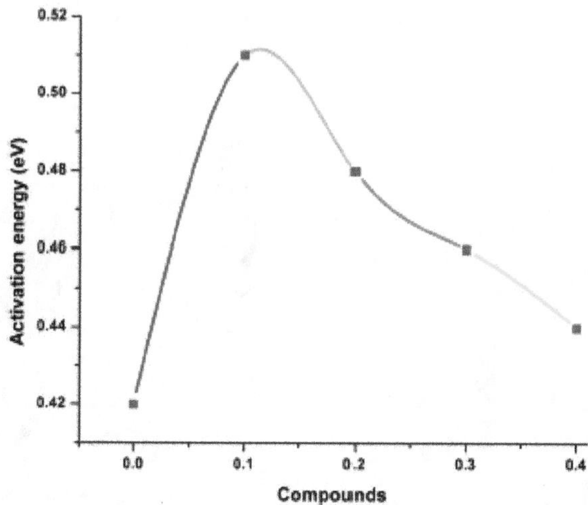

Figure 7: Activation energies of $Ni_{0.5}Zn_{0.5-x}Cd_xFe_2O_4$ nanoferrites (Reprinted from reference [13] with kind permission from Elsevier, Copyright 2020, Elsevier).

Moreover, activation energy increased initially till y= 0.1, afterwards it showed downward trend. A. Bajorek et al. [15] did detailed study of $Ni_{0.5}Zn_{0.5}Fe_2O_4$ ferrite, which was fabricated by co-precipitation method. Crystalline size was found to be 14 nm to 20 nm. FTIR confirmed the occupancy of iron on octahedral and tetrahedral sites. Superparamagnetic property was discovered in the magnetization study. Magnitude of coupling constants derived from mean field theory gave proof of opposite magnetization between tetrahedral interstitial sites and octahedral interstitial sites [15]. M.V.S. Kumar et al. [16] synthesized $Ni_{1-x}Zn_xGd_yFe_{2-y)}O_4$ ($0 \leq x \leq 1$, y = 0.1) samples with low temperature citrate precursor combustion method. In this study, it was observed that Gd ions preferred octahedral sites in crystal structure. Also, all sample found to have cubic spinel structure. FE-SEM revealed the presence of porosity in all ferrite samples. Zn ions enhanced the hopping between octahedral and tetrahedral site, which increase the ac conductivity of ferrite samples. Additionally, DC resistivity & Curie temperature were investigated to be decreased with the addition of Zn ions. Such behaviour described the increased hopping between Fe^{2+} and Fe^{3+}. T. Dippong et al. [17] synthesized $Ni_xZn_{1-x}Fe_2O_4/SiO_2$ ferrites by sol gel method. With the help of thermal analysis, they have confirmed the formation of nickel-zinc and ferrous ions-succinate precursors around 200°C. Crystalline size, lattice parameter, magnetic saturation, coercivity and remanence were studied and they all changed their value with Ni content. Super paramagnetic property was observed in all samples.

Figure 8: FE-SEM micrographs of typical samples (a) x = 0.0, (c) x = 0.4 unirradiated and (b) x = 0.0, (d) x = 0.4; γ-irradiated for $Ni_{1-x}Zn_xFe_{2-x}Cr_xO_4$ NPs (Reprinted from reference [19] with kind permission from Springer, Copyright 2018, Springer).

M.M. Kothawale et al. [18] studied properties of $Ni_xZn_{1-x}Fe_2O_4$ ferrites prepared by Microwave-assisted combustion method. Samples were having large surface to volume ratio with strain value ranges between 9.4×10^{-4} to 3.0×10^{-4}. The crystalline size varies from 51 to 69 nm. In FTIR data absorption bands found at 610 cm^{-1}, 550 cm^{-1}, 450 cm^{-1}, 375 cm^{-1}. The highest value of magnetic saturation was 82.49 emu/g for x = 0.6. A high value of saturation magnetization (Ms) is supposed to be the most vital requirement of magnetic recording, magnetic refrigeration, magnetic fluid, permanent magnet material, magnetic resonance imaging, and the magnetic core material. S. Kumar et al. [19] studied $Ni_{1-x}Zn_xFe_2O_4$ ferrites synthesized through facile solution combustion technique. They analysed XRD, FTIR, VSM of all the samples. In magnetic study, magnetic saturation was found to be maximum at x=0.3 with 38.73 emu/g value, afterwards its value

Materials Research Foundations **112** (2021) 189-217 https://doi.org/10.21741/9781644901595-5

decrease. Also, with increase in concentration of Zn ions, ferrimagnetic behaviour changed into super paramagnetic nature. V.K. Mande et al. [20] prepared $Ni_{1-x}Zn_xFe_{2-x}Cr_xO_4$, (Ni-Zn-Cr) nanoparticles with x=0.0, 0.2, 0.4, 0.6, 0.8 and 1.0 with the help of sol gel auto combustion method. Also, irradiation effects of Co-60 source on each ferrite were investigated.

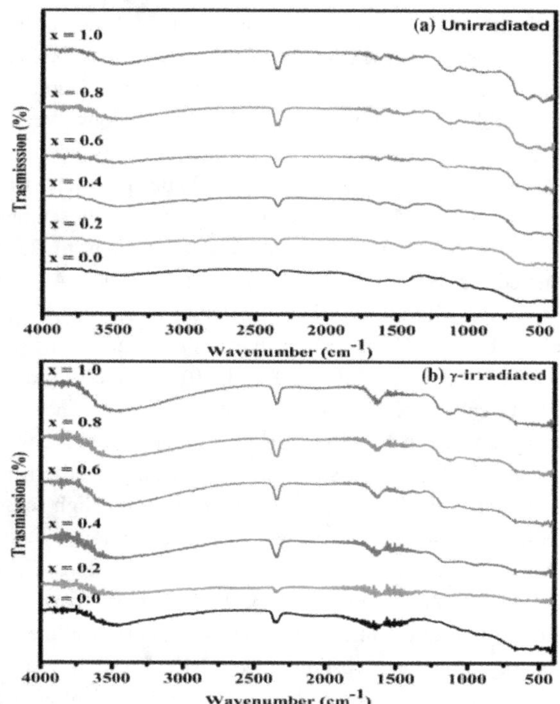

Figure 9: FTIR spectra for (a) unirradiated and (b) γ-irradiated $Ni_{1-x}Zn_xFe_{2-x}Cr_xO_4$ NPs Reprinted from reference [20] with kind permission from Springer, Copyright 2018, Springer).

XRD confirmed the cubic spinel structure of both radiated and irradiated sample. FTIR confirmed the metal oxide bond with its absorption peaks. FE-SEM showed that irradiated samples have some distorted morphology as compared to non-irradiated sample. The dielectric properties were found to decrease with the rise in zinc and Cr^{3+} ions for both the cases i.e. without and with irradiation. S. Ikram et al. [21] prepared $Ni_{0.5}Zn_{0.3}Cd_{0.2}Fe_{2-b}La_bO_4$ (b = 0.0-0.21) cubic spinel ferrite though sol-gel auto combustion synthesis process. Effect of La on the properties of ferrite sample were studied and it was found that lattice constant was increased till b= 0.105 and then decreased. Absorption bands were observed to be at 540.8 cm^{-1} and 490.8 cm^{-1}, which confirmed the tetrahedral interstitial and octahedral interstitial sites in ferrites. Magnetic saturation, remanence has decreased with the addition of La ions. AC conductivity was found to increase at lower frequency and at higher frequency it is found to decrease. Such type of ferrite is suitable for microwave absorbing applications. M. Kuru et al. [22] synthesized $Ca_xNi_{0.75-x}Zn_{0.25}Fe_2O_4$ (x=0, 0.25, 0.5 and 0.75) by adopting chemical co-precipitation synthesis process. In XRD investigation, lattice parameter varied from 8.334 Å to 8.538 Å and crystal structures were cubic spinel for all samples. Afterwards, crystalline size was measured, which varied between 12 nm to 27 nm. Dielectric properties also studied for different temperature and frequencies. AC conductivity was observed to be decreased with the addition of calcium ions. They observed that grain boundaries have profound role in conductivity. Pathania et al. [23] synthesized $Ni_{0.5}Zn_{0.5}W_xFe_{2-x}O_4$ (x = 0.0, 0.2, 0.4, 0.6, 0.8 and 1.0) with the help of co-precipitation method. From SEM, cubic spinal structure was confirmed. They investigated the application of ferrite samples in hydrogen gas sensing at concentration 1000 ppm in a temperature range 80-300 degrees Celsius. Platinum coating has been made on ferrite pallets by sputtering technique which increased the resistivity. High sensitivity, cheaper price, longer stability, high selectivity and quick response at lower temperature proves that such sensor have numerous advantages in industrial applications. K. S. Ramakrishna et al. [24] prepared $Ni_xCu_{0.1}Zn_{0.9-x}Fe_2O_4$ (x =0.5, 0.6, 0.7) ferrite nano particles with co-precipitation method. Exchange interaction and magnetocrystalline anisotropy was observed to increase with the substitution of Ni^{2+} ions. It was observed that magnetocrystalline anisotropy constant was temperature dependant as it decreased with increase in temperature. Jean-Luc Mattei et al. [25] synthesized $Ni_{0.5}Zn_{0.3}Co_{0.2}Fe_yO_{4-\delta}$ where y= 1.98 and y = 2.3 by using chemical co-precipitation method. They analysed its application in antenna. At frequency higher than 700 MHz, both samples were found not suitable for antenna applications. However, frequency lowers than 700 MHZ, they are well suited for such applications as dielectric loss was quite low. M. M. Ismail et al. [26] studied $Ni_{1-a}Zn_{a-2b}Li_bFe_{2+b}O_4$, where ($0 \leq a \leq 0.5$) and (b= 0, 0.01, 0.02, 0.03, and 0.04) ferrites by preparing with sol-gel method. They calculated lattice constant which was observed to be increased with the addition of zinc and decreased with the rise of lithium ions. From SEM, crystalline shape came out to be smooth, spherical and its size came out to be larger than the calculated from XRD data. This suggested the agglomeration in nano particles. Poisson ratio value was calculated and its value was 0.35, this attribute to isotropic theory and also high stability of the materials. Reduction of elastic constant with the addition of zinc and lithium suggested that interatomic bonding was reducing continuously. As Ni with d^8 configuration and Fe with d^5 configuration replaced with Zinc with d^{10} configuration and lithium with s^2 configuration, which described that totally and half-filled shells are more stable and they don't contribute in bond formation whereas partially filled ions such as nickel are good at making strong bond with oxygen. M. V. S. Kumar et al. [27] did research on ($Ni_{0.6}Zn_{0.4}Gd_bFe_{2-b}O_4$; b = 0, 0.1, 0.15, and 0.2) ferrite sample, which was prepared by low-temperature citrate gel auto-combustion synthesis process. Initially lattice constant and crystalline size was reduced with addition of Gd^{+3} substitutions and then it started increasing. Porosity also declined with the addition of Gd^{+3} concentrations, which was depicted by FESEM. Hopping was observed between $Fe^{+3} \leftrightarrow Fe^{+2}$ ions. Such behaviour raised capacitance value. Gd^{+3} replaced the Fe^{+3} ion at B site, while Fe^{+3} occupied A site, which attribute to enhanced dielectric constant. Magnetic saturation value was maximum at b=0.1 with 53.80 emu/g and sample showed super paramagnetic value as remanence was quite low. M.K. Anupama et al. [28] prepared $Ni_{1-x}Zn_xFe_2O_4$ (where, x = 0.0, 0.2, 0.4, 0.6, 0.8 & 1) ferrites with the use of solution combustion method. It was observed that with the addition of Zn ions, ferromagnetic properties have reduced as Zinc is diamagnetic in nature. EPR spectrum examined at room temperature and lands g factor varied from 2.23 to 1.95 with the increase in Zn concentration. This value confirmed that exchange interaction between Fe^{3+}, Ni^{2+} and Zn^{2+} metal ions were not symmetric. T.A. Taha et al. [29] prepared $Ni_{1-x}Zn_xFe_2O_4/C$ (x=0.0, 0.1, 0.3, 0.5 and 0.7) ferrites through self-combustion synthesis process at temperature. All ferrites exhibit cubic spinel structure and lattice constant, crystalline size and density of samples increased as zinc concentration increased. Also, radius of octahedral site was decreasing as zinc concentration was decreasing. However, tetrahedral site radius remained same. A-O-B interaction in spinel ferrites contributed to net magnetic behaviour. When Nickel ion, which is ferromagnetic in nature, was replaced by zinc ions, which is diamagnetic in nature, then magnetic saturation value reduced. Dielectric constant was also decreasing with frequency. Its maximum magnitude was observed for $Ni_{0.5}Zn_{0.5}Fe_2O_4$ sample, which indicate that this ferrite is best suitable for microwave absorbers fabrication. A. Pathania, et al. [30] synthesized $Ni_{0.5}Zn_{0.5}W_bFe_{2-b}O_4$ (b = 0.0, 0.2, 0.4, 0.6, 0.8 & 1.0) ferrites with the help of co-precipitation method and sintered at 850 degree Celsius. Maximum saturation magnetization was found to be 28.05 emu/g and its value was decreased with the replacement of iron ions with non-magnet tungsten ions. Resistivity of samples was found to decline from 2.2×10^5 Ω cm to 1.9×10^5 Ω cm. Activation energy also reduced from magnitude 0.0264 to 0.0221 eV. Semiconductor behaviour was also observed in all the samples. M.A. Islam et al. [31]

fabricated $Li_{0.35-0.5x}Ni_{0.3}Zn_xFe_{2.35-0.5x}O_4$ (x= 0.00-0.40 in steps of 0.10) ferrites with the help of sol- gel auto combustion method, which was sintered at 1100 degree Celsius. In XRD investigation, crystalline size was observed to be 23 nm to 37 nm, which was calculated by Scherrer formula. Lattice constant has also increased with the addition of zinc ions. Other parameters like permeability, magnetization and quality factor also showed increasing trend with the increase in zinc content. The highest value of initial permeability was found to 254 for $Li_{0.15}Ni_{0.3}Zn_{0.4}Fe_{2.15}O_4$ sintered at temperature 1523 K, which was 6 times as compared to parent composition. Dielectric loss also declined six times as compared to parent sample. Moreover, increasing trend in lattice parameter and decreasing trend in Neel temperature showed that there was weakening in A-B interaction with the increase of zinc concentration.

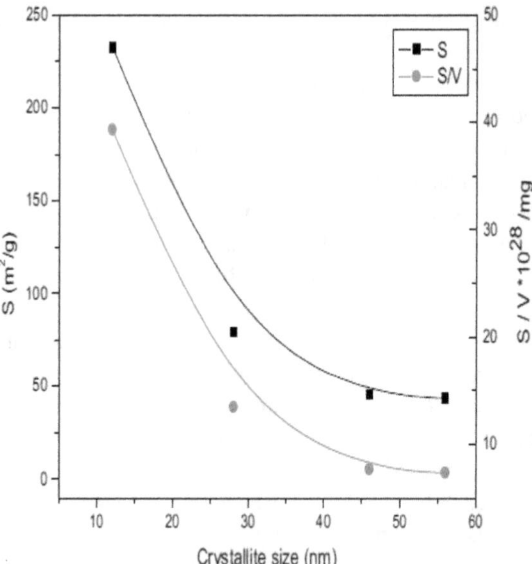

Figure 10: Variation of the specific surface area and specific surface area to volume ratio of $Ni_{0.5}Zn_{0.5}Fe_{2-x}Cr_xO_4$ samples with crystallite size (Reprinted from reference [32] with kind permission from Elsevier, Copyright 2016, Elsevier).

S.P. Dalawai et al. [32] fabricated Nix $Zn_{1+y-x}Fe_{2-2y}Sn_yO_4$ (x = 0, 0.2, 0.4, 0.6, 0.8, 1.0 and y = 0.1, 0.2). In XRD investigation, lattice constant was observed to be similar with parent sample. However, bond lengths at tetrahedral site were decreased due to increase nickel concentration while, bond length for octahedral site remained same. FTIR confirmed the formation of octahedral as well as tetrahedral site. Particle size varied between 20 nm to 60 nm. All samples showed semiconductor behaviour. Moreover, DC resistivity observed to be decrease with nickel content till x=0.6 and an increase was found in the same after x = 0.6.

Activation energy also showed the similar trend till x=0.6 with the increase in nickel ions. M. Ashtar et al. [33] prepared $Ni_{0.5}Zn_{0.5}Fe_{2-x}Cr_xO_4$(x = 0.1, 0.2, 0.3, 0.4) by using sol-gel synthesis process. The specific surface area as well as specific surface area to volume ratio was found to increase by means of decline in the size when the chromium was doped in nickel-zinc nanospinel ferrite. AC conductivity has increased with the increase in frequency. A. Thakur, et al. [34] $Ni_{0.5}Zn_{0.5}Fe_2O_4$ through chemical co-precipitation synthesis process, which was sintered at temperature 800°C and 900°C for 3 hours. The magnitude of saturation magnetization was found to increase from 73.88 to 89.50 emu/g as a variable of sintering temperature telling their potential applications in high frequency devices. M. Veena et al. [35] prepared Ni_{1-x} $Zn_xDy_yFe_{2-y}O_4$ (x=0, 0.2, 0.4 0.6, 0.8 1.0) (y=0.0 and 0.1) nanospinel ferrites by novel combustion technique then all samples were irradiated with ^{60}Co gamma radiation with total exposure dose 31 KGy having rate 6.1 KGy/h. XRD of all samples revealed nano size of ferrites. Presence of dysprosium was observed at lower angle with extra peak before and after irradiation since $DyFe_2O_3$ is present in ferrites.

Also, as Dy^{3+} concentration increases, intensity of peaks reduced. There is a little shift in peaks because of irradiation. Migration of ions in interstitial sites leads into disorder in irradiated samples. Because of irradiation, intensity and broadening of peaks has reduced. As Dy^{3+} with ionic radii 0.092 nm has placed in the position of Fe^{3+} with ionic radii (0.067 nm), which results into raised in lattice parameter from 0.831 nm value to 0.836 nm. Increase in lattice parameters tells that dysprosium ions take octahedral sites. When samples were irradiated, lattice parameter was observed to be enhanced. Dielectric constant was also investigated as function of frequency and concentration for all samples. The value of dielectric constant, loss tangent and ac conductivity was observed to increase with the increase in zinc ions, and then it attained a constant value and found to decrease thereafter. Dielectric constant and ac conductivity both increased in irradiated nano-ferrites.

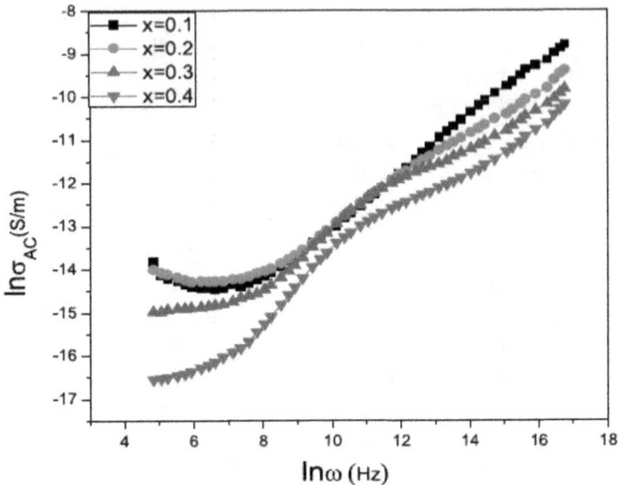

Figure 11: Variation of the ac conductivity as a function of frequency for of Cr^{3+} doped Ni-Zn ferrites (Reprinted from reference [32] with kind permission from Elsevier, Copyright 2016, Elsevier).

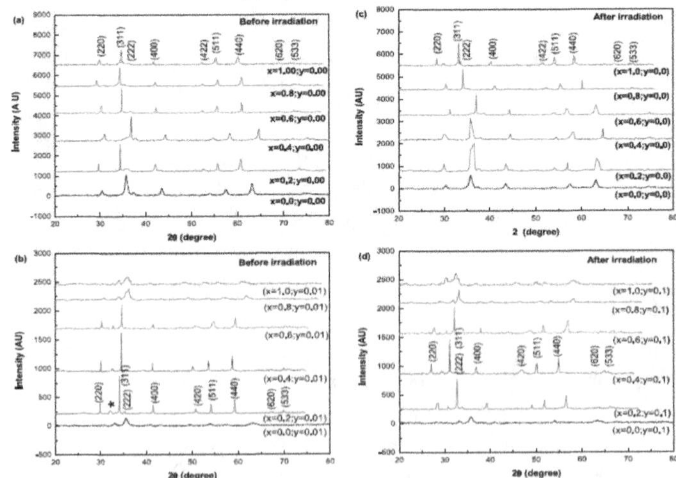

Figure 12: XRD pattern of $Ni_{1-x}Zn_xDy_yFe_{2-y}O_4$ nanoferrites (a–b) before irradiation (c–d) after irradiation (Reprinted from reference [34] with kind permission from Elsevier, Copyright 2016, Elsevier).

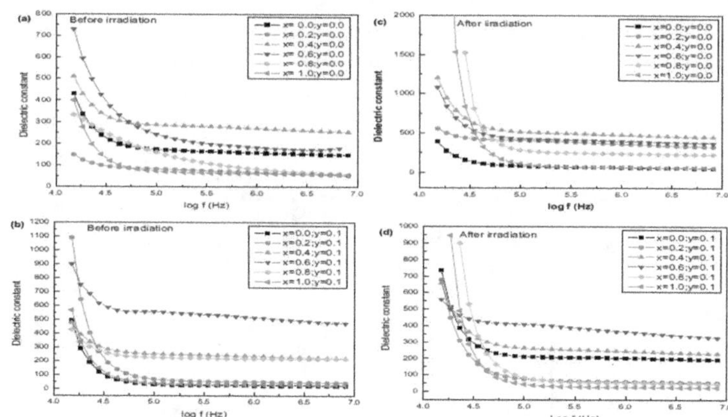

Figure 13: Variation of dielectric constant for $Ni_{1-x}Zn_xDy_yFe_{2-y}O_4$ nanoferrites (a–b) before irradiation (c–d) after irradiation (Reprinted from reference [34] with kind permission from Elsevier, Copyright 2016, Elsevier).

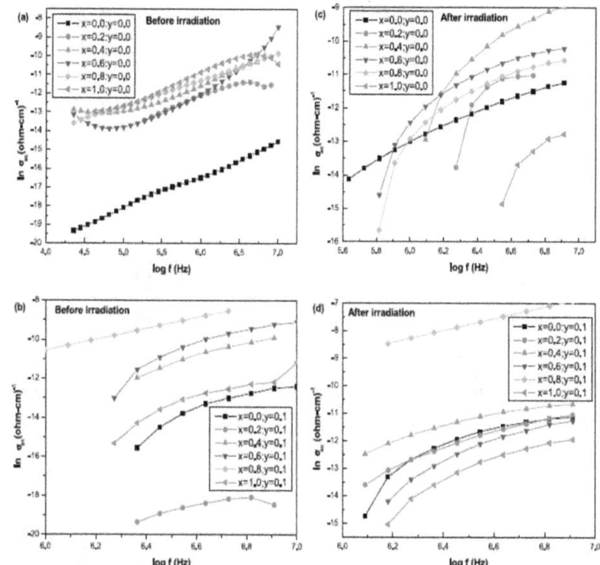

Figure 14: Variation of ac conductivity for $Ni_{1-x}Zn_xDy_yFe_{2-y}O_4$ nanoferrites (a–b) before irradiation (c–d) after irradiation (Reprinted from reference [34] with kind permission from Elsevier, Copyright 2016, Elsevier).

Sarveena et al. [36] prepared $Ni_{0.58}Zn_{0.42}Co_xCu_yFe_{2-x-y}O_4$ through sol-gel method. Particle size was reduced from 20 nm to 18 nm and lattice parameter raised from 8.348 Å to 8.406 Å.

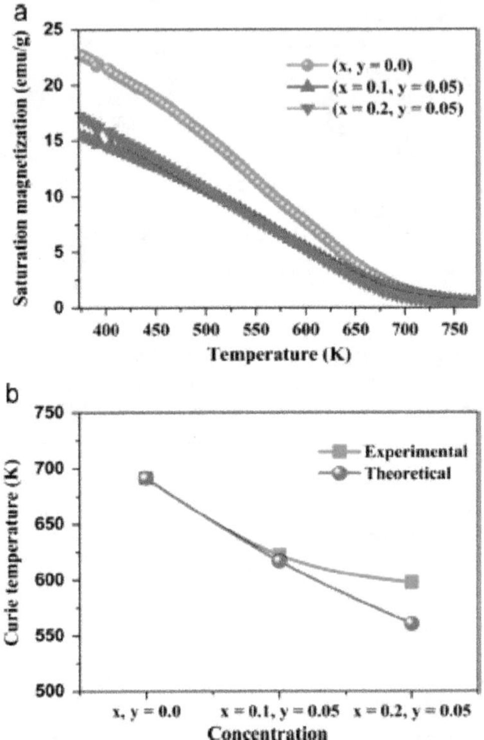

Figure 15: Variation of magnetization with temperature and Curie temperature with composition for $Ni_{0.58}Zn_{0.42}Co_xCu_yFe_{2-x-y}O_4$ ferrites (Reprinted from reference [35] with kind permission from Elsevier, Copyright 2016, Elsevier).

This was in good agreement with theoretical measurements. While studying magnetic behaviour, it was observed that magnetic saturation has reduced as well as in electric behaviour, Curie temperature also reduced with the addition of copper and cobalt ions, which were weakening the tetrahedral and octahedral site interaction. J.S. Ghodake et al. [37] prepared $Ni_{0.60-x}Zn_{0.35}Co_xFe_{2.05}O_4$ by using citrate nitrate combustion method. Lattice constant has increased from 8.09 Å to 8.12 Å as well as crystalline size has increased from 39 nm to 52 nm with the substitution of cobalt ions. Further, in the investigation of magnetic characteristics, increase in retentivity, remanence force and magnetic saturation was observed. Such behaviour was explained on the basis of replacement of Ni^{2+} (low magnetic moment) with Co^{2+} ions (high magnetic moment), which enhanced the A-B interaction. In addition to it, the initial permeability was decreased with the variation of frequency.

Table 3: Summary of structural properties of Ni-Zn ferrites.

Sample	Lattice constant	Crystalline size	Reference
$Ni_{0.5}Mg_xZn_{0.5-x}Fe_2O_4$ ($0.0 \leq x \leq 0.5$)	8.39 to 8.40 Å	8-23 nm	[6]
$Ni_{0.65}Zn_{0.35}Fe_2O_4$	8.3765 Å	43.13 nm and 38.02 nm	[7]
$Zn_{1-x}Ni_xFe_2O_4$ ($x = 0.0$-1.0)	8.33 to 8.43 Å	27.5 to 30.6 nm	[8]
$Ni_{1-x}Zn_xGd_{0.1}Fe_{1.9}O_4$ ($0 \leq x \leq 1$)	8.35 to 8.40 Å	77.6 to 93.2 nm	[9]
$Ni_{0.5}Zn_{0.5-x}Cu_xFe_2O_4$ (where x = 0.05 to 0.25)	8.36 to 8.39 Å		[10]
$Ni_xZn_{1-x}Fe_2O_4$	8.34 to 8.40 Å	15 to 23 nm	[11]
$Ni_{0.5}Zn_{0.5-x}Cd_xFe_2O_4$	8.3894 to 8.4878 Å	42.68 to 40.45 nm	[13]
$Ni_{(1-x)}Zn_xGd_yFe_{(2-y)}O_4$($0 \leq 1 \leq y = 0.1$)	8.354 to 8.404 Å	77.63 to 93.2 nm	[15]
$Ni_{0.5}Zn_{0.3}Cd_{0.2}Fe_{2-y}La_yO_4$ ($y = 0.0$-0.21)	8.439 to 8.465 3 Å		[20]
$Ca_xNi_{0.75-x}Zn_{0.25}Fe_2O_4$ (x=0, 0.25, 0.5 and 0.75)	8.334 and 8.538 Å,	12 nm and 27 nm	[21]
$Ni_xCu_{0.1}Zn_{0.9-x}Fe_2O_4$ (x = 0.5, 0.6, 0.7)	8.429 to 8.399 Å	22.4 to 21.3 nm	[23]
$Ni_{1-x}Zn_{x-2y}Li_yFe_{2+y}O_4$, ($0 \leq x \leq 0.5$) and (y= 0, 0.01, 0.02, 0.03, and 0.04)	0.839 to 0.835 nm	41.9 to 41.935 nm	[25]
($Ni_{0.6}Zn_{0.4}Gd_yFe_{2-y}O_4$; y = 0, 0.1, 0.15, and 0.2)	8.3156 to 8.3259 Å		[26]

$Ni_{1-x}Zn_xFe_2O_4/C$ (x=0.0, 0.1, 0.3, 0.5 and 0.7)	8.260 to 8.397 Å	18.50 to 36.00 nm	[28]
$Ni_{0.5}Zn_{0.5}W_xFe_{2-x}O_4$ (x = 0.0, 0.2, 0.4, 0.6, 0.8 & 1.0)	8.3730 Å		[29]
$Li_{0.35-0.5x}Ni_{0.3}Zn_xFe_{2.35}$-O-0.5 x(4) (x= 0.00-0.40)		23 to 37 nm	[30]
$Ni_{(x)}Zn_{1+y-x}Fe_{2-2y}Sn_{(y)}O_4$ (x = 0, 0.2, 0.4, 0.6, 0.8, 1.0; y = 0.1, 0.2)	8.436 to 8.302 Å	36 nm	[31]
$Ni_{0.5}Zn_{0.5}Fe_{2-x}Cr_xO_4$ (x = 0.1, 0.2, 0.3, 0.4)	8.400 to 8.389 Å	56 to 11 nm	[32]
$Ni_{0.5}Zn_{0.5}Fe_2O_4$		38.36nm to 42.16nm	[33]
$Ni_{1-x}Zn_xDy_yFe_{2-y}O_4$ (x=0, 0.2, 0.4 0.6, 0.8 1.0) (y=0.0 and 0.1)	0.8310 nm to 0.8356 nm		[34]
$Ni_{0.58}Zn_{0.42}Co_xCu_yFe_{2-x-y}O_4$ (x=0.0,0.1,0.2) & (y=0.0,0.05,0.05)	8.348 Å to 8.406 Å		[35]
$Ni_{0.60-x}Zn_{0.35}Co_xFe_{2.05}$ (x=0.0 to 0.4)	8.09 Å to 8.12 Å	39 nm to 52 nm	[36]

Table 4: Magnetic properties of mixed Ni-Zn ferrites.

Sample	Magnetic saturation (Ms) emu/g	Coercivity (H_c) Oe	Remnant (M_r) emu/g	Reference
$Ni_{0.5}Mg_xZn_{0.5-x}Fe_2O_4$ ($0.0 \leq x \leq 0.5$)	44 to 53	-	-	[6]
$Ni_{0.65}Zn_{0.35}Fe_2O_4$	176	26.27	5.20	[7]
$Zn_{1-x}Ni_xFe_2O_4$ (x = 0.0-1.0)	8.21 to 30.48	570.8 to 172.06	-	[8]
$Ni_{0.5}Zn_{0.5-x}Cu_xFe_2O_4$ (where x = 0.05 to 0.25)	45 to 61,	-	-	[10]
$Ni_xZn_{1-x}Fe_2O_4$	44.26 to 14.48	2.94 to 284.53	0.0125 to 11.320	[11]
$Ni_{0.5}Zn_{0.5-x}Cd_xFe_2O_4$	89.51 to	32.12 to	25.36 to	[13]

	71.32,	41.25,	36.48	
$Ni_{0.5}Zn_{0.5}Fe_2O_4$	45.44 to 80.66	0.007 to 0.005	1.93 to 3.43	[14]
$Ni_xZn_{1-x}Fe_2O_4/SiO_2$	0.72 to 29.28	-	-	[16]
$Ni_{0.5}Zn_{0.3}Cd_{0.2}Fe_{2-y}La_yO_4$ (y = 0.0-0.21)	40.31 to 34	71.85 to 54.37	5.2 to 3.25	[20]
$Ni_xCu_{0.1}Zn_{0.9-x}Fe_2O_4$ (x = 0.5, 0.6, 0.7)	23.87 to 36.67	24.09 to 64.99	-	[23]
$Ni_{1-x}Zn_xFe_2O_4/C$ (x=0.0, 0.1, 0.3, 0.5 and 0.7)	44.90 to 47.48	175.71 to 71.38	6.39 to 2.79	[28]
$Ni_{0.5}Zn_{0.5}W_xFe_{2-x}O_4$ (x = 0.0, 0.2, 0.4, 0.6, 0.8 & 1.0)	28.05	-	-	[29]
$Ni_{0.5}Zn_{0.5}Fe_2O_4$	73.88	33.00 to 48.07	89.50	[33]

3. Applications

(a) Inductors

Predominant application of ferrite is used as inductors in several electronic devices such as filters, impedance matching, and low noise amplifiers etc. Recently, ferrite has gain popularity in passive functional devices in which multilayer of ferrite has been used. Such devices have high degree of integration density as well as the ferrite provides large permeability for operation. However, it is necessary to find a method in which density and purity of films can be controlled. Sol gel and spin coating synthesis seemed promising for such requirements [38]. Layered ferrite electronic devices have numerous advantages such as they can easily convert magnetic field into voltage by using magnetostriction, they provide high sensitivity at such small size with almost zero power consumption. One example of this is ($Ni_{1-x}Zn_xFe_2O_4$ with x = 1–0.5)/lead zirconate-titanate ($PbZr_{0.52}Ti_{0.48}O_3$) ferrite [39]. This ferrite was prepared by tape casting with eleven layers.

(b) High frequency applications

Ferrite have been utilizing for telecommunication and radar at higher frequency. Ferrite are far better than metal in terms of penetration of electromagnetic fields [40]. Ferrites application in monitoring, audio-video electronic devices, industrial automation are becoming increasingly popular. Using iron is not an effective option nowadays because of its size and cost. Ferrites provide low eddy current losses, sufficient permeability for power conversion. Microwave absorption when magnetic field is low is getting attention is recognised as LFA. In this application, ferromagnetic material with anisotropy field is required. Results obtained by a research on Ni-Zn ferrite have been quite promising for such applications [41].

(c) Power

Power delivery can also be enhanced by using ferrites. In computers, TV, all sorts of machinery require power for their operations. SMP (Switched mode Power supply) is one of the leading power sources nowadays. By decreasing eddy currents in ferrites, efficient in terms of power can be increased. Recently study of Mn-Zn and Ni-Zn is getting attention for such uses, in which $Mn_xNi_{0.5-x}Zn_{0.5}Fe_2O_4$ has been synthesised with the help of citrate precursor technique. However, its applications at high temperature need to be studied. With increase in Ferrous ions at higher temperature by controlling its state through oxidation, Ni-Zn and Mn-Zn ferrite can also provide numerous power supplies at low as well as high temperature [42]. Miniaturization also required in such area in order to compete with demand of integrated thin films around 200°C. Therefore, a new method for synthesizing transformer can also be given with the help of such devices [43].

(d) Electromagnetic interference suppression

Requirement of high-speed digital devices are increasing drastically in computers, mobile phones, scanners etc. wireless communication is responsible for raising demand of electric and magnetic field. Such communication is known as electromagnetic interference [44]. Noise is one of the biggest issues in electromagnetic interference. Such problems can be reduced by making low pass filters, when frequency becomes greater than circuit frequency. There are many proposed methods to build such devices by using ferrites [45]. Apparently, miniaturization leads to increase in the use of ferrites for suppressors. Ni-Zn ferrites have been using for same application at frequency range 20-200 MHz.

(e) Biosciences

In many living organisms, magnetic particles can easily be found at nano scale. Such magnetic particles have better properties as compared to the magnetic particles that have been prepared in laboratory. Biogenic magnetic nanoparticle possesses perfect crystallography and morphology. These magnetic particles can be utilized for therapeutic applications e.g. MRI also known as magnetic resonance imaging, superparamagnetic behaviour is required as such particles are able to change the speed of proton decay, which is working principle of MRI [46]. Energy coming from hysteresis loss of ferrite have tendency to treat cancer tissues, which is known as hyperthermia. When enzymes, antibodies become immobilized, then these immobilized components can be target and extracted from specific place through external magnetic field [47].

Conclusions

In this chapter, we have tried to provide the basic information of ferrites and then included the recent reported work done so far on the substituted Ni-Zn ferrites along with the applications of these ferrites in various fields. We have proposed Ni-Zn ferrite as suitable and potential candidate for diverse applications.

Conflict of interest

The authors declare that they have no conflict of interest.

Acknowledgment

The authors are grateful to the research and scientific community who have made relentless efforts to bring novel data in the fields and topics of research and development discussed in this review. This work did not receive any funding.

References

[1] A. Hagfeldt, M. Graetzel, Light-induced redox reactions in nanocrystalline systems, Chemical reviews, 95 (1995) 49-68.

[2] T. Abraham, Economics of ceramic magnet, American Ceramic Society Bulletin, 73 (1994) 62-65.

[3] K.K. Kefeni, B.B. Mamba, T.A. Msagati, Application of spinel ferrite nanoparticles in water and wastewater treatment: a review, Separation and Purification Technology, 188 (2017) 399-422.

[4] P. Dhiman, T. Mehta, A. Kumar, G. Sharma, M. Naushad, T. Ahamad, G.T. Mola, $Mg_{0.5}Ni_xZn_{0.5-x}Fe_2O_4$ spinel as a sustainable magnetic nano-photocatalyst with dopant driven band shifting and reduced recombination for visible and solar degradation of Reactive Blue-19, Advanced Powder Technology, (2020).

[5] Ü. Özgür, Y. Alivov, H. Morkoç, Microwave ferrites, part 1: fundamental properties, Journal of Materials Science: Materials in Electronics, 20 (2009) 789-834.

[6] R.C. Pullar, Hexagonal ferrites: a review of the synthesis, properties and applications of hexaferrite ceramics, Progress in Materials Science, 57 (2012) 1191-1334.

[7] F. Hezam, N.O. Khalifa, O. Nur, M. Mustafa, Synthesis and magnetic properties of $Ni_{0.5}Mg_xZn_{0.5-x}Fe_2O_4$ ($0.0\leq x\leq 0.5$) nanocrystalline spinel ferrites, Materials Chemistry and Physics, 257 (2021) 123770.

[8] J.P.K. Chintala, S. Kaushik, M.C. Varma, G. Choudary, K. Rao, An Accurate Low Temperature Cation Distribution of Nano Ni-Zn Ferrite Having a Very High Saturation Magnetization, Journal of Superconductivity and Novel Magnetism, 34 (2021) 149-156.

[9] R. Verma, S. Kane, P. Tiwari, S. Modak, T. Tatarchuk, F. Mazaleyrat, Ni addition induced modification of structural, magnetic properties and antistructural modeling of $Zn_{1-x}Ni_xFe_2O_4$ (x= 0.0-1.0) nanoferrites, Molecular Crystals and Liquid Crystals, 674 (2018) 130-141.

[10] S. Kumar MV, G. Shankarmurthy, E. Melagiriyappa, A. Rao, K. Nagaraja, Cation distribution and magnetic properties of Gd^{+3}-substituted Ni-Zn nano-ferrites, Journal of Superconductivity and Novel Magnetism, 33 (2020) 2821-2827.

[11] D. Venkatesh, B.V. Prasad, K. Ramesh, M. Ramesh, Magnetic Properties of Cu 2+ Substituted Ni–Zn Nano-Crystalline Ferrites Synthesized in Citrate-Gel Route, Journal of Inorganic and Organometallic Polymers and Materials, (2019) 1-10.

[12] S.A. Jadhav, S.B. Somvanshi, M.V. Khedkar, S.R. Patade, K. Jadhav, Magneto-structural and photocatalytic behavior of mixed Ni–Zn nano-spinel ferrites: visible light-enabled active photodegradation of rhodamine B, Journal of Materials Science: Materials in Electronics, 31 (2020) 11352-11365.

[13] X. Zhou, Y. Li, Q. Huang, Preparation of $Ti_3C_2T_x$/NiZn Ferrite Hybrids with Improved Electromagnetic Properties, Materials, 13 (2020) 820.

[14] P.A. Rao, V. Raghavendra, B. Suryanarayana, T. Paulos, N. Murali, P.P. Varma, R.G. Prasad, Y. Ramakrishna, K. Chandramouli, Cadmium substitution effect on structural, electrical and magnetic properties of Ni-Zn nano ferrites, Results in Physics, 19 (2020) 103487.

[15] A. Bajorek, C. Berger, M. Dulski, P. Łopadczak, M. Zubko, K. Prusik, M. Wojtyniak, A. Chrobak, F. Grasset, N. Randrianantoandro, Microstructural and magnetic characterization of $Ni_{0.5}Zn_{0.5}Fe_2O_4$ ferrite nanoparticles, Journal of Physics and Chemistry of Solids, 129 (2019) 1-21.

[16] M.S. Kumar, G. Shankarmurthy, E. Melagiriyappa, K. Nagaraja, H. Jayanna, M. Telenkov, Induced effects of Zn^{+2} on the transport and complex impedance properties of Gadolinium substituted nickel-zinc nano ferrites, Journal of magnetism and Magnetic materials, 478 (2019) 12-19.

[17] T. Dippong, E.-A. Levei, I.G. Deac, F. Goga, O. Cadar, Investigation of structural and magnetic properties of $Ni_xZn_{1-x}Fe_2O_4/SiO_2$ ($0\leq x\leq 1$) spinel-based nanocomposites, Journal of Analytical and Applied Pyrolysis, 144 (2019) 104713.

[18] M.M. Kothawale, R. Tangsali, G. Naik, J. Budkuley, Enhancement of magnetization and tailoring of blocking temperatures of nano-Ni–Zn ferrite powder synthesized using microwave-assisted combustion method, Journal of Superconductivity and Novel Magnetism, 32 (2019) 373-379.

[19] S. Kumar, J. Singh, H. Kaur, H. Singh, H.S. Dosanjh, Microstructural and magnetic properties of Zn substituted nickel ferrite synthesised by facile solution combustion method, Micro & Nano Letters, 14 (2019) 727-731.

[20] V.K. Mande, J.S. Kounsalye, S. Vyawahare, K. Jadhav, Effect of γ-radiation on structural, morphological, magnetic and dielectric properties of Zn–Cr substituted nickel ferrite nanoparticles, Journal of Materials Science: Materials in Electronics, 30 (2019) 56-68.

[21] S. Ikram, J. Jacob, M.I. Arshad, K. Mahmood, A. Ali, N. Sabir, N. Amin, S. Hussain, Tailoring the structural, magnetic and dielectric properties of Ni-Zn-$CdFe_2O_4$ spinel ferrites by the substitution of lanthanum ions, Ceramics International, 45 (2019) 3563-3569.

[22] M. Kuru, T.Ş. Kuru, S. Bağcı, The role of the calcium concentration effect on the structural and dielectric properties of mixed Ni–Zn ferrites, Journal of Materials Science: Materials in Electronics, 30 (2019) 5438-5453.

[23] A. Pathania, P. Thakur, A.V. Trukhanov, S.V. Trukhanov, L.V. Panina, U. Lüders, A. Thakur, Development of tungsten doped Ni-Zn nano-ferrites with fast response and recovery time for hydrogen gas sensing application, Results in Physics, 15 (2019) 102531.

[24] K. Ramakrishna, C. Srinivas, C. Prajapat, S.S. Meena, M. Mehar, D. Potukuchi, D. Sastry, Structural and magnetic investigations: Study of magnetocrystalline anisotropy and magnetic behavior of 0.1% Cu^{2+} substituted Ni–Zn ferrite nanoparticles, Ceramics International, 44 (2018) 1193-1200.

[25] J.-L. Mattei, D. Souriou, A. Chevalier, Magnetic and dielectric properties in the UHF frequency band of half-dense Ni-Zn-Co ferrites ceramics with Fe-excess and Fe-deficiency, Journal of Magnetism and Magnetic Materials, 447 (2018) 9-14.

[26] M.M. Ismail, N.A. Jaber, Structural and elastic properties of nickel–zinc ferrite nano-particles doped with lithium, Journal of the Brazilian Society of Mechanical Sciences and Engineering, 40 (2018) 1-8.

[27] M.S. Kumar, G. Shankarmurthy, E. Melagiriyappa, K. Nagaraja, A.R. Lamani, B. Harish, Dielectric and magnetic properties of high porous Gd+ 3 substituted nickel zinc ferrite nanoparticles, Materials Research Express, 5 (2018) 046109.

[28] M. Anupama, N. Srinatha, S. Matteppanavar, B. Angadi, B. Sahoo, B. Rudraswamy, Effect of Zn substitution on the structural and magnetic properties of nanocrystalline $NiFe_2O_4$ ferrites, Ceramics International, 44 (2018) 4946-4954.

[29] T. Taha, S. Elrabaie, M. Attia, Green synthesis, structural, magnetic, and dielectric characterization of $NiZnFe_2O_4$/C nanocomposite, Journal of Materials Science: Materials in Electronics, 29 (2018) 18493-18501.

[30] A. Pathania, S. Bhardwaj, S.S. Thakur, J.-L. Mattei, P. Queffelec, L.V. Panina, P. Thakur, A. Thakur, Investigation of structural, optical, magnetic and electrical properties of tungsten doped NiZn nano-ferrites, Physica B: Condensed Matter, 531 (2018) 45-50.

[31] M. Islam, M. Hasan, A.A. Hossain, Enhancement of initial permeability and reduction of loss factor in Zn substituted nanocrystalline $Li_{0.35-0.5x}Ni_{0.3}Zn_xFe_{2.35-0.5x}O_4$, Journal of Magnetism and Magnetic Materials, 424 (2017) 108-114.

[32] S. Dalawai, T. Shinde, A. Gadkari, N. Tarwal, J. Jang, P. Vasambekar, Influence of Sn^{4+} on Structural and DC Electrical Resistivity of Ni-Zn Ferrite Thick Films, Journal of Electronic Materials, 46 (2017) 1427-1438.

[33] M. Ashtar, A. Munir, M. Anis-ur-Rehman, A. Maqsood, Effect of chromium substitution on the dielectric properties of mixed Ni-Zn ferrite prepared by WOWS sol–gel technique, Materials Research Bulletin, 79 (2016) 14-21.

[34] A. Thakur, P. Kumar, P. Thakur, K. Rana, A. Chevalier, J.-L. Mattei, P. Queffelec, Enhancement of magnetic properties of Ni0. 5Zn0. 5Fe2O4 nanoparticles prepared by the co-precipitation method, Ceramics International, 42 (2016) 10664-10670.

[35] M. Veena, A. Somashekarappa, G. Shankaramurthy, H. Jayanna, H. Somashekarappa, Effect of 60Co gamma irradiation on dielectric and complex impedance properties of Dy^{3+} substituted Ni–Zn nanoferrites, Journal of Magnetism and Magnetic Materials, 419 (2016) 375-385.

[36] G. Kumar, A. Kumar, R. Kotnala, K.M. Batoo, M. Singh, Investigation of structural, magnetic and Mössbauer properties of Co^{2+} and Cu^{2+} substituted Ni–Zn nanoferrites, Ceramics International, 42 (2016) 4993-5000.

[37] J. Ghodake, R.C. Kambale, T. Shinde, P. Maskar, S. Suryavanshi, Magnetic and microwave absorbing properties of Co2+ substituted nickel–zinc ferrites with the emphasis on initial permeability studies, Journal of Magnetism and Magnetic Materials, 401 (2016) 938-942.

Materials Research Foundations **112** (2021) 189-217 https://doi.org/10.21741/9781644901595-5

[38] F. Liu, C. Yang, T. Ren, L. Liu, H. Feng, A. Wang, H. Long, J. Yu, Fully integrated ferrite-based inductors for RF ICs, in: The 13th International Conference on Solid-State Sensors, Actuators and Microsystems, 2005. Digest of Technical Papers. TRANSDUCERS'05., IEEE, 2005, pp. 895-898.

[39] Y.K. Fetisov, A.A. Bush, K.E. Kamentsev, A.Y. Ostashchenko, G. Srinivasan, Ferrite-piezoelectric multilayers for magnetic field sensors, IEEE Sensors Journal, 6 (2006) 935-938.

[40] M. Pardavi-Horvath, Microwave applications of soft ferrites, Journal of Magnetism and Magnetic Materials, 215 (2000) 171-183.

[41] H. Montiel, G. Alvarez, I. Betancourt, R. Zamorano, R. Valenzuela, Correlations between low-field microwave absorption and magnetoimpedance in Co-based amorphous ribbons, Applied Physics Letters, 86 (2005) 072503.

[42] V. Zaspalis, V. Tsakaloudi, E. Papazoglou, M. Kolenbrander, R. Guenther, P. Van Der Valk, Development of a new MnZn-ferrite soft magnetic material for high temperature power applications, Journal of Electroceramics, 13 (2004) 585-591.

[43] F. Amalou, E.L. Bornand, M.A. Gijs, Batch-type millimeter-size transformers for miniaturized power applications, IEEE transactions on Magnetics, 37 (2001) 2999-3003.

[44] G. Stojanovic, M. Damnjanovic, V. Desnica, L. Zivanov, R. Raghavendra, P. Bellew, N. Mcloughlin, High-performance zig-zag and meander inductors embedded in ferrite material, Journal of Magnetism and Magnetic Materials, 297 (2006) 76-83.

[45] Z. Li, L. Guoqing, L. Chen, W. Yuping, C. Ong, Co 2+ Ti 4+ substituted Z-type barium ferrite with enhanced imaginary permeability and resonance frequency, Journal of applied Physics, 99 (2006) 063905.

[46] T.C. Yeh, W. Zhang, S.T. Ildstad, C. Ho, In vivo dynamic MRI tracking of rat T-cells labeled with superparamagnetic iron-oxide particles, Magnetic resonance in medicine, 33 (1995) 200-208.

[47] T. Matsunaga, S. Kamiya, Use of magnetic particles isolated from magnetotactic bacteria for enzyme immobilization, Applied Microbiology and Biotechnology, 26 (1987) 328-332.

Materials Research Foundations **112** (2021) 218-245 https://doi.org/10.21741/9781644901595-6

Chapter 6

Spinel Ferrite Based Nanomaterials for Water Remediation Application

Rohit Jasrotia* and Jyoti Prakash

School of Physics and Materials Science, Shoolini University, Bajhol, Solan, H.P., India

rohitsinghjasrotia4444@gmail.com

Abstract

The decreasing levels of consumable water on earth have been a serious issue and this issue makes the researchers and scientists develop new technologies for the purification of polluted water. Several reports have been carried on wastewater remediation by utilizing spinel ferrite-based nanoparticles and their composites. The spinel ferrites-based nanoparticles utilized for wastewater treatment was cost effective, chemically stable, easily retrieved and reusable. The present work addresses the various fabrication techniques for the preparation of spinel ferrite-based nanoparticles and their utilization for the removal of organic and inorganic pollutants through the adsorption paths.

Keywords

Spinel Ferrites, Fabrication Techniques, Adsorption Mechanism, Organic and Inorganic Pollutants, Wastewater Remediation

Contents

Materials Research Foundations **112** (2021) 218-245 https://doi.org/10.21741/9781644901595-6

1. Introduction

Water is present in abundant concentration on earth, but the main issue is that only 1% of the water is consumable. From the last decades, consumable water was utilized tremendously for the household works, consumption, industrial work and for many other works which, therefore, makes the level of consumable water decreasing according to the day-to-day life requirements [1]. Moreover, the increasing industrial establishment and population growth will be responsible for the shortage of consumable water on earth. There were various pollutants as organic, inorganic, radiological, and biological, present in water getting from natural resources and thus, making it very difficult to take water for consumption [2,3]. These pollutants are toxic in nature and are dangerous for the human and other living life. Around 80 percent of the diseases are caused in human beings due to polluted water consumption and moreover, 3.5 million people die every year due to water borne diseases. The requirement for clean water increases day by day in the world and for doing this, a number of technologies such as photocatalytic degradation, adsorption, bacterial degradation etc. were taken into practice [4,5]. Out of all these techniques, adsorption and photocatalytic degradation is considered as one of the important approaches for wastewater remediation [6,7]. The adsorption technique is very effective in eliminating pollutants by transferring the polluting reagents from standard media to alterative media in relation to destroy it.

In modern times, spinel ferrites have attracted various researchers and authors from all over the world due to its superior properties such as large surface to volume ratio [8], superparamagnetic nature, high chemical stability [9], high adsorption abilities and moreover, is considered as one of the best photocatalysts for the degradation of pollutants. Based on magnetism, spinel ferrites of chemical formula, MFe_2O_4 where M represents divalent metal ion, are categorized into normal spinel, inverse spinel, and mixed spinel ferrite [10]. A greater attention has been given to the category of spinel ferrite due to its excellent applications such as magnetic storage devices, biomedical, water remediation and many more. The A-A, B-B and A-B exchange interactions were present in the structure of spinel ferrites where A-B interaction was stronger than the B-B interaction and thus, magnetism in spinel ferrites can be observed due to the exchange interactions between the various cations present at the interstitial sites i.e. A-site (tetrahedral site) and B-site (octahedral site) [11,12]. The magnetic property in the spinel ferrites shows its dependence towards the distribution of cations at the interstitial sites and therefore, substitution of dopant ions at the octahedral sites will be responsible for the weakening of A-B exchange interactions [13,14]. Under the cation distribution, in case of normal spinel ferrites, divalent metal ions are distributed among the A-site whereas, ferric ions prefer to occupy the B-site. In inverse spinel ferrites, divalent metal ions occupied the B-site but under mixed spinel ferrites, both the divalent and ferric ions try to occupy the A-site and B-site, respectively. For the water remediation application, spinel ferrite based magnetic nanoparticles were turn out to be very essential for the elimination of pollutants with the utilization of adsorption and photocatalytic degradation technologies [15–19] and these has been extensively utilized in industries to eliminate organic pollutants, dyes, and toxic metallic elements [20–23]. Spinel ferrites are easily retrieved from the resultant mixture and therefore, retrieved material was further utilized for number of reuse cycles [24]. The aim of this book chapter was to address the spinel ferrites and their fabrication approaches. Moreover, to provide a detailed study of spinel ferrites for water remediation application incorporating adsorption.

2. Approaches for the fabrication of Spinel ferrites

There are various approaches for the fabrication of spinel ferrite-based nanoparticles such as sol-gel, auto-combustion, co-precipitation, citrate precursor, solid state reaction and many more [25–27]. To synthesized high quality magnetic nanoparticles, appropriate fabrication technique was taken into practice. The microstructural, optical, magnetic and adsorption properties of spinel ferrites can be improved using incorporation of metal ions at the interstitial sites- Tetrahedral site (A-site) and Octahedral site (B-site). Therefore, the proper selection of fabrication technique will be responsible for deciding the properties of final product [28]. The various approaches for the preparation of nanoparticles were discussed as:

2.1 Solid state reaction technique

The solid-state reaction approach was utilized for the fabrication of spinel ferrite based magnetic nanoparticles by taking two starting solid chemical precursors. For the synthesis, a stoichiometric amount of the starting precursors of AR graded was taken for ball milling around 10 to 12 hours. The chemical reagents are interspersed with distilled water by wet blending in a mortar pestle [29] and then, it was calcinated in a muffle furnace at specified temperature for specified hours. The ball milling of calcinated powder was again taken into practice and most of the times, surfactants were utilized for the reduction of particle size of magnetic nanoparticles. The fabricated magnetic nanoparticles were again pressed into pellets using a hydraulic press and then, these pellets were sintered in a muffle furnace at 1000 °C to 1500 °C for 5-7 hours, respectively. Therefore, the final fabricated nanoparticles were ready for analyzing their properties.

2.2 Sol-gel auto-combustion technique

The sol-gel auto-combustion technique belongs to a category of wet chemical approach utilized to fabricate solid nanoparticles. The metals-based magnetic nanoparticles can be prepared by using technique in which the starting precursors were the metal nitrates and citric acid and here, citric acid acts as a fuel. For the fabrication of nanoparticles, firstly stoichiometric amounts of starting precursors were taken in distilled water and stirred on magnetic stirrer with a hot plate until both the aqueous solutions become clear. Mixed both the aqueous solutions and set their pH~7 by adding few drops of ammonia solution [30]. For the making of gel, add stoichiometric amount of gel precursor called ethylene glycol. Fused out the gel at 100 °C to 150 °C on magnetic stirrer with hot plate to undergo auto-combustion until a loose black material will be obtained [14,31]. The final residue was grinded using mortar pestle to get fine nanoparticles and then, it was calcinated in a muffle furnace at 800 to 1000 °C for 4 to 5 hours. Therefore, the final nanomaterials were ready for their characterization to analyze its various properties. A number of spinel ferrite based nanomaterials have been fabricated with the usage of sol-gel auto-combustion technique [32,33]. Using a hydraulic press, the fabricated powder was pressed in the form of pellets and sintered in a muffle furnace at a specified temperature for some hours, respectively. The Schematic chart of sol-gel auto-combustion technique were provided in Fig. 1.

2.3 Co-precipitation technique

The co-precipitation technique is one of the effective and appropriate techniques for the fabrication of magnetic nanoparticles which incorporates metal chlorides as the starting precursors. Firstly, the stoichiometric weights of metal chlorides are mixed with distilled water and the solution is stirred for 2 hours on a magnetic stirrer. Add few drops of $NH_4(OH)_2$ solution to set the pH of the given solution and stir it continuously. After stirring, the precipitates start inside the solution at the bottom of the beaker and then, these precipitates are separated out by centrifugation, respectively. The precipitates were washed many times for the removal of extra impurities with the utilization of distilled water. The schematic chart of co-precipitation fabrication approach was provided in Fig. 2.

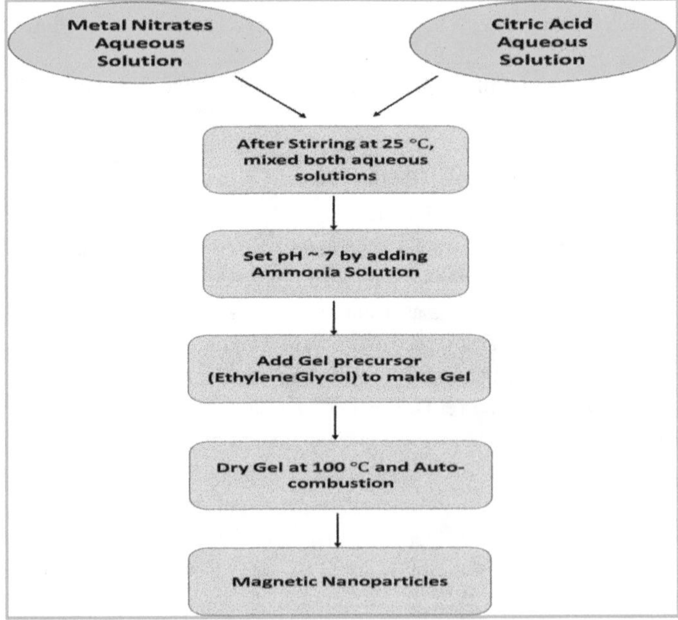

Figure 1: Schematic representation of sol-gel auto-combustion fabrication technique.

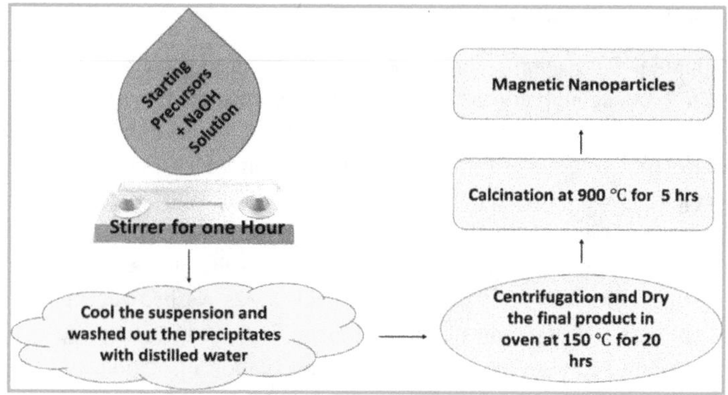

Figure 2: Schematic representation of co-precipitation fabrication technique.

2.4 Citrate precursor technique

The citrate precursor technique is cost effective and a less time consumption fabrication technique which incorporates metal nitrates and citric acid as the starting precursors and it also controls the homogeneity of the fabricated nanoparticles. The stoichiometric amounts of starting precursors were mixed with distilled water. The solution was heated at a temperature of 50-80 °C on magnetic stirrer with hot plate for two hours until a solid residue is obtained and then, this solid residue is crushed into fine powder. The grinded fine powder was calcinated at a temperature of 700-900 °C for five hours, respectively. Therefore, the final nanoparticles were obtained. The Schematic representation of citrate precursor fabrication technique was provided in Fig. 3.

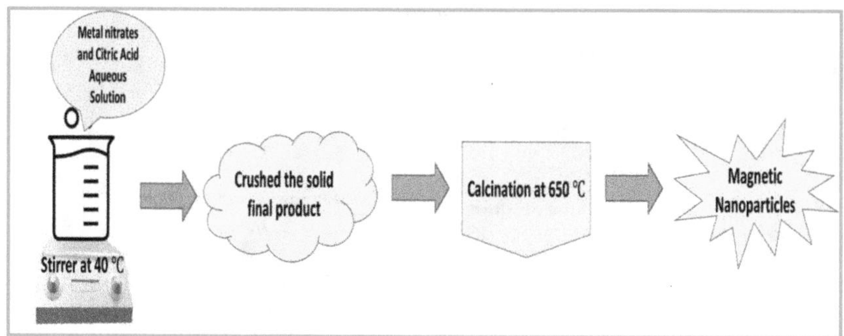

Figure 3: Schematic representation of citrate precursor fabrication technique.

2.5 Hydrothermal technique

The hydrothermal fabrication technique is one of the productive and low temperature approaches to produce high crystallinity and less agglomerated nanoparticles. The whole process was taken out in an autoclave (Teflon lined) at a temperature of 100-170 900 °C. Under this technique, the in a Teflon lined autoclave in which starting precursors were the metal chlorides which were mixed one by one with distilled water and stirred on a magnetic stirrer with hot plate at room temperature [34]. To the stirring solution, KNO_3 or NaOH solution were added and then, pH of the final solution was balanced by adding few drops of ammonia solution. The formed residue was washed several times for the removal of chloride concentration still present in the final product and moreover, washing with ethyl alcohol was also carried out for the prevention of agglomeration [35]. The final residue was then annealed in a Teflon lined autoclave at a temperature of 100-200 °C, respectively. Many times, surfactants were taken into practice for controlling the growth, size, and nucleation of fabricated nanoparticles. The chart of hydrothermal synthesis technique was provided in Fig. 4.

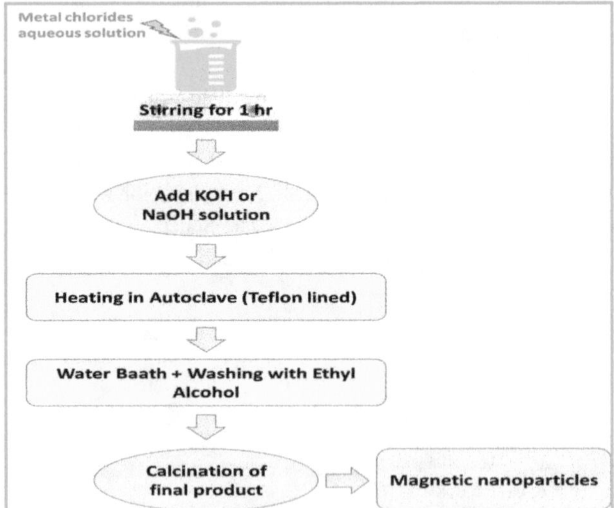

Figure 4: Schematic representation of hydrothermal fabrication technique.

Materials Research Foundations **112** (2021) 218-245 https://doi.org/10.21741/9781644901595-6

3. Spinel ferrites nanoparticles and composites used for wastewater remediation

The comprehensive applications of spinel ferrites nanoparticles and their composites in wastewater remediation have been clearly demonstrated in research. Their use as adsorbent, photocatalyst, sensor and membrane modification etc. Organic dyes, heavy metal ions, pesticides, antibiotics and a large range of aromatic compounds are the most dangerous substances found in wastewater [36]. Natural water unsafe for consumption and even harmful to marine species due to the prevalence of these toxins in water. A variety of methods have been implemented for the remediation of wastewater, with differing degrees of effectiveness. A few of these methods include adsorption strategies focused on filtration, ion exchange, coagulation, anaerobic therapy, aerogels, hydrogels, and non-magnetic metal oxides [37,38]. For the elimination of pollutants from water, these remediation options may be used; but several issues impede their use, such as inadequate pollutant removal, adsorbent recovery complexity, absence of cost-effectiveness of the process, and in some situations the waste sludge is voluminous and requires careful design and a significant amount of disposal space is needed [5,39]. Owing to the above drawbacks, existing adsorbents used in the treatment of wastewater are not suitable. Therefore, we need an adsorbent that is cheap, reliable, reusable and easily retrieved [40,41]. Accordingly, the most suitable materials are spinel ferrites nanoparticles and their composites that can offer alternatives to waste water treatment. Because of their exceptional chemical and physical properties, which allow them to extract a wide variety of pollutants and easily retrieved and reusable. Compared to their bulk counterparts, spinel ferrites nanoparticles and their composites have many benefits, especially in terms of greater adsorption efficiency, cost-effectiveness and pollutants removal capacity [42]. In addition, ease retrieved and ability to reuse retrieved adsorbents for many reuse cycles without losing their adsorption efficiency compared to other non-magnetic nanoparticles-based adsorbents make spinel ferrites nanoparticles and their composites one of the popular materials and recently have been utilized broadly for the remediation of wastewater. According to the literature, spinel ferrites nanoparticles are favored for wastewater remediation due to its huge surface area and large number of active sites for contact with pollutants present in the aqueous solution [43,44]. Moreover, spinel ferrites nanoparticles have superparamagnetic properties that arises from their nanoparticle size, allowing them to be easily removed by external magnetic fields from the aqueous solution. The spinel ferrites nanoparticles might be used directly or as a central material by covering them with sufficient functional groups for the handling of wastewater. For instance, by surface functionalization using an appropriate functional group like amines, trimethoxy silane and phosphoric acid etc. the properties of spinel ferrites nanoparticles are enhanced significantly [45]. By introducing polymers to nanoparticles, such functional groups could be added during the preparation of nanoparticles. Besides the stabilization of spinel ferrites nanoparticles, these functional groups enhance the selectivity in aqueous solution by hyper-branching, ionic or electrostatic interactions. Many isotherms as well as kinetic models are utilized to verify the adsorption potential of various spinel ferrites nanoparticles and their composites [46]. Mainly most frequently utilized models for isotherm are Langmuir and Freundlich isotherm models and for kinetics pseudo first and pseudo second order kinetic models. The adsorption isotherm models help to determine the maximum adsorption capacity of the adsorbent and also explain the interaction between the adsorbent and adsorbate. While the kinetic models help to determine the adsorption rate and behavior of adsorption [47,48].

4. Adsorption pathways

Adsorption process is largely regulated by certain type of interaction between the adsorbent and adsorbate, precisely the surface charge and hydroxyl groups playing a major role in adsorption. The most feasible methods of physicochemical interaction includes ion-exchange, weak forces including van der Waals forces, hydrogen bonding, dipole–dipole and π-π interactions, these are capable of eliminating various contaminants from wastewater by adsorption [49,50]. Furthermore, chemisorption due to the chemical interactions is feasible when there is successful sharing of electrons between the adsorbate and adsorbent and physisorption occurs when there is weak attractive forces between the pollutant and adsorbent's surface as shown in the Fig. 5 [51].

Different mechanisms rely on the form of adsorption and type of contaminant available. There are toxins found in wastewater and the spinel ferrite nanoparticles/spinel ferrite nanoparticle composites are being used to handle wastewater treatment [52]. For more illustration, the ion-exchange process for Pb (II) adsorption on $Co_{0.6}$ $Fe_{2.4}$ O_4 was found to be a significant adsorption mechanism. In this work, when $pH < 7$, the adsorption process is mainly based on ion exchange mechanism and outer-sphere surface complexes, whereas for $pH > 7$, inner-sphere surface complexes are observed [53]. An alternative study showed that adsorption through the method of ion-exchange and electrostatic attraction were the key adsorption mechanisms for copper ion adsorption on the surface of EDTA functionalized silica-coated Fe_3O_4 [54]. Results from the adsorption analysis on many spinel ferrite nanoparticles indicate electrostatic activity and ion-exchange as the key adsorption mechanism for cationic dyes, because surfactant groups are present on these nanoparticles [55,56]. One way to imagine this is to think about extracting MB and Congo red together in the presence of $MnFe_2O_4$ nanoparticles. During this phase, the hydrogen bond association between the dye and the hydroxyl groups found in $MnFe_2O_4$ was discovered to be an important adsorption mechanism [57].

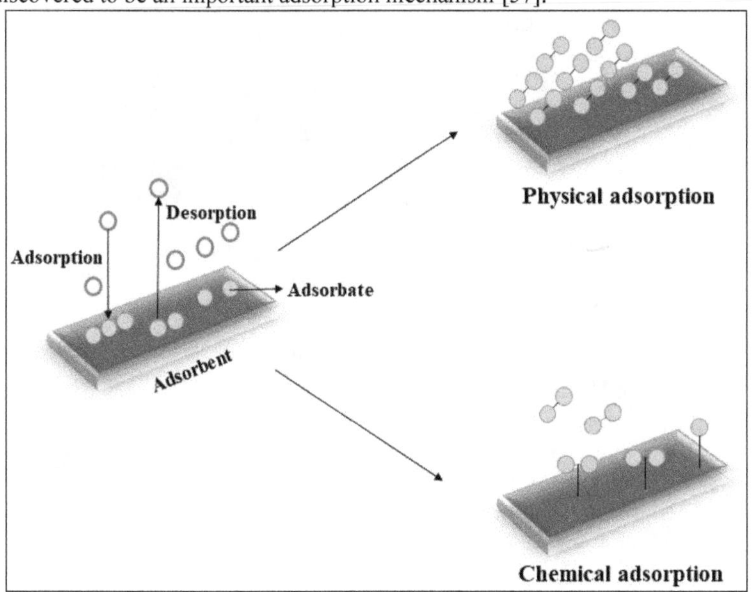

Figure 5: Shows general adsorption mechanism with its basic types.

Similarly, the predominant adsorption mechanism of arsenate on γ-Fe_2O_3 nanoparticles during the removal of arsenate from treated water was found to be inner-sphere complex formation [58]. Because of the various influences on the adsorption mechanism, the nature of the contaminant and the adsorbent such as functional groups, pore structure, oxidation state, and surface charge all significantly affect the outcome.

5. Factors affecting the adsorption ability of spinel ferrite nanoparticles

Multiple factors such as synthesis methods, types and particle size of the adsorbent, dosages, pH value of wastewater, initial concentration of contaminants, temperature, and charge and size of adsorbate all influence the adsorption capacity of spinel ferrite nanoparticles and their corresponding composites. The aforementioned factors (Fig. 6) have already been well studied in regards to adsorption and removal of contaminants [30, 46].

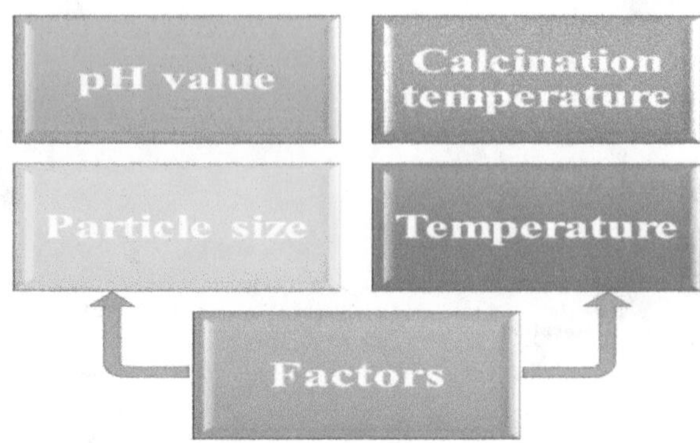

Figure 6: Shows different factors affecting the adsorption ability of spinel ferrite nanoparticles.

5.1 Particle size

Generally, adsorption is related to particle size of nanoparticles, referred to as the nanoscale effect. Larger particle size generally correlates with higher adsorption capacity due to increased surface area- to - volume ratio [60].

5.2 pH value

The pH of the solution determines the volume of content that can be adsorbed, since the surface charge of the spinel ferrite nanoparticles and spinel ferrite nanoparticle composites is influenced by the pH of the solution. For example, spinel ferrite nanoparticles have a greater positive charge at a lower pH and a greater negative charge at a higher pH. This statement supports anion adsorption at lower pH and cation adsorption at higher pH [61]. For example, the best adsorption of Mo(VI) 33.4 mg g^{-1} on γ-Fe_2O_3 nanoparticles between pH 4 and 6. The removal efficacy of $CuFe_2O_4$ for Mo(VI) declined as the pH was raised, with a maximum removal competence at pH 2.75, but almost no removal of Mo(VI) was observed at pH >10. This is related to the growth in attraction between MoO_4^{2-} and positively charged spinel ferrite nanoparticle at declining pH [62]. Related to this, the effect of pH, initial metal concentration, and adsorbent dosage on the removal of Cu(II) were extensively analyzed. As the pH grew from 2.0 to 5.3, the elimination of Cu(II) on Fe_3O_4 NPs improved from 0.73 to 99.1 percent. At a pH of 5.3, the removal decreased up to 66% due to initial solution concentration enhancement, although adsorbent dose from 0.1 to 1 g improved the removal from 41.65% to 97.05% [63].

5.3 Temperature

Under the conditions used in the experiment, adsorption capacities of spinel ferrite nanoparticles mainly influence temperature. Depending on the form of favored chemical reaction, increasing or decreasing temperature determines the accessible free adsorption site. In endothermic reactions with the increase in temperature, adsorption capability increases; as exothermic reactions increase in temperature, adsorption capacity decreases. Pb(II) sorption on $Co_{0.6}Fe_{2.4}O_4$ which was synthesized by a thermal decomposition technique and was found to be endothermic and spontaneous in nature [53,64].

5.4 Calcination temperature

The ability of adsorption is based on the calcination temperature of spinel ferrite nanoparticles. As a result, the calcination temperature and the amount of time used for calcination both affect the size and pore size of nanoparticles, which influences both the surface area and pore size of nanoparticles. In this work process included impregnation and calcination at three separate temperatures with cotton as a template, during which the resulting hollow fibre $NiFe_2O_4$ nanoparticles were used for CR removal. The calcination temperature was increased, but only the highest pore volume was achieved for $NiFe_2O_4$ synthesized at 600°C. Thus, $NiFe_2O_4$ synthesized at 600 degrees Celsius had higher adsorption capacity of CR (89.85 mg g^{-1}) [65]. In order to produce a good technology for complete removal of each targeted contaminant from wastewater, it is absolutely essential to understand the impacts of each parameter.

6. Organic pollutants removal

Spinel ferrites nanoparticles and their composites have demonstrated good adsorption capacity for organic pollutants that are released into the atmosphere from different sources. The organic pollutants are adsorbed on the spinel ferrites nanoparticles and their composites by chemisorption or physisorption [66]. In this section, a brief overview of the current literature dealing with the elimination of organic pollutants present in wastewater.

Naik et al. synthesized the cobalt doped manganese ferrite by combustion method [67]. The structure of prepared sample was confirmed with the help of different characterization techniques. The particle size distribution the prepared samples were confirmed by TEM (Transmission Electron Microscope) analysis and range is 15-30 nm. The variation of magnetic moment and magnetization was analyzed with the help of VSM (Vibrating Sample Magnetometer). At a room temperature antifungal activity of the prepared samples was examined against Rhizopus fungi. Additionally, prepared ferrites

Materials Research Foundations **112** (2021) 218-245 https://doi.org/10.21741/9781644901595-6

were examined for adsorption of methylene blue dye and maximum removal percentage was reported 92.06 % for $Mn_{0.8}Co_{0.2}Fe_2O_4$.

In another study, spinel cobalt ferrite was synthesized successfully through hydrothermal technique [68]. The prepared ferrite was functionalized with the help of humic acid and epichlorohydrin. The prepared ferrite adsorbent was utilized for removal of triarylmethane dyes such as malachite green dye, methyl violet 2B, and basic fuchsin. The maximum removal capacity for malachite green dye, methyl violet 2B, and basic fuchsin were 48.74, 62.62 and 96.49 μmol/g by functionalized spinel cobalt ferrite. And the different adsorption parameters were optimized like effect of pH, dye concentration, ionic strength, temperature, and contact time. The experimental data well fitted for pseudo-second order kinetic model and Langmuir isotherm model. The thermodynamic experiments have shown that the processes of adsorption are spontaneous-endothermic. Moreover, the functionalized cobalt ferrite was examined for ten reuse cycles and findings demonstrated that no significant change after ten reuse cycles.

Yavari et al. synthesized the $CoFe_2O_4$ ferrite nanoparticles and prepared ferrite nanoparticles successfully functionalized by amine groups [69]. The prepared amine group functionalized cobalt ferrite nanoparticle was utilized for removal of acid blue 92, direct green 6 and direct red 80. The maximum reported adsorption capacity for acid blue 92, direct green 6 and direct red 80 were 625, 384.6 and 333.3 mg/g respectively. The different adsorption parameters were optimized such as effect of pH and dose, initial dye concentration and contact time at 25°C. The three different isotherm models such as Langmuir, Freundlich and Tempkin isotherm models were examined and findings show that experimental data best fit for Langmuir isotherm model. Similarly, the three different kinetics models such as pseudo first-order, pseudo second-order and intraparticle diffusion models were examined and findings show that experimental data best fit for pseudo second order kinetic model.

Tatarchuk et al. synthesized the TiO_2 on $CoFe_2O_4$ nanocomposite by the Pechini sol gel technique with the help of chelating agents (citric acid and ethylene glycol) [70]. The prepared sample was utilized for the adsorption of dichromate anions and congo red dye. The reported highest removal percentage for dichromate anions and congo red dye were 83% and 61% respectively. The three different isotherm models such as Langmuir, Freundlich and Dubin-Radushkevich isotherm models were examined and analysis demonstrated that experimental data best fit for Langmuir isotherm model. In another study, Wu et al., synthesized the cobalt ferrite modified with rare earth metals (Ho^{3+}, Tb^{3+}, Sm^{3+}, Pr^{3+}) through hydrothermal technique in the absence of surfactant and template [71]. The prepared cobalt ferrite modified with rare earth metals nanoparticles utilized for the adsorption of Congo red dye. The findings are demonstrated that after doping of rare earth metals adsorption capacity of the ferrite samples increased dramatically. The reported maximum adsorption potential for Congo red dye was 158.0 mg/g and 178.6 mg/g by $CoFe_{1.9}Ho_{0.1}O_4$ and $CoFe_{1.9}Sm_{0.1}O_4$ respectively. The experimental data best fit for pseudo second order kinetics model and Langmuir isotherm model.

Afkhami et al., synthesized the maghemite nanoparticles modified by sodium dodecyl sulfate and prepared sample utilized for the adsorption of cationic dyes such as janus green b, thionine and brilliant cresyl blue from aqueous solution [72]. The reported highest removal capacity for janus green b, thionine and brilliant cresyl blue dyes were 172.4, 166.7 and 200.0 mg/g respectively. In another study, maghemite nanoparticles coated with sodium dodecyl sulfate synthesized by Shahri and Niazi, and maghemite nanoparticles were prepared by co-precipitation technique [73]. The different adsorption parameters were optimized for adsorption of acridine orange dye from aqueous solution. The reported optimized conditions were 30.0 mg/L dye concentration, 0.8 g/L adsorbent dose, pH of dye solution = 5.1 and adsorption time = 43 min. Under the optimized conditions highest adsorption capacity was 285.82mg/g. The Langmuir and Freundlich isotherm models were analyzed and the experiment data best fit for Langmuir isotherm model.

Jiang et al. synthesized nickel ferrite / activated carbon composite prepared by hydrothermal technique [74]. The prepared sample was utilized for the adsorption of methyl orange azo dye and the highest adsorption capacity was reported 182.82 mg/g at 29.85°C. The experimental best fit for Langmuir isotherm model and pseudo second order kinetics model. And thermodynamic analysis examined that methyl orange azo dye adsorption on nickel ferrite / activated carbon composite was endothermic and spontaneous process. Moreover, the summary of organic pollutants removal from wastewater by spinel ferrites nanoparticles and their composites are mentioned in Table 1. As seen in Table 1, the spinel ferrites nanoparticles and their composites are most significant adsorbent for the removal of organic pollutants from aqueous solution.

Materials Research Foundations **112** (2021) 218-245 https://doi.org/10.21741/9781644901595-6

Table 1: The spinel ferrites nanoparticles and their composites are used for organic pollutants removal from aqueous solution with their adsorption capacity.

Sr. No.	Adsorbent name	Organic Pollutants	Adsorption capacity	Reference
1.	$Mn_{0.8}Co_{0.2}Fe_2O_4$	Methylene blue dye	92.06 %	[67]
2.	Epichlorohydrin-crosslinked humic acid with spinel cobalt ferrite	Malachite green dye, Methyl violet 2B, and Basic fuchsin	22.59, 24.67 and 31.25 mg/g	[68]
3.	Spinel cobalt ferrite/ humic acid	Malachite green dye, Methyl violet 2B, and Basic fuchsin	17.30, 18.85 and 21.42 mg/g	[68]
4.	Spinel cobalt ferrite	Malachite green dye, Methyl violet 2B, and Basic fuchsin	2.00, 2.68 and 3.16 mg/g	[68]
5.	Amine group functionalized cobalt ferrite nanoparticle	Acid blue 92, Direct green 6 and Direct red 80	625, 384.6 and 333.3 mg/g	[69]
6.	TiO_2 on $CoFe_2O_4$ nanocomposite	dichromate anions and congo red dye	83% and 61%	[70]
7.	Maghemite nanoparticles coated with sodium dodecyl sulfate	Janus green b, Thionine and Brilliant cresyl blue dyes	172.4, 166.7 and 200.0 mg/g	[72]

8.	Maghemite nanoparticles coated with sodium dodecyl sulfate	Acridine orange dye	285.82 mg/g	[73]
9.	Nickel ferrite/ activated carbon composite	Methyl orange azo dye	182.82 mg/g	[74]
10.	$MgFe_2O_4$ nanoparticle	Basic fuchsin	1.23 mg/g	[75]
11.	$CoFe_{1.9}Sm_{0.1}O_4$	Congo red dye	178.6 mg/g	[71]
12.	$CoFe_{1.9}Ho_{0.1}O_4$	Congo red dye	158.0 mg/g	[71]
13.	$CuFe_2O_4$/drumstick pod biomass composite	Malachite green	952.4 mg/g	[76]

7. Organic pollutants removal

Spinel ferrites nanoparticles and their composites are both economically and technologically very significant in wastewater treatment for the elimination of inorganic pollutants, as they have many advantages such as enhanced adsorption ability, relatively low cost of production, and ease of regeneration and separation. The polluted water is remediated with spinel ferrites nanoparticles and their composites, contaminants could be easily adsorbed to nanoparticles or composites and then separated from a matrix using an external magnetic field. The pure spinel ferrites nanoparticles, ferrites nanoparticles modified with metal oxide or any other appropriate organic functional groups are widely used for water remediation [46]. For example, Asadi et al. synthesized $CoFe_2O_4$ and $MnFe_2O_4$ spinel ferrites nanoparticles prepared by co-precipitation method [77]. The prepared ferrites nanoparticles were utilized for removal of Zn (II) ions from wastewater. The reported magnetization for $CoFe_2O_4$ and $MnFe_2O_4$ spinel ferrites nanoparticles were 37.54 and 61.39 emu/g, respectively. The different adsorption parameters were optimized for adsorption of Zn (II) ions from aqueous solution. Under optimized 6 pH the reported maximum adsorption capacity for $CoFe_2O_4$ and $MnFe_2O_4$ spinel ferrites nanoparticles were 384.6 and 454.5 mg/g respectively. The experimental data best fit for Langmuir isotherm model and pseudo second order kinetics model. From the thermodynamics analysis concluded that Zn (II) ions adsorption was spontaneous and exothermic process. Moreover, the prepared samples were also investigated for adsorption-desorption cycles.

In another study, Tavares et al., synthesized the magnetic nano adsorbent ($CoFe_2O_4$, $MnFe_2O_4$ and Fe_3O_4). The prepared magnetic nano adsorbent utilized for the removal of As (V) ions [78]. Tu et al. synthesized $CuFe_2O_4$ ferrite nanoparticles and prepared ferrites nanoparticles was utilized for the adsorption of As(V) ions [79]. The removal of arsenic (V) by this recycled $CuFe_2O_4$ ferrite nanoparticles exhibited an L-shaped nonlinear isotherm, indicating restricted binding sites on the $CuFe_2O_4$ ferrite surface. The highest arsenic (V) adsorption potential of the $CuFe_2O_4$ ferrite nanoparticle at pH 3.7 was 45.66 mg/g and adsorption capacity reduced with pH due to improved electrostatic repulsion between arsenic (V) and $CuFe_2O_4$ ferrite surface. The findings indicate that the recycled $CuFe_2O_4$ ferrite nanoparticle without surface alteration is an efficient adsorbent for the adsorption of arsenic (V) from aqueous solution.

Because of high toxicity to human and marine environments, the elimination of lead from the wastewater system has drawn substantial public attention. For example, Sreekala et al. synthesized the $CuFe_2O_4$ ferrite nanoparticles through a green technique and its utilized for Pb (II) ions adsorption [80]. The green technique means cost effective, non-toxic and environmentally sustainable procedure utilizing plant extracts. In this research work author used Simarouba glauca leaf extract as a capping and reducing agent. According to the batch adsorption analysis prepared $CuFe_2O_4$ ferrite nanoparticle demonstrated that excellent adsorption capacity for Pb (II) ions from wastewater. The reported maximum adsorption percentage was 80%. In another study, khan et al., synthesized the $CuFe_2O_4$ composite with drumstick pod biomass by co-precipitation method [76]. And prepared composite utilized for lead ions adsorption from aqueous solution. Different adsorption parameters were optimized for Pb (II) ions adsorption and under optimized conditions reported maximum adsorption capacity was 921.1 mg/g. The adsorption experimental data was best fit for Langmuir isotherm models. In another study, Putri et al., synthesized the natural iron sand based Mg_{1-x} $Ni_xFe_2O_4$ ferrite nanoparticles with the help of co-precipitation technique [81]. The prepared nano ferrites used for the removal of Pb and Cu ions from aqueous solution. Under the optimized conditions maximum removal capacity was achieved 190.5 mg/g. Furthermore, the summary of inorganic pollutants removal from wastewater by spinel ferrites nanoparticles and their composites are mentioned in Table 2. As seen in Table 2, the spinel ferrites nanoparticles and their composites are most significant adsorbent for the removal of inorganic pollutants from aqueous solution.

Table 2: The spinel ferrites nanoparticles and their composites are used for inorganic pollutants removal from aqueous solution with their adsorption capacity.

Sr. No.	Adsorbent name	Inorganic Pollutants	Adsorption capacity	Reference
1.	$CoFe_2O_4$	Zn (II)	384.6 mg/g	[77]
2.	$MnFe_2O_4$	Zn (II)	454.5 mg/g	[77]
3.	$CoFe_2O_4$, $MnFe_2O_4$ and Fe_3O_4	As (V)	-	[78]
4.	$CuFe_2O_4$	As (V)	45.66 mg/g	[79]
5.	$CuFe_2O_4$	Pb (II)	80%	[80]
6.	$CuFe_2O_4$/drumstick pod biomass composite	Pb (II)	921.1 mg/g	[76]
7.	$Mg_{1-x}Ni_xFe_2O_4$	Pb and Cu ions	190.5 mg/g	[81]
8.	Rice Straw/ $CoFe_2O_4$ nanocomposite	Fe and Mn	84.25%, 92.45%	[82]
9.	Rice Straw/ $CoFe_2O_4$ nanocomposite	Cu, Cd, Zn, Ni and Pb	100%	[82]
10.	γ-Fe_2O_3	Cu(II), Zn(II) and Pb(II)	111.11, 84.95 and 71.42 mg g^{-1}	[83]

8. Future prospects

The spinel ferrite-based nanoparticles for wastewater remediation needs high quality control for the elimination of pollutants and nanoparticles from the polluted water after remediation i.e., firstly to reduce the cost of nanoparticles by their recovery and reuse after being utilized, and second to overcome the health impact. The application of spinel ferrite-based nanoparticles in fresh water and wastewater remediation was limited and it was one of the biggest gap or issue that has been observed. For wastewater remediation, simulated wastewater was taken for the experiment where cations and anions were present in co-exist situation and practically, simulated wastewater does not consider as real wastewater. Therefore, increasing level of contaminants in the wastewater will be responsible for the increase of active sites in the structure of spinel ferrite-based nanoparticles. This will result in the elimination of anions and cations from the wastewater on the phenomenon of preferential discharge. Thus, the proceeding tests were not effective in the overall elimination of cations and anions from the stimulated wastewater as it does not belong to a real-life issue. So, it was very important to do a systematic way of including all the constituents which were present in the real-life environmental specimens. Moreover, a few studies have been addressed on the effect of spinel ferrite-based nanoparticles on the human health and environment habitat, so, it was very crucial to study their detailed toxicity impact.

Conclusion

Spinel ferrite-based nanoparticles have multifunctional applications in the research areas of magnetic storage devices, recording media, switching devices, high frequency devices, wastewater remediation. Therefore, these ferrites have been tremendously utilized for the elimination of organic as well as inorganic pollutants from polluted water via adsorption. The nanoparticles were highly reactive in nature as a matter of toxicity concern so, it was very important to analyze their toxic behaviour before commercial use or for application concern. Moreover, new technologies were taken into consideration for the fabrication of spinel ferrite-based nanoparticles in regards with environmental and safety research concern. A deep research on the real wastewater remediation have been carried out in which different types of pollutants have been present. The spinel ferrites have excellent capabilities such as high adsorption and disinfection ability in regards with pollutants. Thus, the improvement of remediation technology by utilizing spinel ferrite-based nanoparticles not only lightens the present society issue of poor-quality water but also changes the predictable reagents being utilized for wastewater remediation with better quality amenities and running of safe water quality under sustainable conditions.

References

[1] J. Theron, J.A. Walker, T.E. Cloete, Nanotechnology and water treatment: applications and emerging opportunities, Crit. Rev. Microbiol. 34 (2008) 43–69. https://doi.org/10.1080/10408410701710442

[2] M. Anjum, R. Miandad, M. Waqas, Remediation of wastewater using various nanomaterials. Arabian J of Chem 12: 4897–4919, 2016. https://doi.org/10.1016/j.arabjc.2016.10.004

[3] S. Sharma, A. Bhattacharya, Drinking water contamination and treatment techniques, Appl. Water Sci. 7 (2017) 1043–1067. https://doi.org/10.1007/s13201-016-0455-7

[4] A. Kumar, A. Rana, G. Sharma, M. Naushad, A.H. Al-Muhtaseb, C. Guo, A. Iglesias-Juez, F.J. Stadler, High-performance photocatalytic hydrogen production and degradation of levofloxacin by wide spectrum-responsive Ag/Fe3O4 bridged SrTiO3/g-C3N4 plasmonic nanojunctions: joint effect of Ag and Fe3O4, ACS Appl. Mater. Interfaces. 10 (2018) 40474–40490. https://doi.org/10.1021/acsami.8b12753

[5] A. Verma, S. Thakur, G. Goel, J. Raj, V.K. Gupta, D. Roberts, V.K. Thakur, Bio-based Sustainable Aerogels: New Sensation in CO2 Capture, Curr. Res. Green Sustain. Chem. (2020) 100027. https://doi.org/10.1016/j.crgsc.2020.100027

[6] R. Jasrotia, N. Kumari, R. Kumar, M. Naushad, P. Dhiman, G. Sharma, Photocatalytic degradation of environmental pollutant using nickel and cerium ions substituted Co 0.6 Zn 0.4 Fe 2 O 4 nanoferrites, Earth Syst. Environ. (2021) 1–19. https://doi.org/10.1007/s41748-021-00214-9

[7] S. Kour, R. Jasrotia, P. Puri, A. Verma, B. Sharma, V.P. Singh, R. Kumar, S. Kalia, Improving photocatalytic efficiency of MnFe 2 O 4 ferrites via doping with Zn 2+/La 3+ ions: photocatalytic dye degradation for water remediation, Environ. Sci. Pollut. Res. (2021) 1–16. https://doi.org/10.1007/s11356-021-13147-7

[8] R. Jasrotia, V.P. Singh, R. Kumar, M. Singh, Raman spectra of sol-gel auto-combustion synthesized Mg-Ag-Mn and Ba-Nd-Cd-In ferrite based nanomaterials, Ceram. Int. 46 (2020) 618–621. https://doi.org/10.1016/j.ceramint.2019.09.012

[9] R. Jasrotia, S. Kour, P. Puri, A.D. Jara, B. Singh, C. Bhardwaj, V.P. Singh, R. Kumar, Structural and magnetic investigation of Al3+ and Cr3+ substituted Ni–Co–Cu nanoferrites for potential applications, Solid State Sci. 110 (2020) 106445. https://doi.org/10.1016/j.solidstatesciences.2020.106445

[10] R. Jasrotia, G. Kumar, K.M. Batoo, S.F. Adil, M. Khan, R. Sharma, A. Kumar, V.P. Singh, Synthesis and characterization of Mg-Ag-Mn nano-ferrites for electromagnet applications, Phys. B Condens. Matter. 569 (2019) 1–7. https://doi.org/10.1016/j.physb.2019.05.033

[11] R. Jasrotia, P. Puri, A. Verma, V.P. Singh, Magnetic and electrical traits of sol-gel synthesized Ni-Cu-Zn nanosized spinel ferrites for multi-layer chip inductors application, J. Solid State Chem. (2020) 121462. https://doi.org/10.1016/j.jssc.2020.121462

[12] K. Dulta, G.K. Ağçeli, P. Chauhan, R. Jasrotia, P.K. Chauhan, Ecofriendly Synthesis of Zinc Oxide Nanoparticles by Carica papaya Leaf Extract and Their Applications, J. Clust. Sci. (2021) 1–15. https://doi.org/10.1007/s10876-020-01962-w

[13] N. Murali, S.J. Margarette, G.P. Kumar, B. Sailaja, S.Y. Mulushoa, P. Himakar, B.K. Babu, V. Veeraiah, Effect of Al substitution on the structural and magnetic properties of Co-Zn ferrites, Phys. B Condens. Matter. 522 (2017) 1–6. https://doi.org/10.1016/j.physb.2017.07.043

[14] R. Jasrotia, V.P. Singh, R.K. Sharma, M. Singh, Analysis of optical and magnetic study of silver substituted SrW hexagonal ferrites, in: AIP Conf. Proc., AIP Publishing LLC, 2019: p. 090004. https://doi.org/10.1063/1.5122448

[15] R.D. Ambashta, M. Sillanpää, Water purification using magnetic assistance: a review, J. Hazard. Mater. 180 (2010) 38–49. https://doi.org/10.1016/j.jhazmat.2010.04.105

[16] S. Zeng, S. Duan, R. Tang, L. Li, C. Liu, D. Sun, Magnetically separable Ni0. 6Fe2. 4O4 nanoparticles as an effective adsorbent for dye removal: Synthesis and study on the kinetic and thermodynamic behaviors for dye adsorption, Chem. Eng. J. 258 (2014) 218–228. https://doi.org/10.1016/j.cej.2014.07.093

[17] W. Konicki, D. Sibera, E. Mijowska, Z. Lendzion-Bieluń, U. Narkiewicz, Equilibrium and kinetic studies on acid dye Acid Red 88 adsorption by magnetic ZnFe2O4 spinel ferrite nanoparticles, J. Colloid Interface Sci. 398 (2013) 152–160. https://doi.org/10.1016/j.jcis.2013.02.021

[18] X. Zhang, P. Zhang, Z. Wu, L. Zhang, G. Zeng, C. Zhou, Adsorption of methylene blue onto humic acid-coated Fe3O4 nanoparticles, Colloids Surf. Physicochem. Eng. Asp. 435 (2013) 85–90. https://doi.org/10.1016/j.colsurfa.2012.12.056

[19] R. Jasrotia, V.P. Singh, B. Sharma, A. Verma, P. Puri, R. Sharma, M. Singh, Sol-gel synthesized Ba-Nd-Cd-In nanohexaferrites for high frequency and microwave devices applications, J. Alloys Compd. 830 (2020) 154687. https://doi.org/10.1016/j.jallcom.2020.154687

[20] R. Sivashankar, A.B. Sathya, K. Vasantharaj, V. Sivasubramanian, Magnetic composite an environmental super adsorbent for dye sequestration–A review, Environ. Nanotechnol. Monit. Manag. 1 (2014) 36–49. https://doi.org/10.1016/j.enmm.2014.06.001

[21] S.K. Giri, N.N. Das, G.C. Pradhan, Synthesis and characterization of magnetite nanoparticles using waste iron ore tailings for adsorptive removal of dyes from aqueous solution, Colloids Surf. Physicochem. Eng. Asp. 389 (2011) 43–49. https://doi.org/10.1016/j.colsurfa.2011.08.052

[22] P. Roonasi, A.Y. Nezhad, A comparative study of a series of ferrite nanoparticles as heterogeneous catalysts for phenol removal at neutral pH, Mater. Chem. Phys. 172 (2016) 143–149. https://doi.org/10.1016/j.matchemphys.2016.01.054

[23] Y.-J. Tu, C.-F. You, C.-K. Chang, S.-L. Wang, T.-S. Chan, Arsenate adsorption from water using a novel fabricated copper ferrite, Chem. Eng. J. 198 (2012) 440–448. https://doi.org/10.1016/j.cej.2012.06.006

[24] R.N. Baig, M.N. Nadagouda, R.S. Varma, Magnetically retrievable catalysts for asymmetric synthesis, Coord. Chem. Rev. 287 (2015) 137–156. https://doi.org/10.1016/j.ccr.2014.12.017

[25] R. Jasrotia, V.P. Singh, R. Kumar, R. Verma, A. Chauhan, Effect of Y3+, Sm3+ and Dy3+ ions on the microstructure, morphology, optical and magnetic properties NiCoZn magnetic nanoparticles, Results Phys. 15 (2019) 102544. https://doi.org/10.1016/j.rinp.2019.102544

[26] S. Kour, R.K. Sharma, R. Jasrotia, V.P. Singh, A brief review on the synthesis of maghemite (γ-Fe2O3) for medical diagnostic and solar energy applications, in: AIP Conf. Proc., AIP Publishing LLC, 2019: p. 090007. https://doi.org/10.1063/1.5122451

[27] M. Chandel, V.P. Singh, R. Jasrotia, K. Singha, R. Kumar, A review on structural, electrical and magnetic properties of Y-type hexaferrites synthesized by different techniques for antenna applications and microwave absorbing characteristic materials [J], AIMS Mater. Sci. 7 (2020) 244–268. https://doi.org/10.3934/matersci.2020.3.244

[28] R. Jasrotia, P. Puri, V.P. Singh, R. Kumar, Sol–gel synthesized Mg–Ag–Mn nanoferrites for Power Applications, J. Sol-Gel Sci. Technol. 97 (2021) 205–212. https://doi.org/10.1007/s10971-020-05428-3

[29] Y.-P. Fu, Electrical conductivity and magnetic properties of Li0. 5Fe2. 5- xCrxO4 ferrite, Mater. Chem. Phys. 115 (2009) 334–338. https://doi.org/10.1016/j.matchemphys.2008.12.023

[30] R. Jasrotia, V.P. Singh, R. Kumar, K. Singha, M. Chandel, M. Singh, Analysis of Cd2+ and In3+ ions doping on microstructure, optical, magnetic and mo\" ssbauer spectral properties of sol-gel synthesized BaM hexagonal ferrite based nanomaterials, Results Phys. 12 (2019) 1933–1941. https://doi.org/10.1016/j.rinp.2019.01.088

[31] A. Gatelytė, D. Jasaitis, A. Beganskienė, A. Kareiva, Sol-gel synthesis and characterization of selected transition metal nano-ferrites, Mater. Sci. 17 (2011) 302–307. https://doi.org/10.5755/j01.ms.17.3.598

[32] Z. Yue, J. Zhou, L. Li, X. Wang, Z. Gui, Effect of copper on the electromagnetic properties of Mg–Zn–Cu ferrites prepared by sol–gel auto-combustion method, Mater. Sci. Eng. B. 86 (2001) 64–69. https://doi.org/10.1016/S0921-5107(01)00660-2

[33] G.B. Teh, S. Nagalingam, D.A. Jefferson, Preparation and studies of Co (II) and Co (III)-substituted barium ferrite prepared by sol–gel method, Mater. Chem. Phys. 101 (2007) 158–162. https://doi.org/10.1016/j.matchemphys.2006.03.008

[34] Z. Wang, Y. Xie, P. Wang, Y. Ma, S. Jin, X. Liu, Microwave anneal effect on magnetic properties of Ni0. 6Zn0. 4Fe2O4 nano-particles prepared by conventional hydrothermal method, J. Magn. Magn. Mater. 323 (2011) 3121–3125. https://doi.org/10.1016/j.jmmm.2011.06.068

[35] T. Strachowski, E. Grzanka, W. Lojkowski, A. Presz, M. Godlewski, S. Yatsunenko, H. Matysiak, R.R. Piticescu, C.J. Monty, Morphology and luminescence properties of zinc oxide nanopowders doped with aluminum ions obtained by hydrothermal and vapor condensation methods, J. Appl. Phys. 102 (2007) 073513. https://doi.org/10.1063/1.2786707

[36] K.K. Kefeni, B.B. Mamba, T.A. Msagati, Application of spinel ferrite nanoparticles in water and wastewater treatment: a review, Sep. Purif. Technol. 188 (2017) 399–422. https://doi.org/10.1016/j.seppur.2017.07.015

[37] M.T. Yagub, T.K. Sen, S. Afroze, H.M. Ang, Dye and its removal from aqueous solution by adsorption: a review. Adv Colloid Interfac 209: 172–184, 2014. https://doi.org/10.1016/j.cis.2014.04.002

[38] A. Verma, S. Thakur, G. Mamba, R.K. Gupta, P. Thakur, V.K. Thakur, Graphite modified sodium alginate hydrogel composite for efficient removal of malachite green dye, Int. J. Biol. Macromol. 148 (2020) 1130–1139. https://doi.org/10.1016/j.ijbiomac.2020.01.142

[39] G. Lofrano, M. Carotenuto, G. Libralato, R.F. Domingos, A. Markus, L. Dini, R.K. Gautam, D. Baldantoni, M. Rossi, S.K. Sharma, Polymer functionalized nanocomposites for metals removal from water and wastewater: an overview, Water Res. 92 (2016) 22–37. https://doi.org/10.1016/j.watres.2016.01.033

[40] S. Duan, R. Tang, Z. Xue, X. Zhang, Y. Zhao, W. Zhang, J. Zhang, B. Wang, S. Zeng, D. Sun, Effective removal of Pb (II) using magnetic Co0. 6Fe2. 4O4 micro-particles as the adsorbent: synthesis and study on the kinetic and thermodynamic behaviors for its adsorption, Colloids Surf. Physicochem. Eng. Asp. 469 (2015) 211–223. https://doi.org/10.1016/j.colsurfa.2015.01.029

[41] R. Jasrotia, V.P. Singh, R.K. Sharma, P. Kumar, M. Singh, Analysis of effect of Ag+ ion on microstructure and elemental distribution of strontium W-type hexaferrites, in: AIP Conf. Proc., AIP Publishing LLC, 2019: p. 140004. https://doi.org/10.1063/1.5122517

[42] M. Auffan, J. Rose, O. Proux, D. Borschneck, A. Masion, P. Chaurand, J.-L. Hazemann, C. Chaneac, J.-P. Jolivet, M.R. Wiesner, Enhanced adsorption of arsenic onto maghemites nanoparticles: As (III) as a probe of the surface structure and heterogeneity, Langmuir. 24 (2008) 3215–3222. https://doi.org/10.1021/la702998x

[43] B.J. Kong, A. Kim, S.N. Park, Properties and in vitro drug release of hyaluronic acid-hydroxyethyl cellulose hydrogels for transdermal delivery of isoliquiritigenin, Carbohydr. Polym. 147 (2016) 473–481. https://doi.org/10.1016/j.carbpol.2016.04.021

[44] M. Chandel, V.P. Singh, R. Jasrotia, K. Singha, M. Singh, P. Thakur, S. Kalia, Fabrication of Ni2+ and Dy3+ substituted Y-Type nanohexaferrites: a study of structural and magnetic properties, Phys. B Condens. Matter. 595 (2020) 412378. https://doi.org/10.1016/j.physb.2020.412378

[45] T. Tatarchuk, M. Bououdina, J.J. Vijaya, L.J. Kennedy, Spinel ferrite nanoparticles: synthesis, crystal structure, properties, and perspective applications, in: Int. Conf. Nanotechnol. Nanomater., Springer, 2016: pp. 305–325. https://doi.org/10.1007/978-3-319-56422-7_22

[46] W.A. Khoso, N. Haleem, M.A. Baig, Y. Jamal, Synthesis, characterization and heavy metal removal efficiency of nickel ferrite nanoparticles (NFN's), Sci. Rep. 11 (2021) 1–10. https://doi.org/10.1038/s41598-021-83363-1

[47] A. Ivanets, V. Prozorovich, T. Kouznetsova, T. Dontsova, O. Yanushevska, A. Hosseini-Bandegharaei, V. Srivastava, M. Sillanpää, Effect of Mg2+ ions on competitive metal ions adsorption/desorption on magnesium ferrite: Mechanism, reusability and stability studies, J. Hazard. Mater. 411 (2021) 124902. https://doi.org/10.1016/j.jhazmat.2020.124902

[48] K. Dulta, G.K. Ağçeli, P. Chauhan, R. Jasrotia, P.K. Chauhan, A novel approach of synthesis zinc oxide nanoparticles by bergenia ciliata rhizome extract: antibacterial and anticancer potential, J. Inorg. Organomet. Polym. Mater. 31 (2021) 180–190. https://doi.org/10.1007/s10904-020-01684-6

[49] S. Thakur, B. Sharma, A. Verma, J. Chaudhary, S. Tamulevicius, V.K. Thakur, Recent approaches in guar gum hydrogel synthesis for water purification, Int. J. Polym. Anal. Charact. 23 (2018) 621–632. https://doi.org/10.1080/1023666X.2018.1488661

[50] S.S. Fiyadh, M.A. AlSaadi, W.Z. Jaafar, M.K. AlOmar, S.S. Fayaed, N.S. Mohd, L.S. Hin, A. El-Shafie, Review on heavy metal adsorption processes by carbon nanotubes, J. Clean. Prod. 230 (2019) 783–793. https://doi.org/10.1016/j.jclepro.2019.05.154

[51] W. Sun, W. Pan, F. Wang, N. Xu, Removal of Se (IV) and Se (VI) by MFe2O4 nanoparticles from aqueous solution, Chem. Eng. J. 273 (2015) 353–362. https://doi.org/10.1016/j.cej.2015.03.061

[52] S. Vijayalakshmi, E. Elaiyappillai, P.M. Johnson, I.S. Lydia, Multifunctional magnetic CoFe 2 O 4 nanoparticles for the photocatalytic discoloration of aqueous methyl violet dye and energy storage applications, J. Mater. Sci. Mater. Electron. 31 (2020) 10738–10749. https://doi.org/10.1007/s10854-020-03624-z

[53] S. Duan, R. Tang, Z. Xue, X. Zhang, Y. Zhao, W. Zhang, J. Zhang, B. Wang, S. Zeng, D. Sun, Effective removal of Pb (II) using magnetic Co0. 6Fe2. 4O4 micro-particles as the adsorbent: synthesis and study on the kinetic and thermodynamic behaviors for its adsorption, Colloids Surf. Physicochem. Eng. Asp. 469 (2015) 211–223. https://doi.org/10.1016/j.colsurfa.2015.01.029

[54] N. Neyaz, W.A. Siddiqui, Removal of Cu (II) by modified magnetite nanocomposite as a nanosorbent, Int J Sci Res. 4 (2015) 1868–1873.

[55] C. Vîrlan, R.G. Ciocârlan, T. Roman, D. Gherca, N. Cornei, A. Pui, Studies on adsorption capacity of cationic dyes on several magnetic nanoparticles, Acta Chem. Iasi. 21 (2013) 19–30. https://doi.org/10.2478/achi-2013-0003

[56] K. Singha, R. Jasrotia, V.P. Singh, M. Chandel, R. Kumar, S. Kalia, A study of magnetic properties of Y–Ni–Mn substituted Co 2 Z-type nanohexaferrites via vibrating sample magnetometry, J. Sol-Gel Sci. Technol. (2020) 1–9. https://doi.org/10.1007/s10971-020-05412-x

[57] L. Yang, Y. Zhang, X. Liu, X. Jiang, Z. Zhang, T. Zhang, L. Zhang, The investigation of synergistic and competitive interaction between dye Congo red and methyl blue on magnetic MnFe2O4, Chem. Eng. J. 246 (2014) 88–96. https://doi.org/10.1016/j.cej.2014.02.044

[58] T. Tuutijärvi, J. Lu, M. Sillanpää, G. Chen, Adsorption mechanism of arsenate on crystal γ-Fe 2 O 3 nanoparticles, J. Environ. Eng. 136 (2010) 897–905. https://doi.org/10.1061/(ASCE)EE.1943-7870.0000233

[59] D.H.K. Reddy, Y.-S. Yun, Spinel ferrite magnetic adsorbents: alternative future materials for water purification?, Coord. Chem. Rev. 315 (2016) 90–111. https://doi.org/10.1016/j.ccr.2016.01.012

[60] M. Auffan, J. Rose, O. Proux, D. Borschneck, A. Masion, P. Chaurand, J.-L. Hazemann, C. Chaneac, J.-P. Jolivet, M.R. Wiesner, Enhanced adsorption of arsenic

onto maghemites nanoparticles: As (III) as a probe of the surface structure and heterogeneity, Langmuir. 24 (2008) 3215–3222. https://doi.org/10.1021/la702998x

[61] B. Sharma, S. Thakur, D. Trache, H. Yazdani Nezhad, V.K. Thakur, Microwave-assisted rapid synthesis of reduced graphene oxide-based gum tragacanth hydrogel nanocomposite for heavy metal ions adsorption, Nanomaterials. 10 (2020) 1616. https://doi.org/10.3390/nano10081616

[62] A. Afkhami, R. Norooz-Asl, Removal, preconcentration and determination of Mo (VI) from water and wastewater samples using maghemite nanoparticles, Colloids Surf. Physicochem. Eng. Asp. 346 (2009) 52–57. https://doi.org/10.1016/j.colsurfa.2009.05.024

[63] R. Davarnejad, P. Panahi, Cu (II) removal from aqueous wastewaters by adsorption on the modified Henna with Fe3O4 nanoparticles using response surface methodology, Sep. Purif. Technol. 158 (2016) 286–292. https://doi.org/10.1016/j.seppur.2015.12.018

[64] B. Sharma, S. Thakur, G. Mamba, R.K. Gupta, V.K. Gupta, V.K. Thakur, Titania modified gum tragacanth based hydrogel nanocomposite for water remediation, J. Environ. Chem. Eng. (2020) 104608. https://doi.org/10.1016/j.jece.2020.104608

[65] C. Kong, J. Li, F. Liu, Y. Song, P. Song, Synthesis of NiFe2O4 using degreasing cotton as template and its adsorption capacity for Congo Red, Desalination Water Treat. 57 (2016) 11337–11347. https://doi.org/10.1080/19443994.2015.1043589

[66] P.A. Vinosha, A. Manikandan, A.S.J. Ceicilia, A. Dinesh, G.F. Nirmala, A.C. Preetha, Y. Slimani, M.A. Almessiere, A. Baykal, B. Xavier, Review on recent advances of zinc substituted cobalt ferrite nanoparticles: Synthesis characterization and diverse applications, Ceram. Int. (2021). https://doi.org/10.1016/j.ceramint.2020.12.289

[67] A.B. Naik, P.P. Naik, S.S. Hasolkar, D. Naik, Structural, magnetic and electrical properties along with antifungal activity & adsorption ability of cobalt doped manganese ferrite nanoparticles synthesized using combustion route, Ceram. Int. 46 (2020) 21046–21055. https://doi.org/10.1016/j.ceramint.2020.05.177

[68] S.C.W. Sakti, R.N. Laily, S. Aliyah, N. Indrasari, M.Z. Fahmi, H.V. Lee, Y. Akemoto, S. Tanaka, Re-collectable and recyclable epichlorohydrin-crosslinked humic acid with spinel cobalt ferrite core for simple magnetic removal of cationic triarylmethane dyes in polluted water, J. Environ. Chem. Eng. 8 (2020) 104004. https://doi.org/10.1016/j.jece.2020.104004

[69] S. Yavari, N.M. Mahmodi, P. Teymouri, B. Shahmoradi, A. Maleki, Cobalt ferrite nanoparticles: preparation, characterization and anionic dye removal capability, J. Taiwan Inst. Chem. Eng. 59 (2016) 320–329. https://doi.org/10.1016/j.jtice.2015.08.011

[70] T. Tatarchuk, I. Mironyuk, V. Kotsyubynsky, A. Shyichuk, M. Myslin, V. Boychuk, Structure, morphology and adsorption properties of titania shell immobilized onto cobalt ferrite nanoparticle core, J. Mol. Liq. 297 (2020) 111757. https://doi.org/10.1016/j.molliq.2019.111757

[71] X. Wu, Z. Ding, N. Song, L. Li, W. Wang, Effect of the rare-earth substitution on the structural, magnetic and adsorption properties in cobalt ferrite nanoparticles, Ceram. Int. 42 (2016) 4246–4255. https://doi.org/10.1016/j.ceramint.2015.11.100

[72] A. Afkhami, M. Saber-Tehrani, H. Bagheri, Modified maghemite nanoparticles as an efficient adsorbent for removing some cationic dyes from aqueous solution, Desalination. 263 (2010) 240–248. https://doi.org/10.1016/j.desal.2010.06.065

[73] F.B. Shahri, A. Niazi, Synthesis of modified maghemite nanoparticles and its application for removal of acridine orange from aqueous solutions by using Box-Behnken design, J. Magn. Magn. Mater. 396 (2015) 318–326. https://doi.org/10.1016/j.jmmm.2015.08.054

[74] T. Jiang, Y. Liang, Y. He, Q. Wang, Activated carbon/NiFe2O4 magnetic composite: a magnetic adsorbent for the adsorption of methyl orange, J. Environ. Chem. Eng. 3 (2015) 1740–1751. https://doi.org/10.1016/j.jece.2015.06.020

[75] M. Gao, Z. Wang, C. Yang, J. Ning, Z. Zhou, G. Li, Novel magnetic graphene oxide decorated with persimmon tannins for efficient adsorption of malachite green from aqueous solutions, Colloids Surf. Physicochem. Eng. Asp. 566 (2019) 48–57. https://doi.org/10.1016/j.colsurfa.2019.01.016

[76] M.A. Khan, M. Otero, M. Kazi, A.A. Alqadami, S.M. Wabaidur, M.R. Siddiqui, Z.A. Alothman, S. Sumbul, Unary and binary adsorption studies of lead and malachite green onto a nanomagnetic copper ferrite/drumstick pod biomass composite, J. Hazard. Mater. 365 (2019) 759–770. https://doi.org/10.1016/j.jhazmat.2018.11.072

[77] R. Asadi, H. Abdollahi, M. Gharabaghi, Z. Boroumand, Effective removal of Zn (II) ions from aqueous solution by the magnetic MnFe2O4 and CoFe2O4 spinel ferrite nanoparticles with focuses on synthesis, characterization, adsorption, and desorption, Adv. Powder Technol. 31 (2020) 1480–1489. https://doi.org/10.1016/j.apt.2020.01.028

[78] D.S. Tavares, C.B. Lopes, J.C. Almeida, C. Vale, E. Pereira, T. Trindade, Spinel-type ferrite nanoparticles for removal of arsenic (V) from water, Environ. Sci. Pollut. Res. 27 (2020) 22523–22534. https://doi.org/10.1007/s11356-020-08673-9

[79] Y.-J. Tu, C.-F. You, C.-K. Chang, S.-L. Wang, T.-S. Chan, Arsenate adsorption from water using a novel fabricated copper ferrite, Chem. Eng. J. 198 (2012) 440–448. https://doi.org/10.1016/j.cej.2012.06.006

[80] G. Sreekala, A.F. Beevi, R. Resmi, B. Beena, Removal of lead (II) ions from water using copper ferrite nanoparticles synthesized by green method, Mater. Today Proc. (2020). https://doi.org/10.1016/j.matpr.2020.09.087

[81] W.B.K. Putri, E.A. Setiadi, V. Herika, A.P. Tetuko, P. Sebayang, Natural iron sand-based Mg1-xNixFe2O4 nanoparticles as potential adsorbents for heavy metal removal synthesized by co-precipitation method, in: IOP Conf. Ser. Earth Environ. Sci., IOP Publishing, 2019: p. 012031. https://doi.org/10.1088/1755-1315/277/1/012031

[82] A.E. Alia, W.M. Salema, S.M. Younes, A.Z. Elabdeen, Ferrite Nanocomposite (Rice Straw-CoFe2O4) as New Chemical Modified of for Treatment of Heavy Metal from Waste Water, Hydrol Curr. Res. 10 (2018) 2.

[83] A. Roy, J. Bhattacharya, Removal of Cu (II), Zn (II) and Pb (II) from water using microwave-assisted synthesized maghemite nanotubes, Chem. Eng. J. 211 (2012) 493–500. https://doi.org/10.1016/j.cej.2012.09.097

Ferrite - Nanostructures with Tunable Properties and Diverse Applications Materials Research Forum LLC
Materials Research Foundations **112** (2021) 246-278 https://doi.org/10.21741/9781644901595-7

Chapter 7

Application of Magnetic Nano Particles and their Composites as Adsorbents for Waste Water Treatment: A Brief Review

Meenakshi Dhiman[1], Baljinder Kaur[1] and Balwinder Kaur[2]

[1] Chitkara University Institute of Engineering and Technology, Chitkara University, Punjab, India

[2] Govt Degree College, R S Pura, Jammu, India

meenakshi.dhiman@chitkara.edu.in

Abstract

The present review highlights the different types of nano ferrites and their surface modified composites as an alternative adsorbent in waste water treatment. In this review, the recent progresses and potential applications of SFNPs/SFNCs for the removal of organic and inorganic contaminants through adsorption routes are critically reviewed. There are number of water purification techniques but the adsorption is one of the simplest, effective and economical method for wastewater purification. Adsorption isotherm models, kinetic models, thermodynamic parameters and adsorption mechanism have also been discussed. The present article lists different type of adsorbents and reviews state-of-the-art of the removal of different pollutants from water. The efforts have been made to discuss the sources of contamination and toxicities of pollutants. The possible techniques of recovery and reuse, toxicity, research gaps and the future perspective of SFNPs are also discussed in brief. Based on this review, it is possible to conclude that SFNPs and their derivative composites have unlimited capacity in addressing array of problems encountered in water and wastewater treatment. The present study highlights the future areas of research for waste water treatment.

Keywords

Spinel Ferrite, Spinel Ferrite Composites, Wastewater, Adsorbents, Dyes, Regeneration

Materials Research Foundations **112** (2021) 246-278 https://doi.org/10.21741/9781644901595-7

Contents

1. Introduction

Spinel Ferrites Nanoparticles have interesting aspects in magnetic properties when accompanied with other functional properties like catalytic activities. It is also important to mention here that the magnetic properties of spinel ferrite nanoparticles are strongly dependent on the particle size and shape up to some extent. These are also called as universal and multifunctional due to their excellence magnetic properties, simple chemical composition, and wide applications in several areas which include water and wastewater treatment, biomedical,
catalyst and electronic device.

The structural equation of spinel ferrite is MFe_2O_4, where M is a divalent metal ion. For the last few decades, many researchers studied the physical properties of ferrites substituting M with Ni^{2+}, Zn^{2+}, Mg^{2+},Cu^{2+}, Mn^{2+}, etc.[1,2,3,4]. Spinel ferrites are competitive crystalline materials due to their excellent magnetic properties together with the electrical properties. The magnetic and dielectric properties of ferrites play a crucial role in choosing their applications. Ferrites show an almost semiconducting behaviour that is greatly affected by doping. Magnetic properties of spinel ferrites are sensitive to

Materials Research Foundations **112** (2021) 246-278 https://doi.org/10.21741/9781644901595-7

the type of cations and their distribution amongst the two interstitial sites of the spinel lattice; the tetrahedral and the octahedral sites [5], Changes in the cation distribution between the two interstitial sites alter the spin moment algebraic sum which strongly affects the magnetic and electric properties.

Magnetic properties are due to the super exchange interaction between cations located in tetrahedral and octahedral coordinated sites throughout the surrounding oxygen anions [6]. The doping process at any interstitial site not only impacts the total magnetic moment but also it may control the particle growth and consequently the material properties [7-9], In the formula MFe_2O_4, depending on the position of M(II) and Fe(III) site preference, three possible spinel ferrite structures are known, namely normal, inverse and mixed. In a normal spinel structure of ferrite, M (II) located at tetrahedral sites while Fe (III) at octahedral sites. In an inverse spinel structure of ferrite, Fe (III) equally distributed at both sites while M(II) occupy only at octahedral sties. In a mixed spinel structure of ferrite, both ions randomly occupy the tetrahedral and octahedral sites. $ZnFe_2O_4$is an example of normal spinel ferrite, Fe_3O_4 and $NiFe_2O_4$ are common examples for inverse spinel ferrite while $MnFe_2O_4$ is a good example for mixed spinel ferrite.

Scientific advancement and industrialization have been both beneficial and problematic for the environment and to human populations. Problems associated with drinkable water are particularly notable. The vast advancement in the field of science and technology for making life easy and comfortable, many electronic and consumer products have become essential part of household and office as daily routine. The uncontrollable growth in human population and excessive usage of these products contaminated all the natural resources e.g. water, soil and air essential to sustain life on earth. Now, water pollution is a global problem and concentration of toxic pollutants in the water bodies are well above the designated limits established by world health organization (WHO) and environmental protection agency (EPA). Water quality has become an eminent issue as ground water level has enormously gone down and its contamination by natural contaminants like fluoride and arsenic, man-made chemical pollutants such as dyes, pesticides and insecticides and biological contaminants is rising at an alarming rate due to outburst of population, irrigation and industrialization.

The inconsistency in awareness, socio-economic development, education, poverty, practices and rituals of people in villages has further worsen the problem. This has resulted in the enormous health burden of poor water quality, as ground water is the decentralized source of potable water to about 85% of the rural population in India. It is estimated that annually, around 37.7 million Indians are affected by waterborne diseases; around 1.5 million children are estimated to die of diarrhoea alone. That is why, the government of India has categorized this as one of the major National Missions of India.

And even though a depletion of groundwater resources has been recorded throughout the world, there is an increasing demand for clean water to support human populations, industries, and living organisms [10-12]. For all these reasons, water is no longer considered a free resource. In several developing and developed countries, individuals are paying for clean water and furthermore, some governments have imposed water tariffs [13]. The most significant water problems include: (i) groundwater depletion, (ii) contamination of fresh water resources (rivers, lakes, wells and ponds), and (iii) a high demand for fresh water (due to a rapid increase in population and urbanization) [14, 15] [16]. Polluted water significantly affects the health of the communities, aquatic ecosystems, the sustainability of the natural environment, and the economic and social welfare of society. For example, it has been reported that inadequate water supplies, in terms of both quantity and quality, coupled with poor sanitation globally account for the death of approximately 30 000 people per day. Of these cases, 80% occur in rural areas with the highest percentage occurring among infants[17].

There are other types of contaminations such as heavy metals occurring as trace metals in aqueous systems but their level of contamination is increasing from different activities such as geological, mining, agricultural, acidic rain and industrial waste in aqueous environment. No doubt, there are certain metals like Manganese (Mn), Iron (Fe), Cobalt (Co), Copper (Cu), Vanadium(V) and Zinc (Zn) which are essentials to human body in trace amount called as essential metals. Heavy metals such as arsenic, lead, antimony, mercury, selenium, chromium etc. are very harmful and toxic in nature even in micro level range. Among different heavy metals, Lead (Pb) is the poisonous metal which is widely used in industries dealing in the production of lead acid batteries, solder, alloys, ammunition, refinery and plumbing [18]. The allowed limit set by world health organization(WHO) for lead in drinking water is 0.01 mg/l [19]. The increased concentration of lead in water leads to many serious diseases like inhibition of haemoglobin, kidney damage, loss of memory and reproductive effects in human being [20]. Therefore, removal of Pb (II) ions from contaminated water is necessary for sustainability of environment.

The need for clean, safe and utmost quality water is increasing day by day throughout the world and it ultimately resulted into the need of new and cost-effective methods for water purification. A satisfactory solution to these problems is the recovery of clean and fresh water from contaminated one. In this regard, different approaches to develop highly efficient water purification methods have been undertaken [21-32]. The removal of such toxic elements and compounds which are present in low concentrations is a difficult and challenging task. Various remediation technologies have been developed for the removal of pollutants including toxic heavy metals, dyes, pesticides, fertilizers, organic acids, and

Materials Research Foundations **112** (2021) 246-278 https://doi.org/10.21741/9781644901595-7

halogen attend and phenolic compounds, among others. Techniques such as precipitation, incineration, flocculation, coagulation, ion exchange, reverse osmosis, membrane filtration, electrochemistry, photo electrochemistry, advanced oxidation processes, and biological methods have demonstrated different degrees of remediation efficiency [33]. There were certain drawbacks of the methods like low removal efficiency, high production of sewage sludge, non-cost-and-energy efficiency, generation of toxic by-products, as well as the addition of more potentially toxic chemicals into the environment. Among all these methods, adsorption technique has been the most effective, simplest, economical, less time consuming, environmental friendly and offers high performance also. This method is widely used to remove different types of metal ions from the tainted aqueous system at industrial level [34, 35]. In this process, different types of natural as well as synthetic adsorbents such as clay, mineral oxides, rice husk, fly ash, meso porous materials, nano materials etc. have been reported to remove pollutants from water. The most common adsorbents for removal of different pollutants from aqueous solution are discussed briefly [36].

Researchers from various fields, including environmental, chemical, biotechnological and material science, have come together to develop novel adsorbents to solve wastewater treatment problems [37-47]. Published reports describe the development of a range of adsorbents including: carbon-based three-dimensional architectures [48], three-dimensional graphene-based macrostructures (3D GBM)[49] , magnetic chitosan composites (MCCs) [43], ordered meso-porous materials [50], inorganic nano-adsorbents [51], nano adsorbents [52], agricultural biomasses [53], graphene nano-sheets [54], metal-organic frameworks [55], sludge-derived activated carbons [56], bio sorbents [56, 57], tailored zeolites [58] and bio-derived materials [59]. Activated carbon (AC) is one of the most successful well developed adsorbent materials known for years, and is, still playing a very important role in commercial applications.

But the AC is inappropriate for inorganic pollutant removal because of its non-polar surface and also has low adsorption affinity for low molecular weight polar organic pollutants [60] which further limit its applications. The effectiveness of various adsorbents for the removal of impurities from the aqueous systems have been discussed and Chitosan is found to be most important materials in adsorption applications. Amino and hydroxyl groups present in the molecules contribute to many possible adsorption interactions between chitosan and pollutants (dyes, metals, ions, phenols, pharmaceuticals/drugs, pesticides, herbicides, etc.)[61] [62].

Adsorption is a surface phenomenon and it involves the attachment of dissolved impurity on the surface of the adsorbent. An ideal adsorbent should have(i) high performance, (ii) rapid adsorption, (iii) cost-efficient, (iv) environmentally non-toxic, (v) reusability, and

Materials Research Foundations **112** (2021) 246-278 https://doi.org/10.21741/9781644901595-7

(vi) ease of separation for commercial application. The promising new generation adsorbent materials developed so far are magnetic nano-adsorbents that have these features of high surface area, tunable morphology, ease of separation after sorption, and high efficiency [63, 64].

2. Nano Materials as Adsorbent

Due to their extraordinary physico-chemical properties, the nano-materials have gained much attention in the past two decades in water purification. Among different nano materials, nano ferrites and their composites are showing high potential in waste water purification due to their extra magnetic character which makes their separation easier after treatment. Spinel ferrite magnetic materials are emerging composite metal oxides for removal of aqueous pollutants from the contaminated water due to their size, shape and high surface area [65].

These excellent features of spinel ferrites might give long-lasting and excellent performance in water treatment [66], and aggregation related problems were overcome [67]. Additionally, spinel ferrites and their composites can be prepared in the form of different structure and shapes as per the utilization in various systems, including hierarchical [68], hollow spheres [69] [70], three-dimensional (3D) porous [71], [72] hollow fibres [73], beads [74], and rods [75]. Materials with high porosity and morphology typically exhibit high adsorption properties for a variety of pollutants.

There is vast literature available on spinel ferrites but in the recent years, these spinel ferrites and their composites have gained attraction because of their potential utilization in environmental applications, primarily for the removal of toxic pollutants from water. The reason behind the development of nano composites for purification is their increased affinity, and capacity for selectivity of different pollutants from aqueous system, and particularly due to the following reasons [76-82].

1. Removal of inorganic/organic pollutants at low concentration.
2. High adsorption of hazardous pollutants.
3. Reduction in the toxicity of different pollutants by altering their oxidation state.
4. Providing modifiable reactive surface.
5. Facile removal of adsorbed pollutants from the surface of nano materials.
6. Cost effectiveness and reusability.

The present review work is a combine approach to discuss detailed study of various surface modification methods with their general mechanism and features; applications of different ferrites and their surface modified composites as adsorbent for treatment of organic and inorganic pollutants from waste water; discussion on regeneration of different ferrites and their composites by using various desorption agents and representing the advantages and disadvantages of such types of materials. The past few years has documented a substantial growth in the publication of numerous review articles on different ferrites with focus on synthesis, physical and chemical properties and their applications. In this review, the emphasis has been laid onto the various pros and cons of several synthetic pathways for creating different types of nano ferrite, need for surface modification with general adsorption mechanism for the treatment of organic and inorganic pollutants from water, and their regeneration.

3. Different Synthesis Techniques of SFNP

For the purification of wastewater, which is contaminated with various harmful particles, should be removed efficiently by using various adsorbents. From literature review, it is very well known that spinel ferrite nanoparticle (SFNP) or spinel magnetic particles (SMP) are preferred for wastewater treatment because of its high surface to volume ratio and provide high area to interact with contaminated particles [83, 84]. Most of SFNP shows magnetic behaviour and therefore with the help of external magnetic field, aqueous solution contamination can be removed. Different SFNP synthesised by different synthesis techniques shows different adsorption capacities for the contamination. Even same material synthesised by different techniques shows different adsorption capacity for the same ions as shown in Table 1.

SFNP can be used directly and indirectly for the removal of pollutants from the waste water. Indirectly means coating the SFNP with desirable functional group and SFNP as core material. As core is magnetic these can be used as magnetic separator and coating of such functional group which improve the surface properties, such as phosphoric acid, amine group, trimethoxy silica etc.

Adsorption capacity of various SFNP or SFNC mostly calculated by isothermal equations and pseudo first and second order kinetic equations given by Longmuir and Freundlich models [65, 86, 92] . Adsorption capacity greatly influenced by the distribution of cations that is position of cation is tetrahedral or octahedral sites, because chemical properties of material depend upon the position of cation sites.

Table 1: Data showing variation in adsorption capacity with the method of preparation of different spinel ferrite particles.

Ferrite	Method of preparation	Absorptive removal	Removable capacity(mg/g)	pH value	References
GO-$MnFe_2O_4$	One-pot hydrothermal	Pb^{2+}	621.11	6	[85]
GO-$MnFe_2O_4$ nanohybrids	Coprecipitation	As^{3+}	146		[86]
		As^{5+}	207		
		Pb^{2+}	673		
$MnFe_2O_4$ nanoparticles		As^{3+}	97		
		As^{5+}	136		
		Pb^{2+}	448		
$MnFe_2O_4$-graphene composite	Solvothermal	Pb^{2+}	100	5	[87]
		Cd^{2+}	76.90	7	
$MnFe_2O_4$	Sol-gel	Cr^{4+}	175.43	2	[88]
Co Fe_2O_4	Hydrothermal	CR	244.5	7	[89]
$NiFe_2O_4$ /ZnO	Hydrothermal	CR	221.73	13	[90]
Fe_3O_4 /GO	impregnation	Black 5 dye	164	3	[91]
Fe_3O_4 /GO	Coprecipitation	Black 5 dye	188	3	[91]

Mechanism of the process and factors that effects the adsorption process such as, when surface to volume ratio increases in the nano scale then hydroxyl group is very effective in controlling adsorption [93] . Mostly ion-exchange, inner outer complex, weak interactions hydrogen bond and Π-Π interactions etc. are responsible for contamination removal, which is listed below in Table 2.

Table 2: Data contained ferrite nanoparticles showing mechanism for removal of pollutants.

Particle name	Removal ion	Mechanism	Reference
$MnFe_2O_4$	MB	Hydrogen bound interaction	[94]
γ- Fe_2O_3	Arsenate	Inner sphere-complex formation	[95]
$NiFe_2O_4$	Cr(VI)	Physisorption	[96]

The removal capacity of SFNP and its composites also depends upon the synthesis techniques used, as listed in Table 1. Same spinal ferrites nanoparticles synthesised by different techniques gives different removal capacity for same ions from the aqueous solution [65, 97-99]. Generally, with the decrease in size the removal capacity increases, called nanoscale effect [100, 101] pH also plays significant role in adsorption process as in low pH aqueous solution surface of SFNP is positively charged i.e anions, whereas if pH is high, surface is negatively charged which means it is cation.

Table 3: Data showing removal capacity of ferrites for contamination in specific pH value.

Ferrite	Contamination ion	Removal capacity (mg/g)	pH	References
γFe_3O_4	Mo (VI)	33.4	4-6	[101]
Cu Fe_2O_4	Mo(VI)	30.58	2.75	[102]
Cu Fe_2O_4-rGO	Pb^{2+} Hg^{2+}	299.9 157.9	5.3 4.6	[103]
$MnFe_2O_4$	As^{3+} As^{5+}	93.8 94.4	3 7	[104]
$CoFe_2O_4$	As^{3+} As^{5}	100.3 73.8	3 7	[104]
Fe_3O_4	As^{3+} As^{5}	49.8 44.1	3 7	[104]
$Co_{0.1}Al_{0.03}Fe_{0.17}O_{0.4}$.	MB	~12 ~37 ~81 ~98	3 5 9 11	[105]

4. Types of contamination

The waste water is contaminated by various regions. The contamination in water may be Organic or Inorganic.

4.1 Organic Contaminations

Organic contamination is due to industrial waste which can be effectively removed by the nanoparticles from the literature survey it is found that SFNP is good in this. Organic contamination mostly included dyes, pesticides and pharmaceuticals [106, 107]. All these contaminations are discussed in detail as below:

Dyes: Dyes has complex ring shaped or planar structure with Π- bond in resonance. Therefore, dyes are stable and not easily removed from waste water. When various dyes are discharged from industries it pollutes the water which causes various harmful effect to the environment as well as human life[98] [103, 104]. From literature review, MFe_2O_4 SFNP are effective in removal of harmful contaminations from wastewater and these SFNP are easily removed due to its magnetic properties as well as can be used several times without decrease in capacity of removal of contaminated ions. This is listed in Table 4.
SFNP which are generally used for removal of dyes from waste water are Fe_3O_4 NP, γ-Fe_3O_3 and MFe_2O_4 where M= Mg, Mn, Co, Ni, Zn etc. some of them are discussed below along with their removal capacity which is listed in Table 4.

In Fe_3O_4 NP and NCs, the surface functionalization increases due to its small size, therefore externally useful in removal of dyes from waste water. Some composites of Fe_3O_4 NP are listed below along with removal capacity of dyes from the waste water [86, 108]. From literature review it is noticed that the removal capacity of SFNP is increased when used in composites of SFNP. For example, when Fe_3O_4 NP coated with Humic Acid (HA) absorption capacity for methylene blue increase three times and shows stability in range 3–11 for pH value. When graphene is incorporated with SFNP there is increase in properties such as recovery and removal capacity [91, 109, 110].

Table 4: Data showing removal capacity of various dyes using ferrites with pH and temperature.

Ferrite	Contamination	Removal capacity (mg/g)	pH	Temperature In °C	References
Chitosan-glutamic-Fe_3O_4@SiO_2	MB, CV, CLY	180.1, 375.4, 217.3	7	25	[111]
Chitosan-Fe_3O_4@SiO_2	MB CV CLY	28.8, 78.8, 17.6	7	25	[111]
HA- Fe_3O_4	MB	93.1	8.4	25	[108]
Fe_3O_4/GO	Black 5 dye	164	3	25	[91]
Fe_3O_4/SiO_2- GO	MB	97	7	25	[109]
Fe_3O_4/SiO_2- GO	MB	102.6	7	45	[109]
Fe_3O_4/SiO_2- GO	MB	111.1	7	60	[109]
GO-Fe_3O_4 hybrid composites	MB NR	162.2 171.3	3-7	N/A	[112]
$CoFe_2O_4$@SiO_2-NH_2	Cu^{2+} Cd^{2+} Mn^{2+}	170.829 144.948 110.803	4-6.5	35	[113]
P25-$CoFe_2O_4$-PANI	MO TB BBR	~172 ~160 ~135	N/A	60	[114]
AC- $NiFe_2O_4$	MO	182.8	4.4-8.4	30	[115]
$NiFe_2O_4$	RB5 dye		1	25	[116]
$ZnFe_2O_4$	AR88	111.1	7	30	[117]

Note : Methyl Orange (MO), Trypan Blue (TB), Brilliant Blue R (BBR), Acid Red 88 (AR88), crystal violet (CV).

γ- $Fe_3O_3$is also used largely for removal of organic dyes from biomedical and industrial areas. γ- Fe_3O_3 (Maghemite) coated with dodecyl sulphate improved its adsorption capacity. It removes 285.8 mg/g of acridine orange at room temperature with pH 5.1 [118]. Also, with pH 6 at room temperature the capacity of γ- Fe_3O_3 to remove cationic dyes thionine and Janus green b brilliant crystal blue from the wastewater with removal capacity 166.7, 172.4 and 200.0 mg/g. γ- Fe_3O_3 is also non-toxic as well as has good adsorption capacity. When γ- Fe_3O_3 is made composite with SiO_2 then removal capacity for methyl blue increases to116.09 mg/g by physical adsorption mechanism [119].

Some water pollutants such as direct red (DR), direct green red 6 (DGR6), and acid blue 92 (AB92) dyes can be removed using $CoFe_2O_4$ nanoparticles effectively, when functional group amine is attached with $CoFe_2O_4$ its removal efficiency increases which is shown in the table 4[120]. Whereas $CoFe_2O_4$ composite made with titania- $CoFe_2O_4$ polyaniline, it is concluded anionic dyes are removed effectively than cationic dyes because anionic dyes contribute to electrostatic attractions with positively charged part of polyaniline [114]. Adsorption capacity of this for MO, TB, BBR dyes is given in Table 4. When $NiFe_2O_4$ NPs are used for wastewater treatment then it shows good photo-catalytical properties as well as removal capacity for dyes [71, 90, 121-126].

When $NiFe_2O_4$ made composite activated carbon (AC) then removal capacity for methyl orange dye improved to 182.8 mg/g at 30ºC [115]. For photocatalytic degradation of contamination and antimicrobial, $ZnFe_2O_4$ NPs are used [127-129]. Also, when $ZnFe_2O_4$ prepared by microwave assisted hydrothermal method then acid red 88 dye is effectively removed from wastewater [117]. But with the increase of temperature from 20 to 80 ºC adsorption equilibrium decreases. For acid red dyes absorbed by $ZnFe_2O_4$ at 30ºC with pH 7 is equal to 111.1 mg/g.

Pharmaceuticals are also a major contamination in water [130, 131]. And many of these pharmaceuticals are not removed by generally used wastewater treatment plants, so easily found in surface water or in small amount in ground water, which causes harmful effect on health of humans as well as animals [131-137]. Number of treatments are available in literature, but application of spinel ferrite nanoparticles and nano-composites shows effective removal of pharmaceuticals from polluted water [130, 138]. But using other treatment some of species of lower molecular mass were generated which may be more harmful then of parent pollutant. Form the literature review, treatment for pharmaceuticals and personal care products are effectively done by γ- Fe_2O_4 coated with zeolite which shows 95% of adsorption for the pollutants in wastewater.

Pesticides are used frequently in agriculture these days to enhance their production and to cure various harmful problems in agriculture like insects, fungus and weeds [137]. But these pesticides improve production but on the other side these are also toxic and contaminate the ground water by some amount. Firstly, used pesticides for controlling number of diseases such as malaria and typhus are organochlorine but this is banned by various countries since 1970. Organochlorine is stable and easily available in various

Materials Research Foundations **112** (2021) 246-278 https://doi.org/10.21741/9781644901595-7

ground and surface water resources [138]. These days farmer used mostly organo phosphorus insecticides and triazine herbicides. These pesticides are high in toxicity and high resistance to degradation. Atrazine pesticides easily distribute into environment therefore have low concentration in water, but even in low concentration these are harmful for humans. SFNPs used effectively for removal of pesticides from contaminated water such as Fe_3O_4 NCs.

4.2 Inorganic contamination

Inorganic contamination also plays vital role in water pollution, it occurs naturally or may be caused by human being as their activities in areas such as mining, industries or agriculture. Most commonly occurring inorganic contaminations are caused by the presence of As, Cd, Cr, Pb, Hg, and Se. SFNP /SFNC shows the significant efficiency in the removal of contamination of water because these shows good adsorption capacity and easy to synthesize and can be separated easily due to its magnetic properties as well as cost effective [139, 140]. For waste water treatment, Fe_3O_4 when synthesized with Co – precipitation method, the contaminations in water such as nickel, copper, cadmium and chromium are removed effectively with pH 4 at temperature 20 ºC for 24 hours[141]. Different SFNP or SFNC with different methods have different adsorption capacities whether they may have same surface area. These all contaminations are briefly explained below.

Water polluted by arsenic is highly toxic i.e. 50% more than arsenate, become the causes of cancer. It is easily mixed in environment [92] . Number of metal oxides was used for its removal but after removal, recovery of metal oxides was difficult. This problem is effectively overcome by the use of SFNP/SFNC because of their magnetic properties. With the help of external magnetic field, they are easily recovered and reused. $CuFe_2O_4$ shows good removal capacity for As^{5+} (45.7mg/g) at pH 3.7 but this adsorption deceases with the increasing value of pH due to electrostatic repulsion between AsO_4^{3-} and adsorbent surface [42, 142] $MnFe_2O_4$ shows effective removal of As3 + [143] .

Other pollutant cadmium is harmful when it is present more than recommended limit. It can become the cause of lung cancer, blood pressure, kidney failure etc.[144] Cd^{2+} is effectively removed by the use of SFNP such as $CuFe_2O_4$ when used at temperature 45 ºC with pH 6[145].. However, when Cd^{2+} is removed by using Fe_3O_4 coated with citric acid, its removal capacity become temperature dependent i.e., decrease with decrease in temperature and vice versa. All the water pollutants are toxic as well as harmful for living organisms. Similarly, Chromium (Cr^{6+}) is harmful for humans and animals, also carcinogenic as well as causing health problems if present more than 0.05mg/L in domestic usable water, according to WHO [146] .These are treated efficiently by

Materials Research Foundations **112** (2021) 246-278 https://doi.org/10.21741/9781644901595-7

$MnFe_2O_4$ and after removing chromium ions, these ferrite nanoparticles are retrieved using magnetic field which is provided externally [143]. The removal capacity of Chromium from industrial waste by $MnFe_2O_4$ is 89.16 but when used in the form of composite of γ-Fe_2O_3 with graphene and cellulose synthesized its removal capacity improved and gives exothermic reaction [147, 148]. $CoFe_2O_4$ also shows effective removal of Cr^{6+}. But studies releveled that $MnFe_2O_4$ gives maximum & $CoFe_2O_4$ gives minimum removal of Cr^{6+} ions due to its large surface area whereas it's recovery after reaction is minimum as compared with other SFNP used for chromium removal. Lead is harmful for living organism and causes various deceases, when used above a recommended limit i.e., 0.05 mg/L[97, 146] [149]. It is effectively used in batteries due to its high efficiency but also very toxic and not easy to dispose it off, therefore, banned by many countries. SFNP and its composites shows effective removal of Pb from waste-water, such as by $CoFe_2O_4(NH_2)$ and MWCNTS–$MnFe_2O_4$(NH2) comprise chitosan based SFNP shows high removal capacity of Pb at pH 6 is 370.6 mg/g due to high active sites present on chitosan [150].

Mercury is very toxic for human and aquatic life as it has very strong binding energy. Due to this property, it binds strongly with haemoglobin and proteins which restrict the proper functioning of these. This is strongly toxic which causes hindrance in kidney functioning, chest pain etc. SFNP such as γ- Fe_2O_3 coated with polyrhodanine shows Langmuir adsorption capacity for Hg^{2+} is 179 mg/g. It is also described best from pseudo second order reaction for adsorption of mercury by chemical reaction[151] .

Selenium: Selenite (SeO_3^{2-}) and Selenate (SeO_4^{2-}) are two forms of selenium that exist in oxidation state Se^{4+} and Se^{6+} in inorganic solvent. These are more toxic than the organic solvent, if its value in drinking water is more than subscribed value that is 40 µg/L[152] [153]. Various polymers, composites and oxides are used for the removal of selenium from wastewater but recovery of these adsorbents was difficult. Therefore, SFNP are used for the removal of selenium which gives effective results for removal as well as easily recovered due to its magnetic properties. $CoFe_2O_4$ shows highest removal capacity for Se^{4+} and Se^{6+} whereas $MnFe_2O_4$ gives least removal capacity. γ- Fe_2O_3 also gives significant removal capacity for selenium ions.

5. Adsorption Mechanism

Different adsorbates adsorb on different adsorbents in different ways. Physicochemical interactions of the adsorbate-adsorbent play a key role in an efficient adsorption system; thus, it is essential to understand the adsorption mechanism.

Surfactant coating is a well–known method that protects nano ferrites from oxidation, degradation and enhances their stability by changing structure, particle size, magnetic, electrical and adsorption properties. The process involves either a single layer or double layer formation on the surface of nano ferrites. The surface modification of spinel ferrites and their composites can be easily explained on the basis of electrostatic interaction between surfactant molecules and nano ferrite particles. There are various types of electrostatic interactions like as (i) π–π interaction, (ii) electrostatic interaction, (iii) ion exchange interaction, (iv) hydrogen bonding, (v) surface complex formation etc. exist between the surfactant molecules and nano ferrite particles. The generation of such electrostatic interaction stabilizes the colloidal nano particles and increases their stability with surfactant molecules with modification in their properties [154-156].

Surfactant coating on the surface of ferrites and their interaction toward different pollutants plays important role in adsorption system. The surfactant modified ferrite composite contains different types of functional groups depending on the nature of surfactant and act as active sites for adsorption of different inorganic and organic pollutants in the aqueous solution *[79]* and Carlos *et al[157]* reviewed different types of surfactants used for remodelling iron oxide surface and understand their utility as adsorbent for removal of heavy metal ions from water. The drastic increase in adsorption behaviour of surfactant modified iron was recorded compared to iron oxide. The studies reveal that different types of surfactant coated ferrites have ability to remove numerous types of inorganic and organic pollutants from the waste water[158].

Sorption mechanism of dyes has also been thoroughly described in terms of electrostatic attractions and hydrogen bonding [159]. At a lower pH the surface of $ZnFe_2O_4$ has a net positive charge, and anionic acid red 88 (AR88) adsorption was favourable [114]. With increasing pH, the surface becomes negatively charged and, lower sorption was observed, presumably due to an electrostatic repulsion between the anionic dye and the negatively charged $ZnFe_2O_4$. When methyl blue and CR are adsorbed to $MnFe_2O_4$, hydrogen bond interactions are the major mechanisms attracting dye and hydroxyl groups [94]. Fluorescence spectroscopy was used to further understand the synergetic effect of the interaction between methyl blue and CR. The factors that influence the adsorption performance of SFs and SFCs include (i) cation distribution in SFs, (ii)surface charge, (iii) annealing temperature, and (iv) material used to modify or functionalize SFs.

Adsorption behaviour of metal and their oxide doped nano ferrites can be well explained using ion exchange mechanism. As the dopant concentration increases in nano ferrite sample, it directly affects the crystalline structure and size of nano particles. The difference is due to variation in atomic size of dopant (metal or their oxide particles) ions with the parent metal ions. The variation in atomic size between the parent metal ions and substituted dopant ions in crystal lattice surface directly affect the inter planner spacing, lattice constant between two crystal lattice and increase the concentration of free electrons in the crystal lattice. These variations in crystal lattice structure increase the active site on the doped surface, which involves change in adsorption behaviour of doped nano ferrites. The presence of free electron in crystal lattice surface helps to reduce the high oxidation state of pollutant to lower state, which in turn become less stable and easily adsorb on surface of doped surface nano ferrites. Table 2 shows the different mechanism used for contamination removal in various ferrites.

6. Desorption, Regeneration of Nano Ferrites and their Composites

The removal of the pollutants from aqueous solution using adsorbents is an important step, but desorption of the loaded pollutant from the adsorbent, to recover the adsorbent and to discard the pollutant is also important [160]. As the disposal of pollutant-loaded adsorbent may become a secondary pollution problem. The importance of regeneration in adsorption with respect to economic and environmental factors with highlights on regeneration procedures like thermal, electrical, ultrasonic and chemical methods has been discussed by Kulkarni et al.[161]. It depends upon the selection of desorption agent and operating factors such as temperature, pH and concentration. In particularly, the pH value is a critical parameter that influences the efficiency of desorption of different pollutants from adsorbent surface [161, 162]. Aqueous solutions of strong bases and strong acids are commonly used desorption agents to remove inorganic/organic pollutants from the adsorbent surface.

The following is a list of eluents and processes that have been chosen most often to desorb pollutant from SFs and their composite adsorbents:

(i) Acids (HCl, HNO3, H3PO4)

(ii) Bases (NaOH, NaHCO3)

(iii) Organic solvents (ethanol, methanol, and acetone)

(iv) Chelating agents (EDTA and thiourea)

(v) Thermal regeneration

It is a Challenging task to retrieve these nanoparticles from liquid medium effectively because of their extremely small size. Magnetic Separation can be used to get spinel ferrite-based nanoparticles from the liquid medium. Magnetic field separation is superior to traditional centrifugation and filtration processes because it is more economical, selective and rapid.

7. Future Aspects

The potable water scarcity existing in many parts of the world and the necessity for water treatment confirm the need for efficient adsorbents. Magnetic nanoparticles and their composites with tailored size, composition, magnetic characteristics and structure are currently used for treating polluted water.

Although Fe_3O_4 has traditionally been used for water treatment applications, spinel ferrites like MFe_2O_4 (M=Mn, Cu, Co, Ni, Zn, Mg, Ca etc.) are becoming more widely used. More selective, efficient, inexpensive, reusable and eco-friendly adsorbents should be developed and experiments be conducted with real industrial effluents. Researches are required to have a comparative study of different adsorbents for the removal of different contaminants from water considering the feasibility, effectiveness and cost under similar conditions.

Despite the strong capacity of SFNPs for removing varieties of toxic chemicals from wastewater, studies on the impact of SFNPs on human health and their environmental behaviour and ecological risk as well as the effects of SFNPs ending up as pollutants in the environment, are scant and are not fully addressed. Therefore, an in depth toxicity study is required.

Different type of adsorbents with greater stability (resistance to pH changes and concentrations of chemicals present in contaminated water) and the capacity for the simultaneous removal of multiple contaminants, such as toxic metal ions, organic dyes and bacterial pathogens should be developed. Despite various drawbacks and challenges that currently exist, a widespread and great progress in this area can be expected in the future.

Conclusions

Based on this review, it is possible to conclude that SFNPs/SFNCs have high adsorption and disinfection capacities toward various types of contaminants. It involves detailed discussion on tailored synthesis of different types of nano ferrites, the fine tuning of surface properties, and their application as adsorbents in wastewater treatment. The analysis reveals that the shape and size of different types of nano ferrites directly depend upon synthetic routes. Moreover, the characteristics of different types of nano ferrites are regulated by different factors such as precursor composition, pH of solution and annealing temperature.

Number of methods has been developed for purification of water but adsorption method has been found to be efficient and effective. Various types of adsorbents have been used for removal of different pollutants from wastewater. In recent years the use of nanomaterials in water remediation has increased considerably. Adsorption efficiency is found to depend on the type of adsorbent, adsorbate, pH of solution, concentration of pollutant and temperature. Removal of pollutants has been discussed on the basis of different isotherm and kinetic models.

In addition, the use of SFNPs/SFNCs is an excellent option for water and wastewater treatment because wastewater can be reused, waste generation and environmental pollution is reduced, the treatment technology is simple and easy, and it is cost-effective. This is due to the dose requirement being low, easily recovered by an external magnetic field and used for several cycles.

References

[1] S. Yáñez-Vilar, M. Sánchez-Andújar, C. Gómez-Aguirre, J. Mira, M.A. Señarís-Rodríguez, S. Castro-García, A simple solvothermal synthesis of MFe_2O_4 (M= Mn, Co and Ni) nanoparticles, Journal of Solid State Chemistry, 182 (2009) 2685-2690. https://doi.org/10.1016/j.jssc.2009.07.028

[2] O. Yelenich, S. Solopan, J.-M. Greneche, A. Belous, Synthesis and properties MFe_2O_4 (M= Fe, Co) nanoparticles and core–shell structures, Solid State Sciences, 46 (2015) 19-26. https://doi.org/10.1016/j.solidstatesciences.2015.05.011

[3] W. Wang, Z. Ding, X. Zhao, S. Wu, F. Li, M. Yue, J.P. Liu, Microstructure and magnetic properties of MFe_2O_4 (M= Co, Ni, and Mn) ferrite nanocrystals prepared using colloid mill and hydrothermal method, Journal of Applied Physics, 117 (2015) 17A328. https://doi.org/10.1063/1.4917463

[4] P. Dhiman, N. Dhiman, A. Kumar, G. Sharma, M. Naushad, A.A. Ghfar, Solar active nano-$Zn_{1-x}Mg_xFe_2O_4$ as a magnetically separable sustainable photocatalyst for degradation of sulfadiazine antibiotic, Journal of Molecular Liquids, 294 (2019) 111574. https://doi.org/10.1016/j.molliq.2019.111574

[5] P. Dhiman, T. Mehta, A. Kumar, G. Sharma, M. Naushad, T. Ahamad, G.T. Mola, Mg0. 5NixZn0. 5-xFe2O4 spinel as a sustainable magnetic nano-photocatalyst with dopant driven band shifting and reduced recombination for visible and solar degradation of Reactive Blue-19, Advanced Powder Technology, (2020). https://doi.org/10.1016/j.apt.2020.10.010

[6] E. Santiago, G. Márquez, R. Guillén-Guillén, C. Jaimes, V. Sagredo, G.E. Delgado, caracterización de nanocompósitos de hematita y ferritas mixtas de ni-zn sintetizados mediante el método de coprecipitación, La Revista Latinoamericana de Metalurgia y Materiales, RLMM, 40 (2020).

[7] R. Panda, R. Muduli, G. Jayarao, D. Sanyal, D. Behera, Effect of Cr^{3+} substitution on electric and magnetic properties of cobalt ferrite nanoparticles, Journal of Alloys and Compounds, 669 (2016) 19-28. https://doi.org/10.1016/j.jallcom.2016.01.256

[8] P. Motavallian, B. Abasht, H. Abdollah-Pour, Zr doping dependence of structural and magnetic properties of cobalt ferrite synthesized by sol–gel based Pechini method, Journal of Magnetism and Magnetic Materials, 451 (2018) 577-586. https://doi.org/10.1016/j.jmmm.2017.11.112

[9] B. Abasht, A. Beitollahi, S.M. Mirkazemi, Processing, structure and magnetic properties correlation in co-precipitated Ca-ferrite, Journal of Magnetism and Magnetic Materials, 420 (2016) 263-268. https://doi.org/10.1016/j.jmmm.2016.07.047

[10] D. MacAllister, A. MacDonald, S. Kebede, S. Godfrey, R. Calow, Comparative performance of rural water supplies during drought, Nature communications, 11 (2020) 1-13. https://doi.org/10.1038/s41467-020-14839-3

[11] A. Mark, Science and technology for water purification in the coming decades, Nature, 452 (2008) 20. https://doi.org/10.1038/nature06599

[12] A. Biewald, S. Rolinski, H. Lotze-Campen, C. Schmitz, J.P. Dietrich, Valuing the impact of trade on local blue water, Ecological Economics, 101 (2014) 43-53. https://doi.org/10.1016/j.ecolecon.2014.02.003

[13] B.P. Walsh, S.N. Murray, D. O'Sullivan, The water energy nexus, an ISO50001 water case study and the need for a water value system, Water Resources and Industry, 10 (2015) 15-28. https://doi.org/10.1016/j.wri.2015.02.001

[14] U. Water, WWAP (United Nations World Water Assessment Programme)(2016)(pp. 1–148), Paris, France: The United Nations World Water Development Report, (2016).

[15] K. Feng, K. Hubacek, S. Pfister, Y. Yu, L. Sun, Virtual scarce water in China, Environmental Science & Technology, 48 (2014) 7704-7713. https://doi.org/10.1021/es500502q

[16] A.G. Fane, R. Wang, M.X. Hu, Synthetic membranes for water purification: status and future, Angewandte Chemie International Edition, 54 (2015) 3368-3386. https://doi.org/10.1002/anie.201409783

[17] G.M. Ochieng, E.S. Seanego, O.I. Nkwonta, Impacts of mining on water resources in South Africa: A review, Scientific Research and Essays, 5 (2010) 3351-3357.

[18] G.M. Naja, B. Volesky, 9 Treatment of Metal-Bearing, Heavy Metals in the Environment, (2009) 247. https://doi.org/10.1201/9781420073195.ch9

[19] S. Water, W.H. Organization, Guidelines for drinking-water quality. Vol. 1, Recommendations, (2004).

[20] S. Tong, Y.E.v. Schirnding, T. Prapamontol, Environmental lead exposure: a public health problem of global dimensions, Bulletin of the world health organization, 78 (2000) 1068-1077.

[21] C.A. Quist-Jensen, F. Macedonio, E. Drioli, Membrane technology for water production in agriculture: Desalination and wastewater reuse, Desalination, 364 (2015) 17-32. https://doi.org/10.1016/j.desal.2015.03.001

[22] A. Subramani, J.G. Jacangelo, Emerging desalination technologies for water treatment: a critical review, Water research, 75 (2015) 164-187. https://doi.org/10.1016/j.watres.2015.02.032

[23] A. Cincinelli, T. Martellini, E. Coppini, D. Fibbi, A. Katsoyiannis, Nanotechnologies for removal of pharmaceuticals and personal care products from water and wastewater. A review, Journal of nanoscience and nanotechnology, 15 (2015) 3333-3347. https://doi.org/10.1166/jnn.2015.10036

[24] E. Brillas, C.A. Martínez-Huitle, Decontamination of wastewaters containing synthetic organic dyes by electrochemical methods. An updated review, Applied Catalysis B: Environmental, 166 (2015) 603-643. https://doi.org/10.1016/j.apcatb.2014.11.016

[25] J. Yin, B. Deng, Polymer-matrix nanocomposite membranes for water treatment, Journal of membrane science, 479 (2015) 256-275. https://doi.org/10.1016/j.memsci.2014.11.019

[26] G. Ungureanu, S. Santos, R. Boaventura, C. Botelho, Arsenic and antimony in water and wastewater: overview of removal techniques with special reference to latest advances in adsorption, Journal of environmental management, 151 (2015) 326-342. https://doi.org/10.1016/j.jenvman.2014.12.051

[27] H.M. Saeed, G.A. Husseini, S. Yousef, J. Saif, S. Al-Asheh, A.A. Fara, S. Azzam, R. Khawaga, A. Aidan, Microbial desalination cell technology: a review and a case study, Desalination, 359 (2015) 1-13. https://doi.org/10.1016/j.desal.2014.12.024

[28] N. Wang, X. Zhang, Y. Wang, W. Yu, H.L. Chan, Microfluidic reactors for photocatalytic water purification, Lab on a Chip, 14 (2014) 1074-1082. https://doi.org/10.1039/C3LC51233A

[29] T.A. Kurniawan, G.Y. Chan, W.-H. Lo, S. Babel, Physico–chemical treatment techniques for wastewater laden with heavy metals, Chemical engineering journal, 118 (2006) 83-98. https://doi.org/10.1016/j.cej.2006.01.015

[30] F. Fu, Q. Wang, A review of removal of heavy metal ions from wastewaters, Journal of environmental management, 92 (2011) 407-418. https://doi.org/10.1016/j.jenvman.2010.11.011

[31] M. Barakat, New trends in removing heavy metals from industrial wastewater, Arabian journal of chemistry, 4 (2011) 361-377. https://doi.org/10.1016/j.arabjc.2010.07.019

[32] S.J. Tesh, T.B. Scott, Nano-composites for water remediation: A review, Advanced Materials, 26 (2014) 6056-6068. https://doi.org/10.1002/adma.201401376

[33] H. Shin, D. Tiwari, D.-J. Kim, Phosphate adsorption/desorption kinetics and P bioavailability of Mg-biochar from ground coffee waste, Journal of Water Process Engineering, 37 (2020) 101484. https://doi.org/10.1016/j.jwpe.2020.101484

[34] Y. Anjaneyulu, N.S. Chary, D.S.S. Raj, Decolourization of industrial effluents–available methods and emerging technologies–a review, Reviews in Environmental Science and Bio/Technology, 4 (2005) 245-273. https://doi.org/10.1007/s11157-005-1246-z

[35] Q. Jiuhui, Research progress of novel adsorption processes in water purification: a review, Journal of environmental sciences, 20 (2008) 1-13. https://doi.org/10.1016/S1001-0742(08)60001-7

[36] N.P. Raval, P.U. Shah, N.K. Shah, Adsorptive removal of nickel (II) ions from aqueous environment: A review, Journal of environmental management, 179 (2016) 1-20. https://doi.org/10.1016/j.jenvman.2016.04.045

[37] D.H.K. Reddy, S.-M. Lee, Magnetic biochar composite: facile synthesis, characterization, and application for heavy metal removal, Colloids and Surfaces A: Physicochemical and Engineering Aspects, 454 (2014) 96-103. https://doi.org/10.1016/j.colsurfa.2014.03.105

[38] A. Alsbaiee, B.J. Smith, L. Xiao, Y. Ling, D.E. Helbling, W.R. Dichtel, Rapid removal of organic micropollutants from water by a porous β-cyclodextrin polymer, nature, 529 (2016) 190-194. https://doi.org/10.1038/nature16185

[39] R. Jasrotia, N. Kumari, R. Kumar, M. Naushad, P. Dhiman, G. Sharma, Photocatalytic degradation of environmental pollutant using nickel and cerium ions substituted $Co_{0.6}Zn_{0.4}Fe_2O_4$ nanoferrites, Earth Systems and Environment, (2021). https://doi.org/10.1007/s41748-021-00214-9

[40] H. Zhao, X. Song, H. Zeng, 3D white graphene foam scavengers: vesicant-assisted foaming boosts the gram-level yield and forms hierarchical pores for superstrong pollutant removal applications, NPG Asia Materials, 7 (2015) e168-e168. https://doi.org/10.1038/am.2015.8

[41] P. Dhiman, N. Dhiman, A. Kumar, G. Sharma, M. Naushad, A.A. Ghfar, Solar active nano-$Zn_{1-x}Mg_xFe_2O_4$ as a magnetically separable sustainable photocatalyst for degradation of sulfadiazine antibiotic, Journal of Molecular Liquids, 294 (2019) 111574. https://doi.org/10.1016/j.molliq.2019.111574

[42] Y.-J. Tu, C.-F. You, C.-K. Chang, S.-L. Wang, T.-S. Chan, Arsenate adsorption from water using a novel fabricated copper ferrite, Chemical Engineering Journal, 198-199 (2012) 440-448. https://doi.org/10.1016/j.cej.2012.06.006

[43] D.H.K. Reddy, S.-M. Lee, Synthesis and characterization of a chitosan ligand for the removal of copper from aqueous media, Journal of Applied Polymer Science, 130 (2013) 4542-4550. https://doi.org/10.1002/app.39578

[44] A.J. Howarth, M.J. Katz, T.C. Wang, A.E. Platero-Prats, K.W. Chapman, J.T. Hupp, O.K. Farha, High Efficiency Adsorption and Removal of Selenate and Selenite from Water Using Metal–Organic Frameworks, Journal of the American Chemical Society, 137 (2015) 7488-7494. https://doi.org/10.1021/jacs.5b03904

[45] P. Tian, X.-y. Han, G.-l. Ning, H.-x. Fang, J.-w. Ye, W.-t. Gong, Y. Lin, Synthesis of Porous Hierarchical MgO and Its Superb Adsorption Properties, ACS Applied Materials & Interfaces, 5 (2013) 12411-12418. https://doi.org/10.1021/am403352y

[46] C.J. Madadrang, H.Y. Kim, G. Gao, N. Wang, J. Zhu, H. Feng, M. Gorring, M.L. Kasner, S. Hou, Adsorption Behavior of EDTA-Graphene Oxide for Pb (II) Removal, ACS Applied Materials & Interfaces, 4 (2012) 1186-1193. https://doi.org/10.1021/am201645g

[47] Z.-Y. Sui, Y. Cui, J.-H. Zhu, B.-H. Han, Preparation of Three-Dimensional Graphene Oxide–Polyethylenimine Porous Materials as Dye and Gas Adsorbents, ACS Applied Materials & Interfaces, 5 (2013) 9172-9179. https://doi.org/10.1021/am402661t

[48] B. Chen, Q. Ma, C. Tan, T.-T. Lim, L. Huang, H. Zhang, Carbon-Based Sorbents with Three-Dimensional Architectures for Water Remediation, Small, 11 (2015) 3319-3336. https://doi.org/10.1002/smll.201403729

[49] Y. Shen, Q. Fang, B. Chen, Environmental Applications of Three-Dimensional Graphene-Based Macrostructures: Adsorption, Transformation, and Detection, Environmental Science & Technology, 49 (2015) 67-84. https://doi.org/10.1021/es504421y

[50] Z. Wu, D. Zhao, Ordered mesoporous materials as adsorbents, Chemical Communications, 47 (2011) 3332-3338. https://doi.org/10.1039/c0cc04909c

[51] P.Z. Ray, H.J. Shipley, Inorganic nano-adsorbents for the removal of heavy metals and arsenic: a review, RSC Advances, 5 (2015) 29885-29907. https://doi.org/10.1039/C5RA02714D

[52] M. Khajeh, S. Laurent, K. Dastafkan, Nanoadsorbents: Classification, Preparation, and Applications (with Emphasis on Aqueous Media), Chemical Reviews, 113 (2013) 7728-7768. https://doi.org/10.1021/cr400086v

[53] V.M. Nurchi, I. Villaescusa, Agricultural biomasses as sorbents of some trace metals, Coordination Chemistry Reviews, 252 (2008) 1178-1188. https://doi.org/10.1016/j.ccr.2007.09.023

[54] Z. Guo, C. Xie, P. Zhang, J. Zhang, G. Wang, X. He, Y. Ma, B. Zhao, Z. Zhang, Toxicity and transformation of graphene oxide and reduced graphene oxide in bacteria biofilm,

Science of The Total Environment, 580 (2017) 1300-1308. https://doi.org/10.1016/j.scitotenv.2016.12.093

[55] Z. Hasan, S.H. Jhung, Removal of hazardous organics from water using metal-organic frameworks (MOFs): Plausible mechanisms for selective adsorptions, Journal of Hazardous Materials, 283 (2015) 329-339. https://doi.org/10.1016/j.jhazmat.2014.09.046

[56] S. Aoudj, A. Khelifa, N. Drouiche, R. Belkada, D. Miroud, Simultaneous removal of chromium(VI) and fluoride by electrocoagulation–electroflotation: Application of a hybrid Fe-Al anode, Chemical Engineering Journal, 267 (2015) 153-162. https://doi.org/10.1016/j.cej.2014.12.081

[57] S.W. Won, P. Kotte, W. Wei, A. Lim, Y.-S. Yun, Biosorbents for recovery of precious metals, Bioresource Technology, 160 (2014) 203-212. https://doi.org/10.1016/j.biortech.2014.01.121

[58] H. Figueiredo, C. Quintelas, Tailored zeolites for the removal of metal oxyanions: Overcoming intrinsic limitations of zeolites, Journal of Hazardous Materials, 274 (2014) 287-299. https://doi.org/10.1016/j.jhazmat.2014.04.012

[59] S. Qiu, Z. Lin, Y. Zhou, D. Wang, L. Yuan, Y. Wei, T. Dai, L. Luo, G. Chen, Highly selective colorimetric bacteria sensing based on protein-capped nanoparticles, Analyst, 140 (2015) 1149-1154. https://doi.org/10.1039/C4AN02106A

[60] N.V. Quy, N.D. Hoa, M. An, Y. Cho, D. Kim, A high-performance triode-type carbon nanotube field emitter for mass production, Nanotechnology, 18 (2007) 345201. https://doi.org/10.1088/0957-4484/18/34/345201

[61] M. Vakili, M. Rafatullah, B. Salamatinia, A.Z. Abdullah, M.H. Ibrahim, K.B. Tan, Z. Gholami, P. Amouzgar, Application of chitosan and its derivatives as adsorbents for dye removal from water and wastewater: A review, Carbohydrate Polymers, 113 (2014) 115-130. https://doi.org/10.1016/j.carbpol.2014.07.007

[62] G.Z. Kyzas, D.N. Bikiaris, Recent Modifications of Chitosan for Adsorption Applications: A Critical and Systematic Review, Marine Drugs, 13 (2015). https://doi.org/10.3390/md13010312

[63] J. Gómez-Pastora, E. Bringas, I. Ortiz, Recent progress and future challenges on the use of high performance magnetic nano-adsorbents in environmental applications, Chemical Engineering Journal, 256 (2014) 187-204. https://doi.org/10.1016/j.cej.2014.06.119

[64] D. Peng, B. Wu, H. Tan, S. Hou, M. Liu, H. Tang, J. Yu, H. Xu, Effect of multiple iron-based nanoparticles on availability of lead and iron, and micro-ecology in lead contaminated soil, Chemosphere, 228 (2019) 44-53. https://doi.org/10.1016/j.chemosphere.2019.04.106

[65] D.H.K. Reddy, Y.-S. Yun, Spinel ferrite magnetic adsorbents: Alternative future materials for water purification?, Coordination Chemistry Reviews, 315 (2016) 90-111. https://doi.org/10.1016/j.ccr.2016.01.012

[66] D. Kang, X. Yu, M. Ge, W. Song, One-step fabrication and characterization of hierarchical MgFe2O4 microspheres and their application for lead removal, Microporous and Mesoporous Materials, 207 (2015) 170-178. https://doi.org/10.1016/j.micromeso.2015.01.023

[67] W.-H. Xu, L. Wang, J. Wang, G.-P. Sheng, J.-H. Liu, H.-Q. Yu, X.-J. Huang, Superparamagnetic mesoporous ferrite nanocrystal clusters for efficient removal of arsenite from water, CrystEngComm, 15 (2013) 7895-7903. https://doi.org/10.1039/c3ce40944a

[68] Z. Jia, Q. Qin, J. Liu, H. Shi, X. Zhang, R. Hu, S. Li, R. Zhu, The synthesis of hierarchical $ZnFe_2O_4$ architecture and their application for Cr(VI) adsorption removal from aqueous solution, Superlattices and Microstructures, 82 (2015) 174-187. https://doi.org/10.1016/j.spmi.2015.01.028

[69] Y.-R. He, S.-C. Li, X.-L. Li, Y. Yang, A.-M. Tang, L. Du, Z.-Y. Tan, D. Zhang, H.-B. Chen, Graphene (rGO) hydrogel: A promising material for facile removal of uranium from aqueous solution, Chemical Engineering Journal, 338 (2018) 333-340. https://doi.org/10.1016/j.cej.2018.01.037

[70] R. Rahimi, H. Kerdari, M. Rabbani, M. Shafiee, Synthesis, characterization and adsorbing properties of hollow Zn-Fe_2O_4 nanospheres on removal of Congo red from aqueous solution, Desalination, 280 (2011) 412-418. https://doi.org/10.1016/j.desal.2011.04.073

[71] X. Hou, J. Feng, X. Liu, Y. Ren, Z. Fan, T. Wei, J. Meng, M. Zhang, Synthesis of 3D porous ferromagnetic NiFe2O4 and using as novel adsorbent to treat wastewater, Journal of Colloid and Interface Science, 362 (2011) 477-485. https://doi.org/10.1016/j.jcis.2011.06.070

[72] D.H.K. Reddy, S.-M. Lee, Three-Dimensional Porous Spinel Ferrite as an Adsorbent for Pb(II) Removal from Aqueous Solutions, Industrial & Engineering Chemistry Research, 52 (2013) 15789-15800. https://doi.org/10.1021/ie303359e

[73] T. Dong, X. Zhang, M. Li, P. Wang, P. Yang, Hierarchical flower-like Ni–Co layered double hydroxide nanostructures: synthesis and super performance, Inorganic Chemistry Frontiers, 5 (2018) 3033-3041. https://doi.org/10.1039/C8QI00931G

[74] A.-S.A. Bakr, Y.M. Moustafa, E.A. Motawea, M.M. Yehia, M.M.H. Khalil, Removal of ferrous ions from their aqueous solutions onto $NiFe_2O_4$–alginate composite beads, Journal of Environmental Chemical Engineering, 3 (2015) 1486-1496. https://doi.org/10.1016/j.jece.2015.05.020

[75] Z. Jia, Q. Wang, J. Liu, L. Xu, R. Zhu, Effective removal of phosphate from aqueous solution using mesoporous rodlike NiFe2O4 as magnetically separable adsorbent, Colloids and Surfaces A: Physicochemical and Engineering Aspects, 436 (2013) 495-503. https://doi.org/10.1016/j.colsurfa.2013.07.025

[76] S. Chaturvedi, P.N. Dave, A review on the use of nanometals as catalysts for the thermal decomposition of ammonium perchlorate, Journal of Saudi Chemical Society, 17 (2013) 135-149. https://doi.org/10.1016/j.jscs.2011.05.009

[77] I. Mohmood, C.B. Lopes, I. Lopes, I. Ahmad, A.C. Duarte, E. Pereira, Nanoscale materials and their use in water contaminants removal—a review, Environmental Science and Pollution Research, 20 (2013) 1239-1260. https://doi.org/10.1007/s11356-012-1415-x

[78] S. Thatai, P. Khurana, J. Boken, S. Prasad, D. Kumar, Nanoparticles and core–shell nanocomposite based new generation water remediation materials and analytical techniques: A review, Microchemical Journal, 116 (2014) 62-76. https://doi.org/10.1016/j.microc.2014.04.001

[79] W. Wu, Q. He, C. Jiang, Magnetic Iron Oxide Nanoparticles: Synthesis and Surface Functionalization Strategies, Nanoscale Research Letters, 3 (2008) 397. https://doi.org/10.1007/s11671-008-9174-9

[80] M. Hua, S. Zhang, B. Pan, W. Zhang, L. Lv, Q. Zhang, Heavy metal removal from water/wastewater by nanosized metal oxides: A review, Journal of Hazardous Materials, 211-212 (2012) 317-331. https://doi.org/10.1016/j.jhazmat.2011.10.016

[81] J. Yang, B. Hou, J. Wang, B. Tian, J. Bi, N. Wang, X. Li, X. Huang, Nanomaterials for the Removal of Heavy Metals from Wastewater, Nanomaterials (Basel), 9 (2019) 424. https://doi.org/10.3390/nano9030424

[82] K.K. Kefeni, B.B. Mamba, Photocatalytic application of spinel ferrite nanoparticles and nanocomposites in wastewater treatment: Review, Sustainable Materials and Technologies, 23 (2020) e00140. https://doi.org/10.1016/j.susmat.2019.e00140

[83] F. Moeinpour, A. Alimoradi, M. Kazemi, Efficient removal of Eriochrome black-T from aqueous solution using $NiFe_2O_4$ magnetic nanoparticles, Journal of Environmental Health Science and Engineering, 12 (2014) 112. https://doi.org/10.1186/s40201-014-0112-8

[84] P. Dhiman, M. Patial, A. Kumar, M. Alam, M. Naushad, G. Sharma, D.-V.N. Vo, R. Kumar, Environmental friendly and robust $Mg_{0.5-x}Cu_xZn_{0.5}Fe_2O_4$ spinel nanoparticles for visible light driven degradation of Carbamazepine: Band shift driven by dopants, Materials Letters, 284 (2021) 129005. https://doi.org/10.1016/j.matlet.2020.129005

[85] M. Verma, A. Kumar, K.P. Singh, R. Kumar, V. Kumar, C.M. Srivastava, V. Rawat, G. Rao, S. Kumari, P. Sharma, H. Kim, Graphene oxide-manganese ferrite (GO-$MnFe_2O_4$) nanocomposite: One-pot hydrothermal synthesis and its use for adsorptive removal of Pb^{2+}

ions from aqueous medium, Journal of Molecular Liquids, 315 (2020) 113769. https://doi.org/10.1016/j.molliq.2020.113769

[86] S. Kumar, R.R. Nair, P.B. Pillai, S.N. Gupta, M.A.R. Iyengar, A.K. Sood, Graphene Oxide–MnFe2O4 Magnetic Nanohybrids for Efficient Removal of Lead and Arsenic from Water, ACS Applied Materials & Interfaces, 6 (2014) 17426-17436. https://doi.org/10.1021/am504826q

[87] S. Chella, P. Kollu, E.V.P.R. Komarala, S. Doshi, M. Saranya, S. Felix, R. Ramachandran, P. Saravanan, V.L. Koneru, V. Venugopal, S.K. Jeong, A. Nirmala Grace, Solvothermal synthesis of MnFe2O4-graphene composite—Investigation of its adsorption and antimicrobial properties, Applied Surface Science, 327 (2015) 27-36. https://doi.org/10.1016/j.apsusc.2014.11.096

[88] B. Verma, C. Balomajumder, Magnetic magnesium ferrite–doped multi-walled carbon nanotubes: an advanced treatment of chromium-containing wastewater, Environmental Science and Pollution Research, 27 (2020) 13844-13854. https://doi.org/10.1007/s11356-020-07988-x

[89] L. Wang, J. Li, Y. Wang, L. Zhao, Q. Jiang, Adsorption capability for Congo red on nanocrystalline MFe_2O_4 (M=Mn, Fe, Co, Ni) spinel ferrites, Chemical Engineering Journal, 181-182 (2012) 72-79. https://doi.org/10.1016/j.cej.2011.10.088

[90] H.-Y. Zhu, R. Jiang, Y.-Q. Fu, R.-R. Li, J. Yao, S.-T. Jiang, Novel multifunctional NiFe2O4/ZnO hybrids for dye removal by adsorption, photocatalysis and magnetic separation, Applied Surface Science, 369 (2016) 1-10. https://doi.org/10.1016/j.apsusc.2016.02.025

[91] G.Z. Kyzas, N.A. Travlou, O. Kalogirou, E.A. Deliyanni, Magnetic Graphene Oxide: Effect of Preparation Route on Reactive Black 5 Adsorption, Materials, 6 (2013). https://doi.org/10.3390/ma6041360

[92] M.K. Yadav, A.K. Gupta, P.S. Ghosal, A. Mukherjee, pH mediated facile preparation of hydrotalcite based adsorbent for enhanced arsenite and arsenate removal: Insights on physicochemical properties and adsorption mechanism, Journal of Molecular Liquids, 240 (2017) 240-252. https://doi.org/10.1016/j.molliq.2017.05.082

[93] N. Jordan, A. Ritter, A.C. Scheinost, S. Weiss, D. Schild, R. Hübner, Selenium(IV) Uptake by Maghemite (γ-Fe_2O_3), Environmental Science & Technology, 48 (2014) 1665-1674. https://doi.org/10.1021/es4045852

[94] L. Yang, Y. Zhang, X. Liu, X. Jiang, Z. Zhang, T. Zhang, L. Zhang, The investigation of synergistic and competitive interaction between dye Congo red and methyl blue on magnetic $MnFe_2O_4$, Chemical Engineering Journal, 246 (2014) 88-96. https://doi.org/10.1016/j.cej.2014.02.044

[95] T. Tuutijärvi, J. Lu, M. Sillanpää, G. Chen, Adsorption Mechanism of Arsenate on Crystal γ -$Fe_2 O_3$ Nanoparticles, Journal of Environmental Engineering, 136 (2010) 897-905. https://doi.org/10.1061/(ASCE)EE.1943-7870.0000233

[96] Z. Jia, K. Peng, L. Xu, Preparation, characterization and enhanced adsorption performance for Cr(VI) of mesoporous $NiFe_2O_4$ by twice pore-forming method, Materials Chemistry and Physics, 136 (2012) 512-519. https://doi.org/10.1016/j.matchemphys.2012.07.019

[97] S. Lata, S.R. Samadder, Removal of arsenic from water using nano adsorbents and challenges: A review, Journal of environmental management, 166 (2016) 387-406. https://doi.org/10.1016/j.jenvman.2015.10.039

[98] M.T. Yagub, T.K. Sen, S. Afroze, H.M. Ang, Dye and its removal from aqueous solution by adsorption: A review, Advances in Colloid and Interface Science, 209 (2014) 172-184. https://doi.org/10.1016/j.cis.2014.04.002

[99] G. Lofrano, M. Carotenuto, G. Libralato, R.F. Domingos, A. Markus, L. Dini, R.K. Gautam, D. Baldantoni, M. Rossi, S.K. Sharma, M.C. Chattopadhyaya, M. Giugni, S. Meric, Polymer functionalized nanocomposites for metals removal from water and wastewater: An overview, Water Research, 92 (2016) 22-37. https://doi.org/10.1016/j.watres.2016.01.033

[100] A. Afkhami, R. Norooz-Asl, Removal, preconcentration and determination of Mo(VI) from water and wastewater samples using maghemite nanoparticles, Colloids and Surfaces A: Physicochemical and Engineering Aspects, 346 (2009) 52-57. https://doi.org/10.1016/j.colsurfa.2009.05.024

[101] M. Auffan, J. Rose, O. Proux, D. Borschneck, A. Masion, P. Chaurand, J.-L. Hazemann, C. Chaneac, J.-P. Jolivet, M.R. Wiesner, A. Van Geen, J.-Y. Bottero, Enhanced Adsorption of Arsenic onto Maghemites Nanoparticles: As(III) as a Probe of the Surface Structure and Heterogeneity, Langmuir, 24 (2008) 3215-3222. https://doi.org/10.1021/la702998x

[102] Y.-J. Tu, C.-F. You, C.-K. Chang, T.-S. Chan, S.-H. Li, XANES evidence of molybdenum adsorption onto novel fabricated nano-magnetic $CuFe_2O_4$, Chemical Engineering Journal, 244 (2014) 343-349. https://doi.org/10.1016/j.cej.2014.01.084

[103] Y. Zhang, L. Yan, W. Xu, X. Guo, L. Cui, L. Gao, Q. Wei, B. Du, Adsorption of Pb(II) and Hg(II) from aqueous solution using magnetic $CoFe_2O_4$-reduced graphene oxide, Journal of Molecular Liquids, 191 (2014) 177-182. https://doi.org/10.1016/j.molliq.2013.12.015

[104] S. Zhang, H. Niu, Y. Cai, X. Zhao, Y. Shi, Arsenite and arsenate adsorption on coprecipitated bimetal oxide magnetic nanomaterials: $MnFe_2O_4$ and $CoFe_2O_4$, Chemical Engineering Journal, 158 (2010) 599-607. https://doi.org/10.1016/j.cej.2010.02.013

[105] N. Abbas, N. Rubab, N. Sadiq, S. Manzoor, M.I. Khan, J. Fernandez Garcia, I. Barbosa Aragao, M. Tariq, Z. Akhtar, G. Yasmin, Aluminum-Doped Cobalt Ferrite as an Efficient Photocatalyst for the Abatement of Methylene Blue, Water, 12 (2020). https://doi.org/10.3390/w12082285

[106] G. Sharma, A. Kumar, S. Sharma, M. Naushad, P. Dhiman, D.-V.N. Vo, F.J. Stadler, Fe_3O_4/ZnO/Si_3N_4 nanocomposite based photocatalyst for the degradation of dyes from aqueous solution, Materials Letters, 278 (2020) 128359. https://doi.org/10.1016/j.matlet.2020.128359

[107] P. Dhiman, A. Kumar, M. Shekh, G. Sharma, G. Rana, D.-V.N. Vo, N. AlMasoud, M. Naushad, Z.A. Alothman, Robust magnetic ZnO-Fe_2O_3 Z-scheme hetereojunctions with in-built metal-redox for high performance photo-degradation of sulfamethoxazole and electrochemical dopamine detection, Environmental Research, 197 (2021) 111074. https://doi.org/10.1016/j.envres.2021.111074

[108] X. Zhang, P. Zhang, Z. Wu, L. Zhang, G. Zeng, C. Zhou, Adsorption of methylene blue onto humic acid-coated Fe_3O_4 nanoparticles, Colloids and Surfaces A: Physicochemical and Engineering Aspects, 435 (2013) 85-90. https://doi.org/10.1016/j.colsurfa.2012.12.056

[109] Y. Yao, S. Miao, S. Yu, L.P. Ma, H. Sun, S. Wang, Fabrication of Fe3O4/SiO2 core/shell nanoparticles attached to graphene oxide and its use as an adsorbent, J Colloid Interface Sci, 379 (2012) 20-26. https://doi.org/10.1016/j.jcis.2012.04.030

[110] A. Middea, L.S. Spinelli, F.G. Souza Jr, R. Neumann, T.L.A.P. Fernandes, O.d.F.M. Gomes, Preparation and characterization of an organo-palygorskite-Fe3O4 nanomaterial for removal of anionic dyes from wastewater, Applied Clay Science, 139 (2017) 45-53. https://doi.org/10.1016/j.clay.2017.01.017

[111] H. Yan, H. Li, H. Yang, A. Li, R. Cheng, Removal of various cationic dyes from aqueous solutions using a kind of fully biodegradable magnetic composite microsphere, Chemical Engineering Journal, 223 (2013) 402-411. https://doi.org/10.1016/j.cej.2013.02.113

[112] G. Xie, P. Xi, H. Liu, F. Chen, L. Huang, Y. Shi, F. Hou, Z. Zeng, C. Shao, J. Wang, A facile chemical method to produce superparamagnetic graphene oxide–Fe_3O_4 hybrid composite and its application in the removal of dyes from aqueous solution, Journal of Materials Chemistry, 22 (2012) 1033-1039. https://doi.org/10.1039/C1JM13433G

[113] C. Ren, X. Ding, H. Fu, C. Meng, W. Li, H. Yang, Preparation of amino-functionalized $CoFe_2O_4$@SiO_2 magnetic nanocomposites for potential application in absorbing heavy metal ions, RSC Advances, 6 (2016) 72479-72486. https://doi.org/10.1039/C6RA13304E

[114] P. Xiong, L. Wang, X. Sun, B. Xu, X. Wang, Ternary Titania–Cobalt Ferrite–Polyaniline Nanocomposite: A Magnetically Recyclable Hybrid for Adsorption and

Photodegradation of Dyes under Visible Light, Industrial & Engineering Chemistry Research, 52 (2013) 10105-10113. https://doi.org/10.1021/ie400739e

[115] T. Jiang, Y.-d. Liang, Y.-j. He, Q. Wang, Activated carbon/$NiFe_2O_4$ magnetic composite: A magnetic adsorbent for the adsorption of methyl orange, Journal of Environmental Chemical Engineering, 3 (2015) 1740-1751. https://doi.org/10.1016/j.jece.2015.06.020

[116] I. Khosravi, M. Eftekhar, Characterization and evaluation catalytic efficiency of $NiFe_2O_4$ nano spinel in removal of reactive dye from aqueous solution, Powder Technology, 250 (2013) 147-153. https://doi.org/10.1016/j.powtec.2013.10.021

[117] W. Konicki, D. Sibera, E. Mijowska, Z. Lendzion-Bieluń, U. Narkiewicz, Equilibrium and kinetic studies on acid dye Acid Red 88 adsorption by magnetic $ZnFe_2O_4$ spinel ferrite nanoparticles, Journal of Colloid and Interface Science, 398 (2013) 152-160. https://doi.org/10.1016/j.jcis.2013.02.021

[118] F. Bagheban Shahri, A. Niazi, Synthesis of modified maghemite nanoparticles and its application for removal of Acridine Orange from aqueous solutions by using Box-Behnken design, Journal of Magnetism and Magnetic Materials, 396 (2015) 318-326. https://doi.org/10.1016/j.jmmm.2015.08.054

[119] D. Chen, Z. Zeng, Y. Zeng, F. Zhang, M. Wang, Removal of methylene blue and mechanism on magnetic γ-Fe_2O_3/SiO_2 nanocomposite from aqueous solution, Water Resources and Industry, 15 (2016) 1-13. https://doi.org/10.1016/j.wri.2016.05.003

[120] M.A. Shaker, Adsorption of Co(II), Ni(II) and Cu(II) ions onto chitosan-modified poly(methacrylate) nanoparticles: Dynamics, equilibrium and thermodynamics studies, Journal of the Taiwan Institute of Chemical Engineers, 57 (2015) 111-122. https://doi.org/10.1016/j.jtice.2015.05.027

[121] S.V. Bhosale, N.S. Kanhe, S.V. Bhoraskar, S.K. Bhat, R.N. Bulakhe, J.-J. Shim, V.L. Mathe, Micro-structural analysis of $NiFe_2O_4$ nanoparticles synthesized by thermal plasma route and its suitability for BSA adsorption, Journal of Materials Science: Materials in Medicine, 26 (2015) 216. https://doi.org/10.1007/s10856-015-5547-7

[122] P. Sivakumar, R. Ramesh, A. Ramanand, S. Ponnusamy, C. Muthamizhchelvan, Preparation of sheet like polycrystalline $NiFe_2O_4$ nanostructure with PVA matrices and their properties, Materials Letters, 65 (2011) 1438-1440. https://doi.org/10.1016/j.matlet.2011.02.026

[123] K. Chand Verma, V. Pratap Singh, M. Ram, J. Shah, R.K. Kotnala, Structural, microstructural and magnetic properties of $NiFe_2O4$, $CoFe_2O4$ and $MnFe_2O_4$ nanoferrite thin films, Journal of Magnetism and Magnetic Materials, 323 (2011) 3271-3275. https://doi.org/10.1016/j.jmmm.2011.07.029

[124] A. Ren, C. Liu, Y. Hong, W. Shi, S. Lin, P. Li, Enhanced visible-light-driven photocatalytic activity for antibiotic degradation using magnetic $NiFe_2O_4/Bi_2O_3$ heterostructures, Chemical Engineering Journal, 258 (2014) 301-308. https://doi.org/10.1016/j.cej.2014.07.071

[125] T. Peng, X. Zhang, H. Lv, L. Zan, Preparation of $NiFe_2O_4$ nanoparticles and its visible-light-driven photoactivity for hydrogen production, Catalysis Communications, 28 (2012) 116-119. https://doi.org/10.1016/j.catcom.2012.08.031

[126] L.-X. Yang, R.-C. Jin, Y. Liang, F. Wang, P. Yin, C.-Y. Yi, Preparation and magnetic properties of $NiFe_2O_4$–Fe_2O_3@SnO_2 heterostructures, Materials Letters, 153 (2015) 55-58. https://doi.org/10.1016/j.matlet.2015.04.012

[127] M. Su, C. He, V.K. Sharma, M. Abou Asi, D. Xia, X.Z. Li, H. Deng, Y. Xiong, Mesoporous zinc ferrite: synthesis, characterization, and photocatalytic activity with H_2O_2/visible light, J Hazard Mater, 211-212 (2012) 95-103. https://doi.org/10.1016/j.jhazmat.2011.10.006

[128] N.M. Mahmoodi, Zinc ferrite nanoparticle as a magnetic catalyst: Synthesis and dye degradation, Materials Research Bulletin, 48 (2013) 4255-4260. https://doi.org/10.1016/j.materresbull.2013.06.070

[129] X. Guo, H. Zhu, Q. Li, Visible-light-driven photocatalytic properties of $ZnO/ZnFe_2O_4$ core/shell nanocable arrays, Applied Catalysis B: Environmental, 160-161 (2014) 408-414. https://doi.org/10.1016/j.apcatb.2014.05.047

[130] E.I. Madukasi, X. Dai, C. He, J. Zhou, Potentials of phototrophic bacteria in treating pharmaceutical wastewater, International Journal of Environmental Science & Technology, 7 (2010) 165-174. https://doi.org/10.1007/BF03326128

[131] G. Moussavi, A. Alahabadi, K. Yaghmaeian, M. Eskandari, Preparation, characterization and adsorption potential of the NH4Cl-induced activated carbon for the removal of amoxicillin antibiotic from water, Chemical Engineering Journal, 217 (2013) 119-128. https://doi.org/10.1016/j.cej.2012.11.069

[132] E. Marti, E. Variatza, J.L. Balcazar, The role of aquatic ecosystems as reservoirs of antibiotic resistance, Trends in Microbiology, 22 (2014) 36-41. https://doi.org/10.1016/j.tim.2013.11.001

[133] G.Z. Kyzas, J. Fu, N.K. Lazaridis, D.N. Bikiaris, K.A. Matis, New approaches on the removal of pharmaceuticals from wastewaters with adsorbent materials, Journal of Molecular Liquids, 209 (2015) 87-93. https://doi.org/10.1016/j.molliq.2015.05.025

[134] F.J. Rivas, O. Gimeno, T. Borallho, Aqueous pharmaceutical compounds removal by potassium monopersulfate. Uncatalyzed and catalyzed semicontinuous experiments,

Chemical Engineering Journal, 192 (2012) 326-333. https://doi.org/10.1016/j.cej.2012.03.055

[135] R. Liang, S. Luo, F. Jing, L. Shen, N. Qin, L. Wu, A simple strategy for fabrication of Pd@MIL-100(Fe) nanocomposite as a visible-light-driven photocatalyst for the treatment of pharmaceuticals and personal care products (PPCPs), Applied Catalysis B: Environmental, 176-177 (2015) 240-248. https://doi.org/10.1016/j.apcatb.2015.04.009

[136] T. Chatzimitakos, C. Binellas, K. Maidatsi, C. Stalikas, Magnetic ionic liquid in stirring-assisted drop-breakup microextraction: Proof-of-concept extraction of phenolic endocrine disrupters and acidic pharmaceuticals, Analytica Chimica Acta, 910 (2016) 53-59. https://doi.org/10.1016/j.aca.2016.01.015

[137] Z. Xiong, L. Zhang, R. Zhang, Y. Zhang, J. Chen, W. Zhang, Solid-phase extraction based on magnetic core–shell silica nanoparticles coupled with gas chromatography-mass spectrometry for the determination of low concentration pesticides in aqueous samples, Journal of Separation Science, 35 (2012) 2430-2437. https://doi.org/10.1002/jssc.201200260

[138] Z. He, P. Wang, D. Liu, Z. Zhou, Hydrophilic–lipophilic balanced magnetic nanoparticles: Preparation and application in magnetic solid-phase extraction of organochlorine pesticides and triazine herbicides in environmental water samples, Talanta, 127 (2014) 1-8. https://doi.org/10.1016/j.talanta.2014.03.074

[139] J.S. Suleiman, B. Hu, H. Peng, C. Huang, Separation/preconcentration of trace amounts of Cr, Cu and Pb in environmental samples by magnetic solid-phase extraction with Bismuthiol-II-immobilized magnetic nanoparticles and their determination by ICP-OES, Talanta, 77 (2009) 1579-1583. https://doi.org/10.1016/j.talanta.2008.09.049

[140] G.R. Chaudhary, P. Saharan, A. Kumar, S.K. Mehta, S. Mor, A. Umar, Adsorption Studies of Cationic, Anionic and Azo-Dyes via Monodispersed Fe3O4 Nanoparticles, Journal of Nanoscience and Nanotechnology, 13 (2013) 3240-3245. https://doi.org/10.1166/jnn.2013.7152

[141] K.-S. Lin, K. Dehvari, Y.-J. Liu, H. Kuo, P.-J. Hsu, Synthesis and Characterization of Porous Zero-Valent Iron Nanoparticles for Remediation of Chromium-Contaminated Wastewater, Journal of Nanoscience and Nanotechnology, 13 (2013) 2675-2681. https://doi.org/10.1166/jnn.2013.7381

[142] N.M. Mahmoodi, Manganese ferrite nanoparticle: Synthesis, characterization, and photocatalytic dye degradation ability, Desalination and Water Treatment, 53 (2015) 84-90. https://doi.org/10.1080/19443994.2013.834519

[143] N. Sezgin, A. Yalçın, Y. Köseoğlu, $MnFe_2O_4$ nano spinels as potential sorbent for adsorption of chromium from industrial wastewater, Desalination and Water Treatment, 57 (2016) 16495-16506. https://doi.org/10.1080/19443994.2015.1088808

[144] J. Liao, Z. Wen, X. Ru, J. Chen, H. Wu, C. Wei, Distribution and migration of heavy metals in soil and crops affected by acid mine drainage: Public health implications in Guangdong Province, China, Ecotoxicology and Environmental Safety, 124 (2016) 460-469. https://doi.org/10.1016/j.ecoenv.2015.11.023

[145] Y.-J. Tu, C.-F. You, C.-K. Chang, Kinetics and thermodynamics of adsorption for Cd on green manufactured nano-particles, Journal of Hazardous Materials, 235-236 (2012) 116-122. https://doi.org/10.1016/j.jhazmat.2012.07.030

[146] V. Srivastava, T. Kohout, M. Sillanpää, Potential of cobalt ferrite nanoparticles ($CoFe_2O_4$) for remediation of hexavalent chromium from synthetic and printing press wastewater, Journal of Environmental Chemical Engineering, 4 (2016) 2922-2932. https://doi.org/10.1016/j.jece.2016.06.002

[147] J. Zhu, S. Wei, H. Gu, S.B. Rapole, Q. Wang, Z. Luo, N. Haldolaarachchige, D.P. Young, Z. Guo, One-Pot Synthesis of Magnetic Graphene Nanocomposites Decorated with Core@Double-shell Nanoparticles for Fast Chromium Removal, Environmental Science & Technology, 46 (2012) 977-985. https://doi.org/10.1021/es2014133

[148] B. Qiu, Y. Wang, D. Sun, Q. Wang, X. Zhang, B.L. Weeks, R. O'Connor, X. Huang, S. Wei, Z. Guo, Cr(vi) removal by magnetic carbon nanocomposites derived from cellulose at different carbonization temperatures, Journal of Materials Chemistry A, 3 (2015) 9817-9825. https://doi.org/10.1039/C5TA01227A

[149] S. Lata, P.K. Singh, S.R. Samadder, Regeneration of adsorbents and recovery of heavy metals: a review, International Journal of Environmental Science and Technology, 12 (2015) 1461-1478. https://doi.org/10.1007/s13762-014-0714-9

[150] J.-L. Kuang, L.-N. Huang, L.-X. Chen, Z.-S. Hua, S.-J. Li, M. Hu, J.-T. Li, W.-S. Shu, Contemporary environmental variation determines microbial diversity patterns in acid mine drainage, The ISME Journal, 7 (2013) 1038-1050. https://doi.org/10.1038/ismej.2012.139

[151] J. Song, H. Kong, J. Jang, Adsorption of heavy metal ions from aqueous solution by polyrhodanine-encapsulated magnetic nanoparticles, Journal of Colloid and Interface Science, 359 (2011) 505-511. https://doi.org/10.1016/j.jcis.2011.04.034

[152] N. Jordan, A. Ritter, H. Foerstendorf, A.C. Scheinost, S. Weiß, K. Heim, J. Grenzer, A. Mücklich, H. Reuther, Adsorption mechanism of selenium(VI) onto maghemite, Geochimica et Cosmochimica Acta, 103 (2013) 63-75. https://doi.org/10.1016/j.gca.2012.09.048

[153] M. Vinceti, C.M. Crespi, F. Bonvicini, C. Malagoli, M. Ferrante, S. Marmiroli, S. Stranges, The need for a reassessment of the safe upper limit of selenium in drinking water, Science of The Total Environment, 443 (2013) 633-642. https://doi.org/10.1016/j.scitotenv.2012.11.025

[154] Q. Xia, M. Xu, H. Xia, J. Xie, Nanostructured Iron Oxide/Hydroxide-Based Electrode Materials for Supercapacitors, ChemNanoMat, 2 (2016) 588-600. https://doi.org/10.1002/cnma.201600110

[155] M. Faraji, Y. Yamini, M. Rezaee, Magnetic nanoparticles: Synthesis, stabilization, functionalization, characterization, and applications, Journal of the Iranian Chemical Society, 7 (2010) 1-37. https://doi.org/10.1007/BF03245856

[156] A.-H. Lu, E.L. Salabas, F. Schüth, Magnetic Nanoparticles: Synthesis, Protection, Functionalization, and Application, Angewandte Chemie International Edition, 46 (2007) 1222-1244. https://doi.org/10.1002/anie.200602866

[157] F.S.G. Einschlag, L. Carlos, Waste Water: Treatment Technologies and Recent Analytical Developments, BoD–Books on Demand, 2013. https://doi.org/10.5772/3443

[158] M. Kumar, H. Singh Dosanjh, Sonika, J. Singh, K. Monir, H. Singh, Review on magnetic nanoferrites and their composites as alternatives in waste water treatment: synthesis, modifications and applications, Environmental Science: Water Research & Technology, 6 (2020) 491-514. https://doi.org/10.1039/C9EW00858F

[159] U. Lamdab, K. Wetchakun, W. Kangwansupamonkon, N. Wetchakun, Effect of a pH-controlled co-precipitation process on rhodamine B adsorption of $MnFe_2O_4$ nanoparticles, RSC advances, 8 (2018) 6709-6718. https://doi.org/10.1039/C7RA13570J

[160] D.H.K. Reddy, S.-M. Lee, Application of magnetic chitosan composites for the removal of toxic metal and dyes from aqueous solutions, Advances in Colloid and Interface Science, 201-202 (2013) 68-93. https://doi.org/10.1016/j.cis.2013.10.002

[161] S. Kulkarni, J. Kaware, Regeneration and recovery in adsorption-a review, International Journal of Innovative Science, Engineering & Technology, 1 (2014) 61-64.

[162] G. Busch, Biogas Technology, in: Bioprocessing Technologies in Biorefinery for Sustainable Production of Fuels, Chemicals, and Polymers, pp. 279-292. https://doi.org/10.1002/9781118642047.ch15

Materials Research Foundations **112** (2021) 279-310 https://doi.org/10.21741/9781644901595-8

Chapter 8

Comprehensive Study on Piezo-Phototronic Effect: Way Forward for Ferrite Based Solar Cells

Ritesh Verma[1,2], Ankush Chauhan[1,2], Rajesh Kumar[1,2,3*]

[1]School of Physics and Materials Science, Shoolini University of Biotechnology & Management Sciences, Bajhol-Solan (HP)-173212, India

[2]Himalayan Centre of Excellence in Nanotechnology, Shoolini University of Biotechnology & Management Sciences, Bajhol-Solan (HP)-173212, India

[3]Himalayan Centre of Excellence in Renewable Energy, Shoolini University of Biotechnology & Management Sciences, Bajhol-Solan (HP)-173212, India

*rajeshsharma@shooliniuniversity.com

Abstract

The wurtzite structured materials possess the inner-crystal piezopotential which comprehensively tune or control the charge carrier generation, separation, transportation, and recombination in optoelectronic devices. The piezo-phototronic effect provides a new working principle to the current traditional devices. Its effect on solar cells progresses made by the eminent scientists, and researchers in the field of piezo-phototronic modulated solar cells such as semiconductor-based, perovskite-based, multiple quantum-well (MQW) based and core/shell based. It has been described that the piezo-phototronic effect has improved the power conversion efficiency (PCE) of the solar cells, and the effect of temperature on the piezo-phototronic devices. Ferrite-based materials have emerged as a viable candidate for use in solar cells. As a result of this research, ferrite-based nanomaterials can now be used in solar cell applications, as well as the piezophototronic effect on these materials can be investigated. The piezo-phototronic effect is a novel subject that offers a new platform for material science and electronics to investigate.

Keywords

Wurtzite, Piezopotential, Piezo-Phototronic Effect, Multiple Quantum-Well, Power Conversion Efficiency

Materials Research Foundations **112** (2021) 279-310 https://doi.org/10.21741/9781644901595-8

Contents

1. Introduction

Perovskite structured materials such as Lead zirconate titanate Pb(Zr Ti)O3, Rochelle salt, Topaz, Tourmaline, Cane sugar, Berlinite ($AlPO_4$), Barium Titanate ($BaTiO_3$), Lead Titanate ($PbTiO_3$), Lithium Niobate ($LiNbO_3$) are of great interest in sensors, energy harvesting, and energy generation [1]. Their significant piezoelectric coefficient makes them act as an insulator. Hence, they are studied in the ceramic community [2, 3]. For electronic and photonic applications like energy harvesters, photodetectors, solar cells,

Materials Research Foundations **112** (2021) 279-310 https://doi.org/10.21741/9781644901595-8

etc., Wurtzite structured materials such as ZnO, ZnS, GaN, InN CdS, etc. are gaining much attention, because of their small piezoelectric coefficient and exceptional optical, electrical and piezoelectric properties. Prof. Zhong Lin Wang introduced the concept of coupling between piezoelectric, semiconducting, optical properties of the material, and named it as a piezo-phototronic effect (Figure 1) and its applications in energy science [4, 5]. The piezo-phototronic effect explicitly supports the phenomenon of charge generation, separation, transportation, and recombination, and comprehensively modulates or tunes these processes by the induced inner crystal piezoelectric potential as a gate voltage [6-8]. This effect has propounded the concept of new technology with significant improvement in optoelectronic devices such as solar cells, photodetectors, light emitting diodes, etc. Which are called piezo-phototronic devices? The piezo-phototronic effect has shown much technological advancement for traditional devices by improving their performance. The interest of researchers in the piezo-phototronic effect is growing day by day. Figure 2 provides the data of several publications from 2010 to 2019 (March) related to this stipulated effect. The popularity of the piezo-phototronic effect is extraordinarily increasing and has paved the way to explore new technologies in the field of optoelectronics, renewable energy, electronics, etc. There are various compositions of materials on which the piezo-phototronic effect has been studied for different applications as shown in Table 1. This table gives an idea about the materials which comes under the purview of the piezo-phototronic effect, and the basic parameters that are required to study this effect. Moreover, an effort has been made to summarize almost all the materials with their applications.

This work is split into four extensive sections. First, is the recent progress made by the piezo-phototronic effect, and the different materials used so far. Second, the underlying theory, and fundamental concepts of piezo-phototronic effect. In the third section, we have focused on the fundamentals of piezo-phototronic effect modulated solar cells, the tremendous progress made by various researchers and eminent scientists on different types of solar cells. At last, a brief description to show the dependence of the piezo-phototronic effect on temperature has been conveyed.

2. Piezo-phototronic effect on ferrites

Among all the ferrites, BiFeO3 is one of the most prominent material on which piezo-phototronic effect has been studied so far. BFO has aroused great interest in the past few decades due to its visible light photovoltaic effect, multiferroic properties at room temperature and related potential applications. BFO thin films have the advantage of low cost and accuracy to control the chemical components. As far as other ferrites are concerned, Ba-doped zinc ferrite is again an interesting candidate for photovoltaic

Materials Research Foundations **112** (2021) 279-310 https://doi.org/10.21741/9781644901595-8

applications. But there is no literature available of piezo-phototronic effect on ferrites other then BFO for solar energy applications.

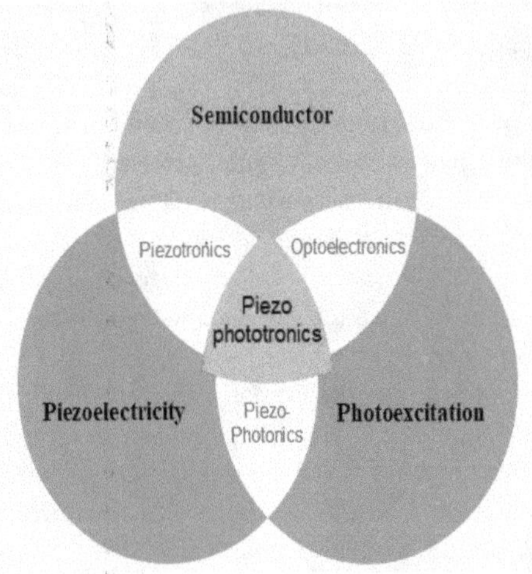

Figure 1: Schematic diagram showing the three-way coupling of piezoelectricity, photoexcitation and semiconducting effect which gives rise to piezo-phototronic [9].

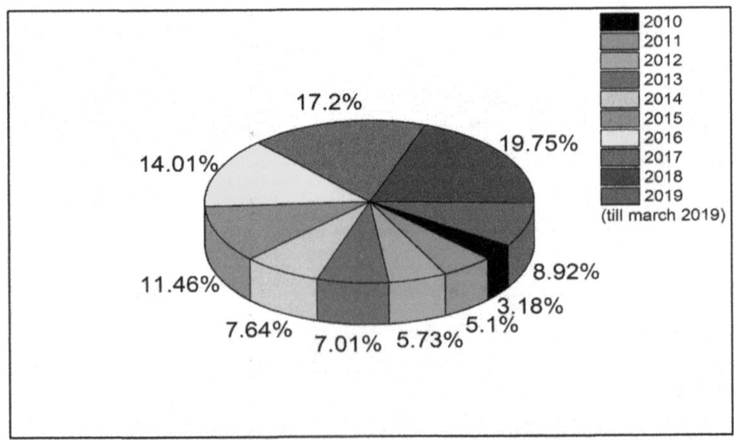

Figure 2: Number of publications piezophototronics has produced from its commencement.

Materials Research Foundations **112** (2021) 279-310 https://doi.org/10.21741/9781644901595-8

Table 1: This table signifies some of the materials on which piezo-phototronic effect has been studied.

S.No	Composition/material	Application	Reference
1.	$CH_3NH_3PbI_3/ZnO$ nanowires	Perovskite solar cells	[10]
2.	GaN/ZnO naowire	The led-based pressure sensor array	[11]
3.	InGaN/GaN SQW heterostructure membrane	Direct lift off and piezo-phototronic study	[12]
4.	ZnO flexible thin film	UV photodetectors	[13]
5.	MoS_2 single layer	NO_2 sensor	[14]
6.	CdTe microdots	Photodetectors	[15]
7.	CdTe nanowires	Photodetectors	[16]
8.	Cu(In, Ga)Se_2(CIGS) heterostructure	Flexibl heterostructured photdetectors	[17]
9.	Si/ZnO/PEDOT: PSS tri-layer heterostructure	Photodetectors using both positive and negative piezo charges	[18]
10.	WSe_2-CdS	Flexible Photodetector	[19]
11.	$TiO_2/BaTiO_3$/Au heterostructure	Photodynamic bacteria killing and wound healing	[8]
12.	$CH_3NH_3PbI_3$ single crystal	Photodetectors	[20]
13.	Ga_2O_3-ZnO heterojunction microwire	Deep UV light photodetectors	[21]
14.	MoS_2/WSe_2	Strain tunable van der waals photdiodes	[22]
15.	ZnO based Cu(In, Ga)Se_2 heterostructure	Flexible solar cells	[23]
16.	Si/CdS	Self-powered flexible photodetectors	[24]
17.	Ag/ZnO nanotetrapods	Piezo photocatalytic activity	[25]
18.	Si microwires/ZnO nanofilms	Flexible light emitting diodes	[26]
19.	n-ZnO/p-NiO films	Photodetectors	[27]
20.	GaN microwires	UV Light Emitting Diodes	[28]
21.	p-Si/n-ZnO	Visible NIR broadband photodiodes	[29]
22.	CdS_xSe_{1-x} nanowires	Photosensors	[30]
23.	CdS/P3HT microwires	Self-powered UV/Visible/NIR photodetectors	[31]

24. ZnO/ZnS core/shell nanowires	Electrical transport and photo sensing application	[32]
25. p-CuO/n-MoS_2 heterostructure	Photodetector with enhanced photoresponse	[33]
26. ZnO/P3HT	Solar cells	[34]
27. CdSe nanowires	Photodetectors	[35]
28. $BiFeO_3$/ZnO	Photovoltaics	[36]
29. $BiFeO_3$/TiO_2	Photocatalytic activity	[37]
30. Ni-(Ba,Gd)FeO_3	Dye sensitized solar cells	[38]

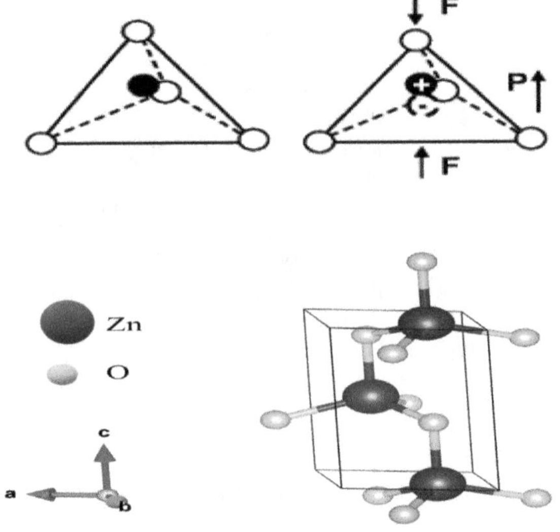

Figure 3: Piezopotential in Wurtzite crystal: atomic model of ZnO crystal [42].

3. Theory of piezophototronics

To discuss the structure and piezopotential generated in Wurtzite family we have used the structure of ZnO. It is an II-VI compound semiconductor. Its ionicity resides at the borderline between the covalent and ionic semiconductors. It is non-centrosymmetric and has a hexagonal structure with a large anisotropy in c-axis or perpendicular to c-axis, which gives rise to piezoelectricity. In a single unit cell of ZnO, Zn^{2+} and O^{2-} ions are tetrahedrally co-ordinated at the corners of a tetrahedron with lattice parameters a and c in ratio $\frac{c}{a}=1.633$, and the centre of two ions overlap with each other, due to which it is non-polarized in stress-free condition [39]. When we apply the stress on the apex of the unit cell, the center of two charges gets separated, results in the dipole moment as shown in Figure 3. Valuable addition of all these tiny dipoles results in a potential macroscopic drop in the direction of applied stress. This is piezoelectric potential or piezopotential [40, 41]. This piezopotential in the crystal is created by the non-mobile and non-annihilative ionic charges, and its magnitude depends upon the stress applied.

Lippman theory has been used to calculate the distribution of piezopotential in deformed ZnO crystal [43]. Following formula has been used to calculate the piezopotential on the surface of the whole material, i.e. nanowire.

$$\phi_{max}(T,C) = \pm\frac{3}{4(K_0+K_\perp)}\left[e_{33}-2(1+\upsilon)e_{15}-2\upsilon e_{15}\right]\frac{a^3}{l^3}\upsilon_{max} \quad (1)$$

Where υ_{max} is the maximum deflection of the nanowire tip; υ is the Poisson's ratio, $K_{11}=K_{22}=K_\perp$, K_0 is the permittivity in a vacuum, T is the tensile side, C is the compressive side. Also, the Finite Element Analysis (FEA) has been used to investigate the piezoelectric and to depict the motion of charge carriers [44]. Two typical configurations have been used to investigate the distribution of piezopotential; transversely deflected nanowires, and the axially strained nanowires. The transversely deflected nanowires usually are utilized in energy harvesting applications [45], while axially strained nanowires are used in piezotronic applications on flexible substrates [46].

The piezophototronic effect is predominantly described by the semiconductor and piezoelectricity in the form of electrostatic equations, current density equations, continuity equations, and piezoelectric equations. The electrostatic behaviour of the charges in the piezoelectric semiconductors is described by Poisson's equation [47].

$$\nabla^2\psi_i = \frac{\rho(r)}{\varepsilon_s} \quad (2)$$

Where ψ_i is the electric potential distribution, $\rho(r)$ is the charge density distribution and ε_s is the permittivity of the material. The drift and diffusion current density equations that correlate the local fields charge densities and local currents are given by the following equations [47].

$$\begin{aligned} J_n &= q\mu_n nE + qD_n\nabla n \\ J_p &= q\mu_p pE - qD_p\nabla p \end{aligned} \quad (3)$$

Materials Research Foundations **112** (2021) 279-310 https://doi.org/10.21741/9781644901595-8

moreover, the total current density of electrons and holes is given by,

$$J = J_n + J_p \tag{4}$$

Where, μ_n and μ_p, D_n and D_p are the charge mobility, diffusion coefficient of electrons and holes respectively. The charge transport under the driving of a field is described by the continuity equations [47],

$$\frac{\partial n}{\partial t} = G_n - U_n + \frac{1}{q}\nabla . J_n$$
$$\frac{\partial p}{\partial t} = G_p - U_p - \frac{1}{q}\nabla . J_p \tag{5}$$

Where, G_n and G_p are the generation rates of electrons and holes. U_n Moreover, U_p are the recombination rates of electrons and holes respectively. Also, we know that the piezoelectricity is the predominant factor in piezo-phototronic, so the relationship between polarization and mechanical strain is given by [48] $(P)_i = (e)_{ijk}(S)_{jk}$ where $(e)_{ijk}$ is the third order piezoelectric tensor, and mechanical strain is given by $(S)_{jk}$. These basic equations will be used for discussing the p-n junction or Metal-semiconductor junction with applied compressive stress or tensile strain.

4. Bandgap modulation by the piezo-phototronic effect

The p-n junction within the same crystal forms the homojunction, whereas two different crystals form heterojunction [49]. This heterojunction may form p-n junction or metal-semiconductor contact. This signifies that bandgap alignment becomes more critical [50]. In the upcoming section, a brief description has been given to elaborate on how piezopotential or a piezo-phototronic effect modulates or control the band gap of p-n junction or metal-semiconductor contact.

4.1 Modulation of p-n junction

A p-type and n-type semiconductor formed a junction where the holes of p- side, and electrons of n- side, diffused across the junction and redistribute to balance the local potential. They tend to reach thermal equilibrium. The inter-diffusion and recombination of the electrons and holes formed a charge depletion zone which was carrier free [51]. This carrier free zone significantly increases the piezoelectric effect, because of the preservation of piezo charges. Let n-side is piezoelectric then local net negative piezo-charges remained preserved at the junction when the compressive strain was applied on it. The piezopotential on the piezoelectric side (n-side) tended to raise the local band and introduced a gradual slope to the band structure (shoulder like structure) [9]. Further, the direction of the strain was reversed, the piezo charges at the interface created the downward bending in the local band. The modification of the local band at the junction introduced a significant change in the local band structure. This modification in the local band was effective for trapping the holes so that the e^-- h^+ separation rate was effectively enhanced. Furthermore, the slow slop tends to change the mobility of the carriers moving towards the junction [51]. For a p-n junction of two distinct materials with different band gaps, local piezo charges have affected the band profile and the transportation of charge carriers across the interface [52]. For p-n junctions, Shockley theory gave the fundamentals of I-V characteristics [53]. Schematic energy band diagram of the p-n junction has been shown in Figure 4.

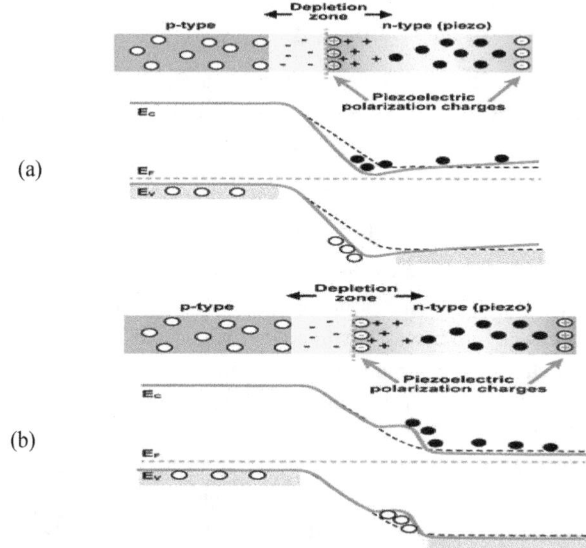

Figure 4: Schematic energy band diagram of the p-n junction. The red curve shows the deformation of the energy band under strain. (a) during the compression a gradual slope can be observed, (b) during the tensile strain sharp decline in the energy band can be observed [54].

The n-type is considered piezoelectric and p-type as non-piezoelectric. Then the built-in-potential is given by the expression,

Materials Research Foundations **112** (2021) 279-310 https://doi.org/10.21741/9781644901595-8

$$\psi_{bi} = q.(2\varepsilon_S)^{-1}[N_A W_{Dp}^2 + \rho_{piezo} W_{piezo}^2 + N_D W_{Dn}^2] \quad (6)$$

Where, N_D and N_A are the donor and acceptor concentration respectively W_{Dp} and W_{Dn} are the width of the depletion layer in p-side and n-side respectively, ρ_{piezo} is the density of polarization charge, W_{piezo} is the width of distribution zone of the piezo charges at the interface, ε_S is the relative dielectric constant [55]. It was observed from the above-stated equation that piezo-charges has a precise effect on built-in-potential. Thus, it can be modulated by induced strain. The total current density of p-n junction, where the n-type side has an abrupt junction with the donor concentration N_D, is given by,

$$J = J_{C0}.\exp[q^2 \rho_{piezo} W_{piezo}^2 (2kT\varepsilon_S)^{-1}].\exp[qV.(kT)^{-1} - 1] \quad (7)$$

J_{C0} Is the saturation current density with the absence of piezopotential [55]. Thus, it was concluded that current transport across the p-n junction largely depends upon the exponential of piezo-charges so current transportation can be modulated by the piezopotential that is the magnitude and polarity of the strain.

4.2 Modulation of metal-semiconductor contact

For semiconductor electronics and optoelectronics, the Metal-Semiconductor contact is a basic structure [56]. When metal –semiconductor (M-S) contact takes place, redistribution of charges occurs across the interface due to the overlapping of wave functions of both the metal and the semiconductor [57, 58]. At the equilibrium, the Fermi level gets aligned on each side of the interface which results in the introduction of an energy barrier called a Schottky barrier with barrier height $e\varphi_{Bn}$. This Schottky barrier between a metal electrode and semiconductor is a result of the difference between the metal-vacuum work function and the semiconductor-vacuum electron affinity [47]. For n-type semiconductor, the Schottky barrier height can be determined by $e\varphi_{Bn} = e(\varphi_m - \chi)$. Schottky barrier dictates the transport of charge carriers across the M-S interface. For n-type semiconductor, current can only pass through the interface if the external positive bias is higher than the threshold value φ_i. For semiconducting piezoelectric material, the negative piezopotential got induced at the semiconducting side with the application of external strain and increased the local Schottky band height. On the other hand, the local Schottky band height decreases with the reversal of polarity of applied strain [59, 60]. Thus, it has been observed that the piezo-charges and piezopotential present in the semiconducting side have the significant effect on the Schottky barrier height due to the crystallographic orientation of material and the polarity of applied strain [45, 59, 61-64].

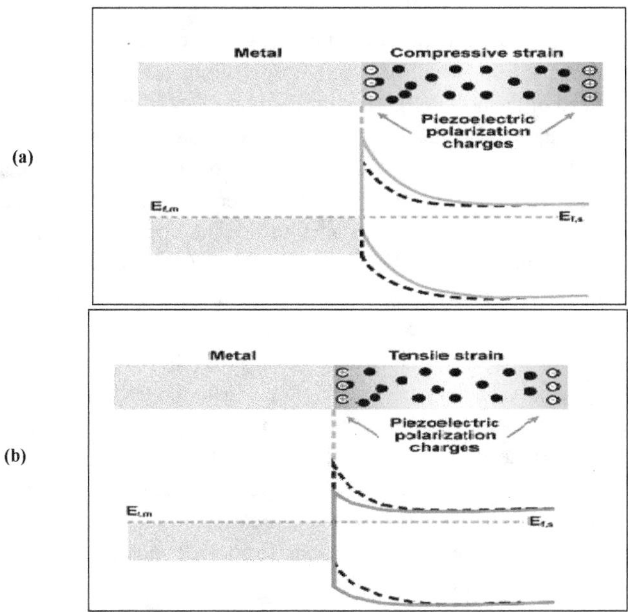

Figure 5: Schematic energy band diagram of metal-semiconductor contact (a) Compressive strain increases the local Schottky barrier height, shown by raised by the blue curve. (b) Tensile strain decreases the local Schottky barrier height shown by the declining red curve [54].

During the excitation of the M-S contact with a photon having energy more significant than the band gap of the semiconductor, electron-hole pair got generated around the contact. The presence of free carriers at the interface can effectively reduce the Schottky barrier height [65].

Piezo-phototronic effect has applications in solar cells [10], photodetectors [66], LEDs [26] etc. However, solar cells are gaining most of the attraction, because of the increasing demand for energy. In the following discussion, an effort has been made to ascribe how piezo-phototronic effect has controlled the efficiency of various types of solar cells.

Materials Research Foundations **112** (2021) 279-310 https://doi.org/10.21741/9781644901595-8

5. Piezo-phototronic effect modulated solar cells

One of the greatest challenges for the development is environmental pollution and to fulfil the demand of energy along with [67, 68]. To fulfil our energy needs sustainable development is the need of the hour [69]. The sources like solar, wind, biomass, etc. have attracted the scientists, and environmentalists to circumvent or to mitigate these burning issues [70-74]. As we know, the sun is a ubiquitous source of energy, and solar cells have shown a mesmerizing effect on the scientific world in order to address some of the issues. The basic working principle of the solar cells is to use the large electric field in the depletion region to assist the separation of electron-hole pair generated by the incident photon. However, the piezoelectric charges created at the junction under strain has found a significant effect on the performance of solar cells, and piezopotential significantly modifies the band structure at the interface controlling the process from charge generation to recombination. The key parameters to evaluate the performance of a solar cell: short circuit current, open circuit voltage, maximum output power, fill factor, and ideal power conversion efficiency [75-78]. The light absorption, electron-hole generation, surface interface recombination, and surface barrier are the processes that constitute the basis of power conversion efficiency (PCE) and the external quantum efficiency [79]. The EQE is the ratio of photo-generated electrons to the per incident monochromatic photon which is given by the relation,

$$EQE(\%) = \left(\frac{1240 J_{SC}}{\lambda I_0} \right) \times 100 \tag{8}$$

Where I_o is the incident intensity of the light source, J_{SC} is the short circuit current density and λ is the radiation length [26]. The PCE is the general efficiency of the solar cell, which is defined as the ratio of generated electricity to the incoming light and given by the relation,

$$PCE = \left(\frac{J_{solar} V_{0C} FF}{P_{in}} \right) \tag{9}$$

Where P_{in} is the power and FF is the fill factor. The fill factor is defined as the ratio of the maximum output power to the product of short circuit current and open circuit voltage, and given by the relation,

$$FF = \frac{P_m}{J_{solar} V_{0C}} = \frac{J_m V_m}{J_{solar} V_{0C}} \tag{10}$$

Here J_m is the current density at maximum output power, V_m is the voltage at maximum output power and P_m is the maximum output power. These are p-n junction and M-S contact based solar cells on which we are discussing in further sections.

5.1 p-n junction based solar cells

In 2012, Zhang et al. proposed an analytical and theoretical calculations to n-type ZnO nanowire solar cell which was based on the p-n junction where n-type, i.e., ZnO was piezoelectric and p-type was non-piezoelectric [80]. Thus, under the compressive stress or tensile strain, it was observed that the induced piezopotential could modify the band structure at the interface. This ultimately increases or decreases the e^--h^+ pair separation based on the polarity of piezopotential, which resulted in the increased or decreased performance of solar cells. Further, it can be elaborated as an ideal p-n junction having two typical types of effects; one is piezo resistance, and another is piezotronic effect [9, 81]. These effects males the output of the solar cell depends on the polarity of the crystal once it is subjected to an external strain, suggesting the key role played by the piezo-phototronic effect on solar cell output [82, 83]. To analyze the piezo-phototronic effect on solar cells, let us assume that the generation rates of e^--h^+ are constant [84],

$$G_n = G_p = J_{solar} \cdot [q(L_n + L_p)]^{-1} \tag{11}$$

Where, J_{solar} is short circuit current density, L_n and L_p represents the e^- and h^+ diffusion lengths respectively. Moreover, $U_n = U_p = 0$, which means there is no photon emission. For ideal p-n junction piezoelectric solar cells, the total current density can be written as [85],

$$J = J_{pn}[\exp\left(\frac{qV}{kT}\right) - 1] - J_{solar} \tag{12}$$

Where, $J_{pn} = J_{pn0} \cdot \exp[q^2 \rho_{piezo} W_{piezo}^2 (2kT\varepsilon_S)^{-1}]$. Therefore, the total current density is,

$$J = J_{pn0} \cdot \exp[q^2 \rho_{piezo} W_{piezo}^2 (2kT\varepsilon_S)^{-1}].[\exp\left(\frac{qV}{kT}\right) - 1] - J_{solar} \tag{13}$$

The above equation gives that J_{pn} (saturation current density) will decrease exponentially with the generation of piezo charges at the interface. The open circuit voltage of piezoelectric solar cells could be obtained as,

$$V_{0C} = \frac{kT}{q}.\ln\left[\frac{J_{solar}}{J_{pn}} + 1\right] \text{For J=0} \tag{14}$$

Here, $J_{solar} >> J_{pn}$

Therefore,

$$V_{0C} = \frac{kT}{q}.\ln\left[\frac{J_{solar}}{J_{pn}}\right] = \frac{kT}{q}.\ln\left[\frac{J_{solar}}{J_{pn}}\right] + q^2 \rho_{piezo} W_{piezo}^2 (2kT\varepsilon_S)^{-1} \tag{15}$$

Thus, the above equation suggests that the open circuit voltage V_{0C} is not only affected by the magnitude of stress, but also the polarity of the stress applied. This effect can be applied on nanowires, thin films or bulk solar cells.

5.2 M-S contact based solar cells

For M-S structure, let the semiconducting side be n-type ZnO nanowire. Thus, the total current density across the M-S interface could be written as [85],

$$J = J_{MS}\cdot\left[\exp\left(\frac{qV}{kT}\right) - 1\right] - J_{solar} \tag{16}$$

Where, $J_{MS} = J_{MS0}.\exp[q^2 \rho_{piezo} W_{piezo}^2 (2kT\varepsilon_S)^{-1}$ and,

$$J_{MS0} \approx q^2 D_n N_C (kT)^{-1}.\sqrt{[2qN_D(\psi_{bi0} - V)\varepsilon_S^{-1}}.\exp[-qe\varphi_{Bn0}.(kT)^{-1}] \tag{17}$$

J_{MS} represents the saturation current density in the case with piezo-charges, J_{MS0} represents the saturation current density without piezo charges ψ_{bi0} and $e\varphi_{Bn0}$ is the built-in potential and Schottky barrier height without piezo charges. The open circuit voltage of an M-S contact can be expressed as,

$$V_{0C} \approx \frac{kT}{q}\left[\ln\left(\frac{J_{solar}}{J_{MS0}}\right)\right] - q^2 \rho_{piezo} W_{piezo}^2 (2kT\varepsilon_S)^{-1} \tag{18}$$

This equation suggests that tensile strain induces negative piezopotential at the interface which increases the barrier height and hence decreases the saturation current J_{MS}. It increases the open circuit voltage V_{0C}. The opposite phenomenon will occur under compressive stress. These theoretical simulations using the piezo-phototronic effect not only provide a basic understanding of solar cells but also assist in designing these solar cells [86]. Also, it was observed that light absorption and photon excitation are the two important factors for the efficiency of solar cells. To enhance these factors, the possible ways are light trapping and anti-reflective coating [87-90]. In recent studies, it was observed that the piezo-phototronic effect could provide the benefit of anti-reflective and light trapping properties in solar cells and hence eliminates the need for anti-reflective coatings [91-94].

6. Effect on solar cells

This concept has provided a clear idea about how the piezo-phototronic effect is going to modulate the solar cells based on p-n junction and metal-semiconductor contact. The following discussion will present the effect of piezo-phototronic on the power conversion efficiency (PCE) of different type of solar cells.

6.1 Heterostructure based solar cells

Zhang et al. (2020) [36] investigated the piezophototronic effect on La-doped BFO/ZnO heterojunction and observed that rise time of ZnO significantly modulated from 153.7 ms to 61.28 ms. In this study they used la-doped BFO as p-type film and ZnO as n-type film. This work not only offers a feasible and common approach for improving the heterojunction based on ferrite film by piezophototronic effect but also facilitates the development of human machine interaction systems investigated with photoexcitation.
In 2013, Wen *et. al.* [34] observed the enhancement in solar cell performance with the use of the piezo-phototronic effect. They observed that the efficiency along with the fill factor for solar cells varies with the variation in the intensity of solar radiation, because of their limited charge separation ability. Thus, to improve this, they demonstrated the piezo-phototronic effect in the ZnO/P3HT solar cell system with four groups depending upon the growth rate and mechanics. The highest current density was 0.258 mA/cm^2 and PCE was 0.203% under 0.32% tensile strain. However, with compressive strain results were found to decrease, which indicates that the piezo-phototronic effect is direction dependent. It was observed that the alignment of grown ZnO nanowires plays an important role to define the parameters of the piezo-phototronic effect.

Zhu. L *et al.*, [95] investigated and designed the silicon-based nano-hetero-structure (p$^+$-Si/p-Si/n$^+$-Si (and n-Si)/n-ZnO nanowire array) solar cell with increased PCE. Due to the enhanced light absorption efficiency through nanowire arrays, it was found that the performance of the p+-Si/p-Si/n+-Si/n-ZnO nanowire array solar cells has been enhanced significantly when subjected to 800 KPa static compressive pressure. With the vast area, it is easy for nanowires to collect the exciting photon carriers. It means the size of the electrode plays a vital role in the performance of the solar device. Larger the area, higher will be the performance of the device. Although it was observed that the nanowire array reduces the collection ability, the increase in the light collection ability of nanowires can circumvent or tackle this problem easily.

$Cu(In,Ga)Se_2$ (CIGS)-based heterostructures have been considered to be the most prospective thin film solar cells as they exhibit high PCE, low price, sizeable optical absorption coefficient ($\geq 10^5 cm^{-1}$), tunable direct band gap (1.04eV-1.67eV), long term thermal and electrical stability with tremendous environmental benefits[1-3]. However, Qiao *et al.,* [23] coupled piezo-phototronic effect on CIGS solar cells. ZnO nanowires were used on CIGS as a replacement of ZnO thin film and observed that PCE quickly improved to 11.40% from 9.8%. By replacing the substrate from ITO glass to steel substrate improvement in PCE was observed from 5.43% to 5.96%. They also observed the effect of the length of nanowires on the efficiency and found that longer nanowires have less ability to transport and collect the photo-excited carriers for ZnO nanowires which suggests that moderate length nanowire sample, i.e., 850nm, have shown best PV properties.

6.2 Multiple Quantum Well (MQW) solar cells

$In_xGa_{1-x}N$ is one of the fascinating materials for designing the multiple quantum-well solar cells, because of the wide tunable bandgap range, i.e., from 0.7eV to 3.4eV for GaN [99-101]. It has been observed that InGaN based solar cells can exhibit the efficiency of more than 50% when it has 40% of the In content [102]. Jiang *et al.,* [103] successfully analysed the piezo-phototronic effect on InGaN/GaN MQW solar cell. It was observed that the parameters of solar cells were enhanced as conversion efficiency improved from 1.12% to 1.24% and fill factor decreased slightly from 57.3% to 56.7% under external stress of 0.134%. It indicated that the piezo-phototronic modulation of solar cells is highly efficient technology to enhance the numerical power conversion efficiency of InGaN/GaN MQW solar cell. The numerical model to compare the experimental and theoretical results was also established. Both results, when compared, shown considerable enhancement is photon conversion efficiency of MQW solar cell using piezo-phototronic effect.

Jiang *et al.* presented the design of InGaN/GaN MQW solar cell coupling with Ag nanoparticles, i.e. the coupling of piezo-phototronic effect with plasmonic effect, where plasmonic nanoparticles are those whose electron density can be coupled with electromagnetic radiations of a wavelength that are fare larger than the particles [104]. Power conversion efficiency increases from 0.75% to 0.98% without strain because of Ag (silver) nanoparticles as they enhance the light absorption efficiency and plasmons couples the incident light into guided modes that propagate through MQW region. When

0.152% of the strain was applied, then the PCE further improved from 0.98% to 1.25% with an overall increment of 66%.

Thus, this new coupling concept has shown effective, recoverable and straightforward method in developing the high-efficiency solar cells.

6.3 Perovskite solar cells

Perovskite materials like $CH_3NH_3PbX_3$ (X= Cl, F, I) have become the center of attraction in fabricating large scale commercial solar cells due to their low-cost fabrication, high absorbance and low carrier recombination rate with a high-power conversion efficiency of more than 22% [105-109]. Gu *et al.*,[110] analysed the effect of the piezo-phototronic effect on $CH_3NH_3PbI_3$ thin films based on simulation results. The PCE improved from 4.85% to 5.05% under the stress range from -1% to 1% and the fill factor improved from 78.5% to 79%. The high piezoelectric coefficient and tiny dielectric constant of perovskite material are suitable for PV application by comparing the $CH_3NH_3PbI_3$ with $BaTiO_3$. $BaTiO_3$ has high dielectric constant, i.e., 600, which is responsible for its less use in the PV industry. Thus, the piezo-phototronic effect has significantly improved the PCE of perovskite solar cells.

Wang *et al.*, [111] successfully described the effect of the piezo-phototronic effect on perovskite/ GaN solar cell and observed that the piezo-phototronic effect has comprehensively improved the efficiency of this solar cell. The maximum efficiency was around 25.46%, and the fill factor was 84.79% under the pressure range from -1% to 1%. The performance of GaN-perovskite solar cells with different types of solar cells was also compared and found that GaN-based perovskite solar cell has gained the highest fill factor, PCE and modulation ratio, i.e., 3.66.

In another study, Sun *et al.*, [10], fabricated a perovskite solar cell with ZnO nanowires as en electron transfer layer and observed the parameters of ZnO-Perovskite Solar cell (ZPSC) using piezo-phototronic effect. A significant enhancement in PCE with the inclination from 9.3% to 12.8% under the tensile strain of 1.88% has been found. This enhancement is due to the reason that ZnO has facilitated the separation of e- generated from $CH_3NH_3PbI_3$. It also suggests that when stress is applied on the ZPSC, the piezopotential created at the interface of ZnO, and perovskite material has significantly improved the energy level structure.

6.4 Core/Shell based solar cells

The core/shell geometry have found great attention, because of their high efficiency of charge collection by shortening the path travelled by minority carriers and high optical quality of material [112-115]. Pan *et al.,* reported the enhanced performance of n-CdS/p-Cu_2S by a factor of 70% using the piezo-phototronic effect, which offers a new concept for improving solar energy conversion efficiency. The improved efficiency has been gained under the -0.41% applied strain, but as the strain was further increased the relative PCE decreased by a factor of 15%. This decrease in relative PCE suggests that by introducing the piezo-phototronic effect, the decrease can be made minimal.

Zhu *et al.,* [116], reported the enhanced performance of the n-ZnO/p-SnS core/shell-based PV device. SnS was used as a shell, which shows p-type conductivity and easily forms the Sn vacancy defects [117-119]. Moreover, SnS is a polycrystalline material. Piezopotential was created by ZnO only. It was observed that under no stress condition the PCE was 1.2% and the fill factor was 0.43 under AM1.5 condition. When the external strain of 320 KPa was applied, the PCE increased to 1.644% and fill factor improved to 0.48. Here the enhancement is due to the piezopotential created by ZnO which supports the charge separation and band modulation at the interface of ZnO/SnS.

In another study, Liu et al. [37] investigated the catalytic activity of BFO/TiO_2 core shell nanocomposite for degradation of organic dye molecule through piezophototronic effect. They observed that BFO nanoparticles can effectively reduce the recombination rate of carriers by built in electric field. It is observed that reaction rate constant (K) of piezophototronic activity is about 560% and 388% higher than that of under the sole piezo-catalytic and photocatalytic process, respectively. Now, this trend is observed when p-type BFO and n-type TiO_2 form a heterojunction structure at the interface area, which establishes the internal electric fields on both sides to enhance the separation of electron and hole pairs under the light irradiation. Without the assistance of piezoelectric polarization in the BFO/TiO_2 nanocomposite the carriers could migrate to either side.

Ni-doped (Bi,Gd)FeO_3 ceramics have been investigated and it is observed that enhanced PV effects are attributed to the reduced bandgap, polarization modulated Schottky barriers, increased O 2p-Fe 3d orbital hybridization and nucleation of domain walls to improve the transportation of charge carriers [38]. Also, Ni-doped (Bi,Gd)FeO_3 can be potential candidate for self-powered UV photodetector.

The power conversion efficiency of the solar cell is the most important parameter which represents how much incident light is being converted into sufficient output power. To improve power conversion efficiency is the most critical issue.

7. Temperature dependence of Piezo-phototronic effect

To understand the basic physics of any phenomenon, it is essential to understand its temperature dependence [120-123]. Various devices were investigated, and it was observed that with the decrease in temperature from room temperature to low temperature, the width of the depletion region increases [124-126]. GaN nanobelts were used for investigation under 0.28% tensile strain at 77K [126].

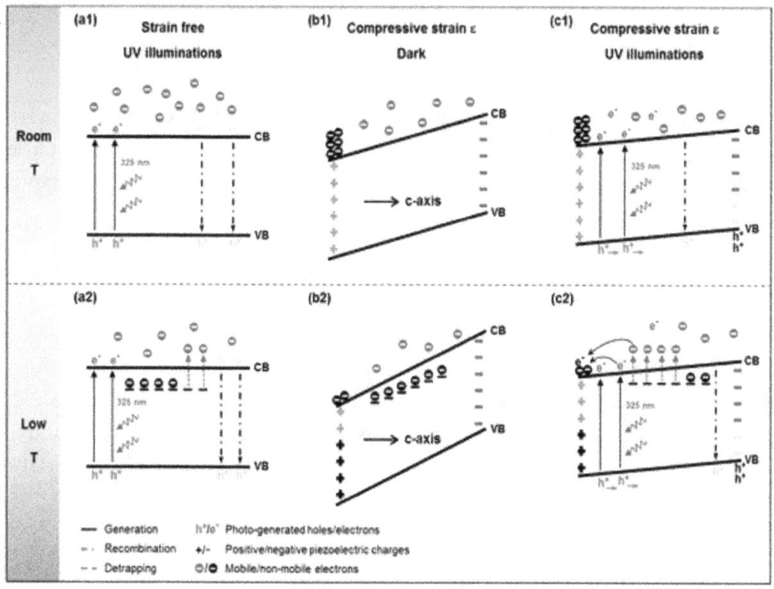

Figure 6: Temperature dependence of piezo-phototronic effect describing how piezopotential reacts at room temperature (T) and lower temperature than room temperature (T) [126, 127].

The figure shows how piezopotential affects the photocurrent with a change in temperature. At room temperature, the compressive strain at dark condition, shown in (b1), induce the piezopotential, and causes the screening effect which decreases the current. When UV illumination takes place under the same compressive strain condition, shown in (c1) the e-h pair generated by illumination of UV are readily attracted by the negative piezoelectric polarization charges. Hence, reduction in recombination occurs which increases the photocurrent. At lower temperature than room temperature, reduction of the carrier in conduction band occurs, because free carriers got trapped in sallow donor energy state, shown in (b2). Under compressive strain with UV illumination, shown in (c2), the increase and decrease of photocurrent simultaneously take place [127]. The decrease in free mobile charges with the decrease in temperature gives the freeze out effect [123, 128] and due to this, the free electrons get trapped in the narrow impurity centres which can be easily activated by the incident UV photon [129, 130]. Now for thermally activated transport, the conductivity can be expressed as [52],

$$G(T) = G_0 \exp\left[\frac{E_D}{k_B T}\right] \quad (19)$$

Where E_D is the thermal activation energy of the donor state. If the material can have polarization, then to check how the polarization charges change with the temperature, it is necessary to calculate the effective surface charges (M-S interface) with the expression [131, 132],

$$\rho_S = \sqrt{\left\{2\varepsilon_r\varepsilon_0 kT\left[N_D\left(\exp\frac{e\varphi_{Bn}}{kT}\right) + n_{bulk}\left(\exp\frac{e\varphi_0}{kT} - 1\right)\right]\right\}} \quad (20)$$

Where ρ_S is the total surface charge, N_D is the donor dopant concentration, n_{bulk} is the temperature dependent bulk carrier density which is equal to the ionized donor dopant concentration. Here, $N_D\left(\exp\frac{e\varphi_{Bn}}{kT}\right)$, represents the screening of surface charge by fixed ionized dopants and $n_{bulk}\left(\exp\frac{e\varphi_0}{kT} - 1\right)$ represents the screening by free electrons. Thus, the above numerical methods show how the temperature variation effectively compensates the piezopotential induced charges [129].

Conclusion

In this chapter, we have outlined the fundamentals of the piezo-phototronic effect, fundamentals of solar cells and recent progress made by the piezophototronic effect on solar cells. This chapter also explores the growth of the piezo-phototronic effect in terms of publications. It also describes the various materials or compositions used so far to study this effect.

As the piezo-phototronic effect is the three-way coupling and predominantly modulates or controls the free charge carrier generation, separation, transportation, and separation at the interface or junction in optoelectronic processes. These controllable or tunning characteristics make it a very powerful concept to introduce in the widespread traditional devices with new working principle.

Ferrite based materials have come up as a potential candidate for solar cell application. Thus, study paves a way for the ferrite-based nanomaterials for the use of solar cell applications as well to investigate the piezophototronic effect on these materials. The piezo-phototronic effect is an emerging field and provides a new platform for material science and electronics and explore the new area to work upon. This concept needs a multidisciplinary contribution so that every field of society could be benefitted.

Acknowledgement
Authors would like to acknowledge the Shoolini University, Solan, Himachal Pradesh, India.

References

[1] C. N. Kumar, Energy collection via Piezoelectricity, IOP Conference Series J. Phys. 1 (2015) 012031. https://doi.org/10.1088/1757-899X/73/1/012031

[2] A. I. Kingon, S. Srinivasan, Lead zirconate titanate thin films directly on copper electrodes for ferroelectric, dielectric and piezoelectric applications, Nat. Mater. 4(3) (2005) 233. https://doi.org/10.1038/nmat1334

[3] R. Ferren, Advances in polymeric piezoelectric transducers, Nature. 350(6319) (1991) 26-27.

[4] Q. Yang, W. Wang, S. Xu, Z. L. Wang, Enhancing light emission of ZnO microwire-based diodes by piezo-phototronic effect, Nano Lett. 11(9) (2011) 4012-4017. https://doi.org/10.1021/nl202619d

[5] Q. Yang, X. Guo, W. Wang, Y. Zhang, S. Xu, D. H. Lien, Z. L. Wang, Enhancing sensitivity of a single ZnO micro-/nanowire photodetector by piezo-phototronic effect, ACS Nano 4(10) (2010) 6285-6291. https://doi.org/10.1021/nn1022878

[6] J. M. Wu, C. Xu, Y. Zhang, Z. L. Wang, Lead-free nanogenerator made from single ZnSnO3 microbelt, ACS Nano 6(5) (2012) 4335-4340. https://doi.org/10.1021/nn300951d

[7] J. M. Wu, C. Xu, Y. Zhang, Y. Yang, Y. Zhou, Z. L. Wang, Flexible and transparent nanogenerators based on a composite of lead-free ZnSnO3 triangular-belts, Adv. Mater. 24(45) (2012) 6094-6099. https://doi.org/10.1002/adma.201202445

[8] J. M. Wu, Y. Chen, Y. Zhang, K. H. Chen, Y. Yang, Y. Hu, J. H. He, Z. L. Wang, Ultrahigh sensitive piezotronic strain sensors based on a ZnSnO3 nanowire/microwire, ACS Nano 6(5) (2012) 4369-4374. https://doi.org/10.1021/nn3010558

[9] Z. L. Wang, Piezopotential gated nanowire devices: Piezotronics and piezo-phototronics, Nano Today 5(6) (2010) 540-552. https://doi.org/10.1016/j.nantod.2010.10.008

[10] J. Sun, Q. Hua, R. Zhou, D. Li, W. Guo, X. Li, G. Hu, C. Shan, Q. Meng, L. Dong, Piezo-Phototronic Effect Enhanced Efficient Flexible Perovskite Solar Cells, ACS Nano 13 (2019) 4507-4513. https://doi.org/10.1021/acsnano.9b00125

[11] Y. Peng, M. Que, H. E. Lee, R. Bao, X. Wang, J. Lu, Z. Yuan, X. Li, J. Tao, J. Sun, Achieving High-resolution Pressure Mapping via Flexible GaN/ZnO Nanowire LEDs Array by Piezo-phototronic Effect. Nano Energy 58 (2019) 633-640. https://doi.org/10.1016/j.nanoen.2019.01.076

[12] Jiang, J., Wang, Q., Wang, B., Dong, J., Li, Z., Li, X., Zi, Y., Li, S., Wang, X.: Direct lift-off and the piezo-phototronic study of InGaN/GaN heterostructure membrane. Nano Ener. 59 (2019) 545-552. https://doi.org/10.1016/j.nanoen.2019.02.066

[13] W. Zhang, D. Jiang, M. Zhao, Y. Duan, X. Zhou, X. Yang, C. Shan, J. Qin, S. Gao, Q. Liang, Piezo-phototronic effect for enhanced sensitivity and response range of ZnO thin film flexible UV photodetectors, J. Appl. Phys. 125(2) (2019) 024502. https://doi.org/10.1063/1.5057371

[14] J. Guo, R. Wen, J. Zhai, Z. L. Wang, Enhanced NO2 gas sensing of a single-layer MoS2 by photogating and piezo-phototronic effects, Sci. Bull. 64(2) (2019) 128-135. https://doi.org/10.1016/j.scib.2018.12.009

[15] D. J. Lee, G. Mohan Kumar, P. Ilanchezhiyan, F. Xiao, S. U. Yuldashev, Y. D. Woo, D. Y. Kim, T. W. Kang, Arrayed CdTeMicrodots and Their Enhanced Photodetectivity via Piezo-Phototronic Effect, Nanomater. 9(2) (2019) 178. https://doi.org/10.3390/nano9020178

[16] X. Wang, G. Dai, Y. Chen, X. Mo, X. Li, W. Huang, J. Sun, J. Yang, Piezo-phototronic enhanced photoresponsivity based on single CdTe nanowire photodetector, J. Appl. Phys. 125(9) (2019) 094505. https://doi.org/10.1063/1.5067371

[17] S. Qiao, J. Liu, X. Niu, B. Liang, G. Fu, Z. Li, S. Wang, K. Ren, C. Pan, Piezophototronic Effect Enhanced Photoresponse of the Flexible Cu (In, Ga) Se2 (CIGS) Heterojunction Photodetectors, Adv. Funct. Mater. 28(19) (2018) 1707311. https://doi.org/10.1002/adfm.201707311

[18] F. Li, W. Peng, Z. Pan, Y. He, Optimization of Si/ZnO/PEDOT: PSS tri-layer heterojunction photodetector by piezo-phototronic effect using both positive and negative piezoelectric charges, Nano Ener. 48 (2018) 27-34. https://doi.org/10.1016/j.nanoen.2018.03.025

[19] P. Lin, L. Zhu, D. Li, L. Xu, Z. L. Wang, Tunable WSe2-CdS mixed-dimensional van der Waals heterojunction with a piezo-phototronic effect for an enhanced flexible photodetector, Nanoscale 10(30) (2018) 14472-14479. https://doi.org/10.1039/C8NR04376K

[20] Q. Lai, L. Zhu, Y. Pang, L. Xu, J. Chen, Z. Ren, J. Luo, L. Wang, L. Chen, K. Han, Piezo-phototronic Effect Enhanced Photodetector Based on CH3NH3PbI3 Single

Crystals, ACS Nano 12(10) (2018) 10501-10508. https://doi.org/10.1021/acsnano.8b06243

[21] M. Chen, B. Zhao, G. Hu, X. Fang, H. Wang, L. Wang, J. Luo, X. Han, X. Wang, C. Pan, Piezo-Phototronic Effect Modulated Deep UV Photodetector Based on ZnO-Ga2O3 Heterojuction Microwire, Adv. Funct. Mater. 28(14) (2018) 1706379. https://doi.org/10.1002/adfm.201706379

[22] P. Lin, L. Zhu, D. Li, L. Xu, C. Pan, Z. Wang, Piezo-Phototronic Effect for Enhanced Flexible MoS2/WSe2 van der Waals Photodiodes, Adv. Funct. Mater. 28(35) (2018) 1802849. https://doi.org/10.1002/adfm.201802849

[23] S. Qiao, J. Liu, G. Fu, K. Ren, Z. Li, S. Wang, C. Pan, ZnO nanowire based CIGS solar cell and its efficiency enhancement by the piezo-phototronic effect, Nano Ener. 49 (2018) 508-514. https://doi.org/10.1016/j.nanoen.2018.04.070

[24] Y. Dai, X. Wang, W. Peng, C. Wu, Y. Ding, K. Dong, Z. L. Wang, Enhanced performances of Si/CdS heterojunction near-infrared photodetector by the piezo-phototronic effect, Nano Ener. 44 (2018) 311-318. https://doi.org/10.1016/j.nanoen.2017.11.076

[25] L. Zhang, D. Zhu, H. He, Q. Wang, L. Xing, X. Xue, Enhanced piezo/solar-photocatalytic activity of Ag/ZnO nanotetrapods arising from the coupling of surface plasmon resonance and piezophototronic effect, J. Phys. Chem. Solids 102 (2017) 27-33. https://doi.org/10.1016/j.jpcs.2016.11.009

[26] X. Li, R. Liang, J. Tao, Z. Peng, Q. Xu, X. Han, X. Wang, C. Wang, J. Zhu, C. Pan, Flexible light emission diode arrays made of transferred Si microwires-ZnO nanofilm with piezo-phototronic effect enhanced lighting, ACS Nano 11(4) (2017) 3883-3889. https://doi.org/10.1021/acsnano.7b00272

[27] J. Sun, P. Li, R. Gao, X. Lu, C. Li, Y. Lang, X. Zhang, J. Bian, Piezo-phototronic effect enhanced photo-detector based on ZnO nano-arrays/NiO structure, Appl. Surf. Sci. 427 (2018) 613-619. https://doi.org/10.1016/j.apsusc.2017.09.023

[28] X. Wang, W. Peng, R. Yu, H. Zou, Y. Dai, Y. Zi, C. Wu, S. Li, Z. L. Wang, Simultaneously enhancing light emission and suppressing efficiency droop in GaN microwire-based ultraviolet light-emitting diode by the piezo-phototronic effect, Nano Lett. 17(6) (2017) 3718-3724. https://doi.org/10.1021/acs.nanolett.7b01004

[29] H. Zou, X. Li, W. Peng, W. Wu, R. Yu, C. Wu, W. Ding, F. Hu, R. Liu, Y. Zi, Piezo-Phototronic Effect on Selective Electron or Hole Transport through Depletion Region of Vis-NIR Broadband Photodiode, Adv. Mater. 29(29) (2017) 1701412. https://doi.org/10.1002/adma.201701412

[30] G. Dai, H. Zou, X. Wang, Y. Zhou, P. Wang, Y. Ding, Y. Zhang, J. Yang, Z. L. Wang, Piezo-phototronic Effect Enhanced Responsivity of Photon Sensor Based on Composition-Tunable Ternary CdSxSe1-x Nanowires, ACS Photo. 4(10) (2017) 2495-2503. https://doi.org/10.1021/acsphotonics.7b00724

[31] C. Takahashi, N. Matsubara, Y. Akachi, N. Ogawa, G. Kalita, T. Asaka, M. Tanemura, Y. Kawashima, H. Yamamoto, Visualization of silver-decorated poly (DL-lactide-co-glycolide) nanoparticles and their efficacy against Staphylococcus epidermidis, Mater. Sci. Eng. C 72 (2017) 143-149. https://doi.org/10.1016/j.msec.2016.11.051

[32] S. Jeong, M. W. Kim, Y. R. Jo, T. Y. Kim, Y. C. Leem, S. W. Kim, B. J. Kim, S. J. Park, Crystal-Structure-Dependent Piezotronic and Piezo-Phototronic Effects of ZnO/ZnS Core/Shell Nanowires for Enhanced Electrical Transport and Photosensing Performance, ACS Appl. Mater. Inter. 10(34) (2018) 28736-28744. https://doi.org/10.1021/acsami.8b06192

[33] K. Zhang, M. Peng, W. Wu, J. Guo, G. Gao, Y. Liu, J. Kou, R. Wen, Y. Lei, A. Yu, A flexible p-CuO/n-MoS 2 heterojunction photodetector with enhanced photoresponse by the piezo-phototronic effect, Mater. Hor. 4(2) (2017) 274-280. https://doi.org/10.1039/C6MH00568C

[34] X. Wen, W. Wu, Z. L. Wang, Effective piezo-phototronic enhancement of solar cell performance by tuning material properties, Nano Ener. 2(6) (2013) 1093-1100. https://doi.org/10.1016/j.nanoen.2013.08.008

[35] L. Dong, S. Niu, C. Pan, R. Yu, Y. Zhang, Z. L. Wang, Piezo-Phototronic Effect of CdSe Nanowires, Adv. Mater. 24(40) (2012) 5470-5475. https://doi.org/10.1002/adma.201201385

[36] Y. Zhang, L. Yang, Y. Zhang, Z. Dig, M. Wu, Y. Zhou, C. Diao, H. Zhang, X. Wang, Z. L. Wang, Enhanced photovoltaic performances of La-doped Bismuth ferrite/zinc oxide heterojunction by coupling piezo-phototronic effect and ferroelectricity, ASC Nano 14 (2020) 10723-10732. https://doi.org/10.1021/acsnano.0c05398

[37] Y. L. Liu, J. M. Wu, Synergistically catalytic activities of BiFeO3/TiO2 core shell nanocomposites for degredation of roganic dye molecule through piezophototronic effect, Nano Ener. 56 (2019) 74-81. https://doi.org/10.1016/j.nanoen.2018.11.028

[38] Y. H. Hsu, P. Y. Chen, C. S. Tu, Anthoniappen, Polarization enhanced potovoltaic response and mehanism in Ni-doped (Ba0.93Gd0.07)FeO3 ceramics for self-powered photodetector, J. Eur. Ceram. Soc. 41 (2021) 1934-1944. https://doi.org/10.1016/j.jeurceramsoc.2020.10.037

[39] H. Morkoç, U. Özgür, Zinc oxide: fundamentals, materials and device technology, John Wiley & Sons, 2008. https://doi.org/10.1002/9783527623945

[40] Z. L. Wang, Nanopiezotronics, Adv. Mater. 19(6) (2007) 889-892. https://doi.org/10.1002/adma.200602918

[41] Z. L. Wang, Piezotronic and piezophototronic effects, J. Phys. Chem. Lett. 1(9) (2010) 1388-1393. https://doi.org/10.1021/jz100330j

[42] J. G. Webster, Wiley encyclopedia of electrical and electronics engineering, John Wiley New York, 1999. https://doi.org/10.1002/047134608X

[43] Y. Gao, Z. L. Wang, Electrostatic potential in a bent piezoelectric nanowire. The fundamental theory of nanogenerator and nanopiezotronics, Nano Lett. 7(8) (2007) 2499-2505. https://doi.org/10.1021/nl071310j

[44] H. Hao, K. Jenkins, X. Huang, Y. Xu, J. Huang, R. Yang, Piezoelectric potential in single-crystalline ZnO nanohelices based on finite element analysis, Nanomater. 7(12) (2017) 430. https://doi.org/10.3390/nano7120430

[45] Z. L. Wang, J. Song, Piezoelectric nanogenerators based on zinc oxide nanowire arrays, Sci. 312(5771) (2006) 242-246. https://doi.org/10.1126/science.1124005

[46] J. Zhou, P. Fei, Y. Gu, W. Mai, Y. Gao, R. Yang, G. Bao, Z. L. Wang, Piezoelectric-potential-controlled polarity-reversible Schottky diodes and switches of ZnO wires, Nano Lett. 8(11) (2008) 3973-3977. https://doi.org/10.1021/nl802497e

[47] S. M. Sze, K. K. Ng, Physics of semiconductor devices, John wiley & Sons, 2006. https://doi.org/10.1002/0470068329

[48] T. Ikeda, Fundamentals of piezoelectricity, Oxford university press, 1996.

[49] U. Muhammad, U. Mushtaq, D. Zheng, D. P. Han, N. Muhammad, Investigation of optoelectronic characteristics of indium composition in InGaN-based light-emitting diodes, Mater. Res. Express 6 (2019) 045909. https://doi.org/10.1088/2053-1591/aaff15

[50] S. O. Kasap, R. K. Sinha, Optoelectronics and photonics: principles and practices, Prentice Hall New Jersey, 2001.

[51] Z. L. Wang, Progress in piezotronics and piezo-phototronics, Adv. Mater. 24(34) (2012) 4632-4646. https://doi.org/10.1002/adma.201104365

[52] Z. L. Wang, W. Wu, Piezotronics and piezo-phototronics: fundamentals and applications, Nat. Sci. Rev. 1(1) (2013) 62-90. https://doi.org/10.1093/nsr/nwt002

[53] S. Sze, Physics of Semiconductor Devices. John Wiley, New York (1981) 122-129.

[54] Z. L. Wang, Introduction of Piezotronics and Piezo-Phototronics. In: Piezotronics and Piezo-Phototronics, Springer, (2012) 1-17. https://doi.org/10.1007/978-3-642-34237-0_1

[55] Y. Zhang, Y. Liu, Z. L. Wang, Fundamental theory of piezotronics, Adv. Mater. 23(27) (2011) 3004-3013. https://doi.org/10.1002/adma.201100906

[56] R. T. Tung, Recent advances in Schottky barrier concepts, Mater. Sci. Eng. R. 35 (2001) 1-138. https://doi.org/10.1016/S0927-796X(01)00037-7

[57] E. H. Rhoderick, R. H. Williams, MS Contacts, Clarendon, Oxford, (1988) 1-70.

[58] L. J. Brillson, Y. Lu, ZnO Schottky barriers and Ohmic contacts, J. Appl. Phys. 109(12) (2011) 8. https://doi.org/10.1063/1.3581173

[59] Z. L. Wang, Piezopotential gated nanowire devices: Piezotronics and piezo-phototronics, Nano Today 5 (2010) 540-552. https://doi.org/10.1016/j.nantod.2010.10.008

[60] Z. L. Wang, Preface to the special section on piezotronics, Adv. Mater. 24(34) (2012) 4630-4631. https://doi.org/10.1002/adma.201202888

[61] J. H. He, C. L. Hsin, J. Liu, L. J. Chen, Z. L. Wang, Piezoelectric gated diode of a single ZnO nanowire, Adv. Mater. 19(6) (2007) 781-784. https://doi.org/10.1002/adma.200601908

[62] Z. Gao, J. Zhou, Y. Gu, P. Fei, Y. Hao, G. Bao, Z. L. Wang, Effects of piezoelectric potential on the transport characteristics of metal-ZnO nanowire-metal field effect transistor, J. Appl. Phys. 105(11) (2009) 113707. https://doi.org/10.1063/1.3125449

[63] Z. L. Wang, The new field of nanopiezotronics, Mater. Today 10(5) (2007) 20-28. https://doi.org/10.1016/S1369-7021(07)70076-7

[64] J. Zhou, Y. Gu, P. Fei, W. Mai, Y. Gao, R. Yang, G. Bao, Z. L. Wang, Flexible piezotronic strain sensor, Nano Lett. 8(9) (2008) 3035-3040. https://doi.org/10.1021/nl802367t

[65] Y. Hu, Y. Chang, P. Fei, R. L Snyder, Z. L. Wang, Designing the electric transport characteristics of ZnO micro/nanowire devices by coupling piezoelectric and photoexcitation effects, ACS Nano 4(2) (2010) 1234-1240. https://doi.org/10.1021/nn901805g

[66] D. Xiong, W. Deng, G. Tian, Y. Gao, X. Chu, C. Yan, L. Jin, Y. Su, W. Yan, W. Yang, Piezo-phototronic Enhanced Serrate-structured ZnO-based Heterojunction Photodetector for Optical Communication, Nanoscale 11 (2019) 3021-3027. https://doi.org/10.1039/C8NR09418G

[67] S. M. Lélé, Sustainable development: a critical review, World Devp. 19(6) (1991) 607-621. https://doi.org/10.1016/0305-750X(91)90197-P

[68] H. Meyar-Naimi, S. Vaez-Zadeh, Sustainable development based energy policy making frameworks, a critical review, Energy Policy 43 (2012) 351-361. https://doi.org/10.1016/j.enpol.2012.01.012

[69] L. Zheng, Z. H. Lin, G. Cheng, W. Wu, X. Wen, S. Lee, Z. L. Wang, Silicon-based hybrid cell for harvesting solar energy and raindrop electrostatic energy, Nano Ener. 9 (2014) 291-300. https://doi.org/10.1016/j.nanoen.2014.07.024

[70] B. Tian, X. Zheng, T. J. Kempa, Y. Fang, N. Yu, G. Yu, J. Huang, C. M. Lieber, Coaxial silicon nanowires as solar cells and nanoelectronic power sources, Nature 449(7164) (2007) 885. https://doi.org/10.1038/nature06181

[71] P. Devine-Wright, Beyond NIMBYism: towards an integrated framework for understanding public perceptions of wind energy, Wind Ener. 8(2) (2005) 125-139. https://doi.org/10.1002/we.124

[72] M. Hoogwijk, A. Faaij, R. Van Den Broek, G. Berndes, D. Gielen, W. Turkenburg, Exploration of the ranges of the global potential of biomass for energy, Biomass and Bioener.25(2) (2003) 119-133. https://doi.org/10.1016/S0961-9534(02)00191-5

[73] Y. Yang, W. Guo, K. C. Pradel, G. Zhu, Y. Zhou, Y. Zhang, Y. Hu, L. Lin, Z. L. Wang, Pyroelectric nanogenerators for harvesting thermoelectric energy, Nano Lett. 12(6) (2012) 2833-2838. https://doi.org/10.1021/nl3003039

[74] W. Liu, X. Yan, G. Chen, Z. Ren, Recent advances in thermoelectric nanocomposites, Nano Ener. 1(1) (2012) 42-56. https://doi.org/10.1016/j.nanoen.2011.10.001

[75] M. A. Green, The path to 25% silicon solar cell efficiency: history of silicon cell evolution, Progress in Photovoltaics: Res. Appl. 17(3) (2009) 183-189. https://doi.org/10.1002/pip.892

[76] A. Polman, M. Knight, E. C. Garnett, B. Ehrler, W. C. Sinke, Photovoltaic materials: Present efficiencies and future challenges, Sci. 352(6283) (2016) aad4424. https://doi.org/10.1126/science.aad4424

[77] A. Rohatgi, D. Meier, Developing novel low-cost, high-throughput processing techniques for 20%-efficient monocrystalline silicon solar cells, Photovol. Int. 10 (2010) 87-93.

[78] M. L. Brongersma, Y. Cui, S. Fan, Light management for photovoltaics using high-index nanostructures, Nat. Mater. 13(5) (2014) 451. https://doi.org/10.1038/nmat3921

[79] M. Girtan, New Trends in Solar Cells Research: Future Solar Energy Devices, Springer, (2018) 45-47. https://doi.org/10.1007/978-3-319-67337-0_3

[80] T. Someya, T. Sekitani, S. Iba, Y. Kato, H. Kawaguchi, T. Sakurai, A large-area, flexible pressure sensor matrix with organic field-effect transistors for artificial skin applications, Proceedings of the National Academy of Sciences of the USA 101(27) (2004) 9966-9970. https://doi.org/10.1073/pnas.0401918101

[81] A. Bykhovski, V. Kaminski, M. Shur, Q. Chen, M. A. Khan, Piezoresistive effect in wurtzite n-type GaN. Applied physics letters 68(6) (1996) 818-819. https://doi.org/10.1063/1.116543

[82] R. Gaska, M. Shur, A. Bykhovski, J. Yang, M. Khan, V. Kaminski, S. Soloviov, Piezoresistive effect in metal-semiconductor-metal structures on p-type GaN, Appl. Phys. Lett. 76(26) (2000) 3956-3958. https://doi.org/10.1063/1.126833

[83] K. Vandewal, K. Tvingstedt, A. Gadisa, O. Inganäs, J. V. Manca, On the origin of the open-circuit voltage of polymer-fullerene solar cells, Nat. Mater. 8(11) (2009) 904. https://doi.org/10.1038/nmat2548

[84] C. Chen, Y. Shi, Y. S. Zhang, J. Zhu, Y. Yan, Size dependence of Young's modulus in ZnO nanowires, Phys. Rev. Chem. 96(7) (2006) 075505. https://doi.org/10.1103/PhysRevLett.96.075505

[85] Y. Zhang, Y. Yang, Z. L. Wang, Piezo-phototronics effect on nano/microwire solar cells, Ener. Environm. Sci. 5(5) (2012) 6850-6856. https://doi.org/10.1039/c2ee00057a

[86] F. Boxberg, N. Søndergaard, H. Xu, Photovoltaics with piezoelectric core– shell nanowires, Nano Lett. 10(4) (2010) 1108-1112. https://doi.org/10.1021/nl9040934

[87] A. Rahman, A. Ashraf, H. Xin, X. Tong, P. Sutter, M. D. Eisaman, C. T. Black, Sub-50-nm self-assembled nanotextures for enhanced broadband antireflection in silicon solar cells, Nat. Comm. 6 (2015) 5963. https://doi.org/10.1038/ncomms6963

[88] J. Oh, H. C. Yuan, H. M. Branz, An 18.2%-efficient black-silicon solar cell achieved through control of carrier recombination in nanostructures, Nat. Nanotech. 7(11) (2012) 743. https://doi.org/10.1038/nnano.2012.166

[89] C. Battaglia, C. M. Hsu, K. Söderström, J. Escarre, F. J. Haug, M. Charrière, M. Boccard, M. Despeisse, D. T. Alexander, M. Cantoni, Light trapping in solar cells: can periodic beat random, ACS Nano 6(3) (2012) 2790-2797. https://doi.org/10.1021/nn300287j

[90] P. H. Wang, R. E. Nowak, S. Geißendörfer, M. Vehse, N. Reininghaus, O. Sergeev, K. von Maydell, A. G. Brolo, C, Agert, Cost-effective nanostructured thin-film solar cell with enhanced absorption, Appl. Phys. Lett. 105(18) (2014) 183106. https://doi.org/10.1063/1.4901167

[91] L. Aé, D. Kieven, J. Chen, R. Klenk, T. Rissom, Y. Tang, M. C. Lux-Steiner, ZnO nanorod arrays as an antireflective coating for Cu (In, Ga) Se2 thin film solar cells,

Progress in Photovoltaics: Res. Appl. 18(3) (2010) 209-213. https://doi.org/10.1002/pip.946

[92] Y. J. Lee, D. S. Ruby, D. W. Peters, B. B. McKenzie, J. W. Hsu, ZnO nanostructures as efficient antireflection layers in solar cells, Nano Lett. 8(5) (2008) 1501-1505. https://doi.org/10.1021/nl080659j

[93] R. E. Nowak, M. Vehse, O. Sergeev, K. von Maydell, C. Agert, ZnO nanorod arrays as light trapping structures in amorphous silicon thin-film solar cells, Solar Ener. Mater. Solar Cells 125 (2014) 305-309. https://doi.org/10.1016/j.solmat.2013.12.025

[94] R. E. Nowak, M. Vehse, O. Sergeev, T. Voss, M. Seyfried, K. von Maydell, C. Agert, ZnO Nanorods with Broadband Antireflective Properties for Improved Light Management in Silicon Thin-Film Solar Cells, Adv. Opt. Mater. 2(1) (2014) 94-99. https://doi.org/10.1002/adom.201300455

[95] L. Zhu, L. Wang, C. Pan, L. Chen, F. Xue, B. Chen, L. Yang, L. Su, Z. L. Wang, Enhancing the efficiency of silicon-based solar cells by the piezo-phototronic effect, ACS Nano 11(2) (2017) 1894-1900. https://doi.org/10.1021/acsnano.6b07960

[96] W. Metzger, I. Repins, M. Romero, P. Dippo, M. Contreras, R. Noufi, D. Levi, Recombination kinetics and stability in polycrystalline Cu (In, Ga) Se2 solar cells, Thin Solid Films 517(7) (2009) 2360-2364. https://doi.org/10.1016/j.tsf.2008.11.050

[97] C. Xu, H. Zhang, J. Parry, S. Perera, G. Long, H. Zeng, A single source three-stage evaporation approach to CIGS absorber layer for thin film solar cells, Solar Ener. Mater. Solar Cells 117 (2013) 357-362. https://doi.org/10.1016/j.solmat.2013.06.006

[98] C. Persson, Anisotropic hole-mass tensor of CuIn 1− x Ga x (S, Se) 2: Presence of free carriers narrows the energy gap, Appl. Phys. Lett. 93(7) (2008) 072106. https://doi.org/10.1063/1.2969467

[99] O. Jani, C. Honsberg, Y. Huang, J. O. Song, I. Ferguson, G. Namkoong, E. Trybus, A. Doolittle, S. Kurtz, Design, growth, fabrication and characterization of high-band gap InGaN/GaN solar cells, IEEE 4th World Conference on Photovoltaic Energy Conference (2006) 20-25. https://doi.org/10.1109/WCPEC.2006.279337

[100] R. Dahal, J. Li, K. Aryal, J. Lin, H. Jiang, InGaN/GaN multiple quantum well concentrator solar cells, Appl. Phys. Lett. 97(7) (2010) 073115. https://doi.org/10.1063/1.3481424

[101] S. Lee, Y. Honda, H. Amano, Effect of piezoelectric field on carrier dynamics in InGaN-based solar cells, J. Phys. D: Appl. Phys. 49(2) (2015) 025103. https://doi.org/10.1088/0022-3727/49/2/025103

[102] R. Dahal, B. Pantha, J. Li, J. Lin, H. Jiang, InGaN/GaN multiple quantum well solar cells with long operating wavelengths, Appl. Phys. Lett. 94(6) (2009) 063505. https://doi.org/10.1063/1.3081123

[103] C. Jiang, L. Jing, X. Huang, M. Liu, C. Du, T. Liu, X. Pu, W. Hu, Z. L. Wang, Enhanced Solar Cell Conversion Efficiency of InGaN/GaN Multiple Quantum Wells by Piezo-Phototronic Effect, ACS Nano 11(9) (2017) 9405-9412. https://doi.org/10.1021/acsnano.7b04935

[104] K. Nakayama, K. Tanabe, H. A. Atwater, Plasmonic nanoparticle enhanced light absorption in GaAs solar cells, Appl. Phys. Lett. 93(12) (2008) 121904. https://doi.org/10.1063/1.2988288

[105] S. Albrecht, B. Rech, Perovskite solar cells: On top of commercial photovoltaics, Nat. Ener. 2(1) (2017) 16196. https://doi.org/10.1038/nenergy.2016.196

[106] G. E. Eperon, M. T. Hörantner, H. J. Snaith, Metal halide perovskite tandem and multiple-junction photovoltaics, Nat. Rev. Chem. 1(12) (2017) 0095. https://doi.org/10.1038/s41570-017-0095

[107] Y. Yang, J. You, Make perovskite solar cells stable, Nat. News 544(7649) (2017) 155. https://doi.org/10.1038/544155a

[108] J. Huang, Y. Yuan, Y. Shao, Y. Yan, Understanding the physical properties of hybrid perovskites for photovoltaic applications, Nat. Rev. Mater. 2(7) (2017) 17042. https://doi.org/10.1038/natrevmats.2017.42

[109] T. Jiang, W. Fu, Improved performance and stability of perovskite solar cells with bilayer electron-transporting layers, RSC Adv. 8(11) (2018) 5897-5901. https://doi.org/10.1039/C8RA00248G

[110] K. Gu, D. Zheng, L. Li, Y. Zhang, High-efficiency and stable piezo-phototronic organic perovskite solar cell, RSC Adv. 8(16) (2018) 8694-8698. https://doi.org/10.1039/C8RA00520F

[111] Y. Wang, D. Zheng, L. Li, Y. Zhang, Enhanced Efficiency of Flexible GaN/Perovskite Solar Cells Based on the Piezo-Phototronic Effect, ACS Appl. Ener. Mater. 1(7) (2018) 3063-3069. https://doi.org/10.1021/acsaem.8b00713

[112] D. R. Kim, C. H. Lee, P. M. Rao, I. S. Cho, X. Zheng, Hybrid Si microwire and planar solar cells: passivation and characterization, Nano Lett. 11(7) (2011) 2704-2708. https://doi.org/10.1021/nl2009636

[113] J. A. Czaban, D. A. Thompson, R. R. LaPierre, GaAs core− shell nanowires for photovoltaic applications, Nano Lett. 9(1) (2008) 148-154. https://doi.org/10.1021/nl802700u

[114] N. Sköld, L. S. Karlsson, M. W. Larsson, M.E. Pistol, W. Seifert, J. Trägårdh, L. Samuelson, Growth and Optical Properties of Strained GaAs− Ga x In1-x P Core−Shell Nanowires, Nano Lett. 5(10) (2005) 1943-1947. https://doi.org/10.1021/nl051304s

[115] B. Hua, J. Motohisa, Y. Kobayashi, S. Hara, T. Fukui, Single GaAs/GaAsP coaxial core− shell nanowire lasers, Nano Lett. 9(1) (2008) 112-116. https://doi.org/10.1021/nl802636b

[116] L. Zhu, L. Wang, F. Xue, L. Chen, J. Fu, X. Feng, T. Li, Z. L. Wang, Piezo-Phototronic Effect Enhanced Flexible Solar Cells Based on n-ZnO/p-SnS Core-Shell Nanowire Array, Adv. Sci. 4(1) (2017) 1600185. https://doi.org/10.1002/advs.201600185

[117] J. Vidal, S. Lany, M. d'Avezac, A. Zunger, A. Zakutayev, J. Francis, J. Tate, Band-structure, optical properties, and defect physics of the photovoltaic semiconductor SnS, Appl. Phys. Lett. 100(3) (2012) 032104. https://doi.org/10.1063/1.3675880

[118] H. Noguchi, A. Setiyadi, H. Tanamura, T. Nagatomo, O. Omoto, Characterization of vacuum-evaporated tin sulfide film for solar cell materials, Solar Ener. Mater. Solar Cells 35 (1994) 325-331. https://doi.org/10.1016/0927-0248(94)90158-9

[119] W. Albers, C. Haas, H. Vink, J. Wasscher, Investigations on SnS, J. Appl. Phys. 32(10) (1961) 2220-2225. https://doi.org/10.1063/1.1777047

[120] A. Bosman, E. Havinga, Temperature dependence of dielectric constants of cubic ionic compounds, Phys. Rev. 129(4) (1963) 1593. https://doi.org/10.1103/PhysRev.129.1593

[121] Y. P. Varshni, Temperature dependence of the energy gap in semiconductors, Physica 34(1) (1967) 149-154. https://doi.org/10.1016/0031-8914(67)90062-6

[122] C. de Julián Fernández, Influence of the temperature dependence of anisotropy on the magnetic behavior of nanoparticles, Phys. Rev. B 72(5) (2005) 054438. https://doi.org/10.1103/PhysRevB.72.054438

[123] T. Sjostrom, F. E. Harris, S. Trickey, Temperature-dependent behavior of confined many-electron systems in the Hartree-Fock approximation, Phys. Rev. B 85(4) (2012) 045125. https://doi.org/10.1103/PhysRevB.85.045125

[124] Y. Hu, B. D. Klein, Y. Su, S. Niu, Y. Liu, Z. L. Wang, Temperature dependence of the piezotronic effect in ZnO nanowires, Nano Lett. 13(11) (2013) 5026-5032. https://doi.org/10.1021/nl401702g

[125] X. Wang, W. Peng, C. Pan, Z. L. Wang, Piezotronics and piezo-phototronics based on a-axis nano/microwires: fundamentals and applications, Semicond. Sci. Tech. 32(4) (2017) 043005. https://doi.org/10.1088/1361-6641/aa5fcd

[126] X. Wang, R. Yu, W. Peng, W. Wu, S. Li, Z. L. Wang, Temperature dependence of the piezotronic and piezophototronic effects in a-axis GaN nanobelts, Adv. Mater. 27(48) (2015) 8067-8074. https://doi.org/10.1002/adma.201504534

[127] W. Sha, J. Zhang, S. Tan, X. Luo, W. Hu, III-Nitrides piezotronic/piezo-phototronic materials and devices, J. Phys. D: Appl. Phys. 52 (2019) 213003. https://doi.org/10.1088/1361-6463/ab04d6

[128] G. A. Slack, L. J. Schowalter, D. Morelli, J. A. Freitas Jr, Some effects of oxygen impurities on AlN and GaN, J. Cryst. Grow. 246 (2002), 287-298. https://doi.org/10.1016/S0022-0248(02)01753-0

[129] W. Moore, J. Jr. Freitas G. Braga, R. Molnar, S. Lee, K. Lee, I. Song, Identification of Si and O donors in hydride-vapor-phase epitaxial GaN, Appl. Phys. Lett. 79(16) (2001) 570-2572. https://doi.org/10.1063/1.1411985

[130] R. Molnar, T. Lei, T. Moustakas, Electron transport mechanism in gallium nitride, Appl. Phys. Lett. 62(1) (1993) 72-74. https://doi.org/10.1063/1.108823

[131] W. Mönch, Semiconductor surfaces and interfaces, Vol. 26, Springer Science & Business Media, 2013.

[132] S. Chevtchenko, X. Ni, Q. Fan, A. Baski, H. Morkoç, Surface band bending of a-plane GaN studied by scanning Kelvin probe microscopy, Appl. Phys. Lett. 88(12) (2006) 122104. https://doi.org/10.1063/1.2188589

Ferrite - Nanostructures with Tunable Properties and Diverse Applications Materials Research Forum LLC
Materials Research Foundations **112** (2021) 311-335 https://doi.org/10.21741/9781644901595-9

Chapter 9

Multiferroicity in Aurivillius Based $Bi_4Ti_3O_{12}$ Ceramics: An Overview, Future Prospective and Comparison with Ferrites

Sumit Bhardwaj[1*], Joginder Paul[2], Gagan Kumar[1], Pankaj Sharma[3], Ravi Kumar[4]

[1]Department of Physics, Chandigarh University, Gharuan, Mohali, Punjab, 140413, India

[2]Government (Girls) Senior Secondary School, Hamirpur, Himachal Pradesh, 177001, India

[3]Department of Applied Sciences, National Institute of Technical Teachers Training and Research, Sector 26, Chandigarh, 160019, India

[4]Department of Materials Science and Engineering, National Institute of Technology, Hamirpur, Himachal Pradesh, 177005, India

* sumit.bhardwaj4@gmail.com (Sumit Bhardwaj)

Abstract

The growing modern society demands more new generation devices to fulfil their requirements. This has forced the scientific community to develop multifunctional smart devices. Aurivillius based $Bi_4Ti_3O_{12}$ ceramics are one of the leading families of oxide materials, which attract immense attention due to their electrical, ferroelectric, optical, and dielectric properties. These materials have gained special attention due to their numerous device applications such as magnetic recording, sensors, read head technology, spintronic devices, switching devices, data storage devices and multiple state memory devices etc. Multiferroic are the materials in which two or more than two ferroic orders exist simultaneously. This chapter focuses on the possibility of existence of multiferroic behaviour in Aurivillius based compounds specially $Bi_4Ti_3O_{12}$. Firstly, we have discussed the basics of multiferroics and their types and the magnetoelectric effect. The effect of different dopants in originating the multiferroism in $Bi_4Ti_3O_{12}$ have been reviewed and discuss in detail. At the end comparison of multiferroic and ferrite materials and their future perspective have been discussed.

Keywords

Multiferroics, $Bi_4Ti_3O_{12}$, Ferroelectrics, Ferromagnetic

Materials Research Foundations **112** (2021) 311-335 https://doi.org/10.21741/9781644901595-9

Contents

1. Introduction

Functional electronics and magnetic materials are playing a significant role in modern technology. Ferroelectric materials are used in computers for manufacturing of ferroelectric random-access memory (Fe-RAM), whereas, magnetic materials are generally used for recording and data storage. Current technology is following a trend towards device miniaturization. This trendy behaviour has led to the growing interest in combining the ferroelectric and magnetic properties in multifunctional devices.

Multiferroics are materials in which two or more primary ferroic orders exists simultaneously [1-2]. The term 'multiferroic' was introduced by Schmid in 1994 to define materials in which two or more than two types of ferroic order occurs simultaneously in the same phase as shown in Figure 1.

One of the most appealing aspects of multiferroics is called magnetoelectric coupling. There are very small group of materials that can show both magnetic and electric orders simultaneously [3].

In these materials, the magnetic field can not only control magnetization but also give rise to ferroelectricty, or electric field can reorient polarization as well as spin arrangements. This property offers an extra degree of freedom and provides the possibility to create new devices based on these materials.

The existence of multiferroicity is determined by number of factors, including crystal structure, symmetry, electronic properties, and chemical properties. There are only thirteen-point groups that can gives rise to multiferroic behaviour [3-4]. The origins of ferromagnetism and ferroelectricity are different. Ferromagnetism needs transition metals with unpaired *3d* electrons and unfilled *3d* orbitals; while ferroelectric polarization needs transition metals with filled *3d* orbitals. Due to different origins, the single phase multiferroic compounds are few in number especially magnetoelectric multiferroics [5-6].

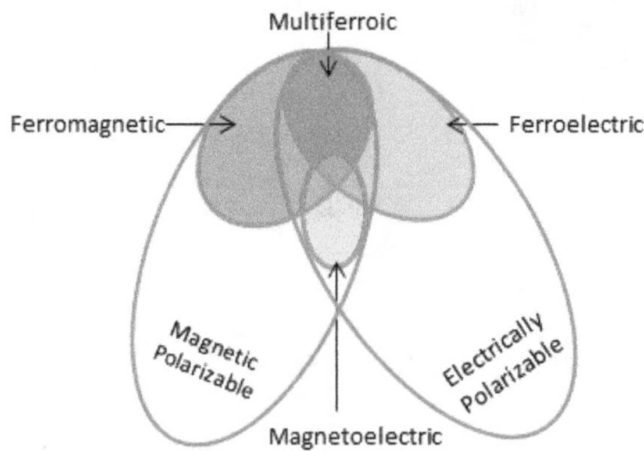

Figure 1: Relationship between ferroelectric, ferromagnetic, multiferroic and magnetoelectric materials [1].

Materials Research Foundations **112** (2021) 311-335 https://doi.org/10.21741/9781644901595-9

2. Types of multiferroic materials

Multiferroic materials are classified into two types:

i. Single phase multiferroics

ii. Composite multiferroics

Single Phase Multiferroics: Single phase multiferroics are those materials that show both ferroelectric and ferromagnetic order [7-11]. Recently, various multiferroic materials have been discovered in which a polar state is induced by different types of ordering, these materials are also known as "improper ferroelectrics". Khomskii classified single-phase multiferroics into two groups and other subgroups, according to the physical mechanism behind ferroelectricity [12]. The grouping of single-phase multiferroics is as given below:

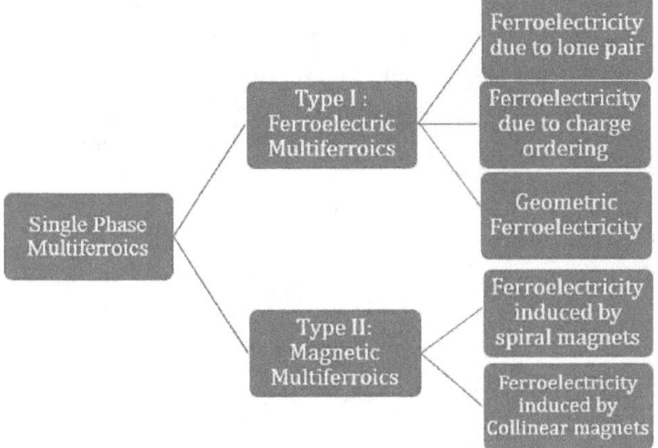

Figure2: Grouping of single phase multi-ferroic materials.

Type-I multiferroics are those materials in which ferroelectricity and magnetism have different sources; usually they show large polarization values and ferroelectricity appears at much higher temperatures than magnetism. This difference in transition temperatures reveals that both orders involve different energy scales and mechanism, which provokes to occurrence of weak magnetoelectric coupling. In this class of multiferroic materials multiferroicity is not due to ferroelectricity, because the magnetic and ferroelectric order have two different sources. In this kind of materials, the origin of ferroelectricity is from different sources such as:

i. Ferroelectricity due to lone pair

ii. Ferroelectricity due to charge ordering

iii. Geometric Ferroelectricity

On the other hand, type-II multiferroics corresponds to the material, in which magnetism causes ferroelectricty, due to which there is a strong coupling between the two orders. These materials show smaller electric polarization values and ferro-electricity always appears at lower temperature than magnetic order. These materials are in turns separated into two subgroups.

i. Ferroelectricity induced by spiral magnets

ii. Ferroelectricity induced by collinear magnets

3. Magnetoelectric effect

Multiferroic materials exhibit very interesting features; they show ferroelectric properties as well as magnetic properties, which points to multifunctional applications. Moreover, multiferroic likely to exhibit the magnetoelectric effect (ME), which is the phenomenon of tuning of magnetic properties with the application of an external electric field and electric properties with magnetic field. In other words, we can control magnetic properties by electric field and electric properties by magnetic field [13-15]. The ME effect provides an additional degree of freedom for devices fabrication such as multiple state memories in which data can be written magnetically and read electrically and vice versa [16]. The ME effect can be linear or / and non-linear with respect to external fields. This effect was first observed by Rontgen in 1888 and Pierre Curie in 1894 in two independent studies. Rontgen found that, when a dielectric material is moved in an electric field, it become magnetized [17, 18]. In 1894, the possibility of intrinsic coupling between magnetic and electric properties was introduced by Curie on the basis of magnetic symmetry [5]. Dzyaloshinskii predicted this effect in the single phase antiferromagnet Cr_2O_3 based on the symmetry consideration and Astrov confirmed this prediction experimentally in 1960 [19-21]. This was a breakthrough in magnetoelectric effect and afterward, the research was carried on various materials. Till now, many materials that exhibit the magnetoelectric effect has been discovered and synthesized.

Restraining our discussion to oxide materials, especially perovskite compounds, it was proven that ferroelectricty and magnetism have different origins. The ferroelectricity originates from the B-site d^0 electrons. Whereas, the magnetic properties require d^k electrons ($k \neq 0$). Moreover, the existence of polarization requires distorted asymmetric crystal structure, whereas ferromagnetism requires a symmetric structure. This concept has been proven experimentally on multiferroics.

In case of perovskite materials, it is realized that these materials do not have the same mechanism of ferroelectricity. For example, in $BaTiO_3$, the central Ti ions play the key role, whereas in $PbTiO_3$, the loan-pair Pb ion play the leading role [22-24]. On the other hand, in $BiFeO_3$ ferroelectricity is mostly dominated by the $6s^2$ lone-pair of Bi ions. Thus, the polarization in $BiFeO_3$ comes from the A-site (Bi^{3+}) ions, while magnetization comes from B-site (Fe^{3+}) ions. The same concept has been applied in designing and fabrication of new class of oxide materials.

4. Aurivillius Based Layer-Structured Ferroelectrics

The most promising ferroelectric material for Random access memories (RAMs) has been considered as Lead zirconium titanate (PZT). But PZT has some problematic issues such as fatigue, non-eco-friendly nature of lead, high operating voltage and low Curie temperature [25]. In comparison with non-layered perovskite ferroelectrics such as PZT and strontium bismuth tantalate (SBT), the bismuth layer-structured ferroelectrics are the promising candidates. These materials have several advantages such as fatigue-free, lead free, low operating voltage and high curie temperature. This makes the study and understanding of these materials interesting, and hence there is hunt for new materials with improved properties.

Bengt Aurivillius has discovered the family of layered structured materials, in which layers of bismuth oxide are sandwiched between perovskite structure layers [26-28]. The unit cell of Aurivillius compounds may be described as an assembling of bismuth oxide layers $(Bi_2O_2)^{2+}$ interwoven with pseudo-perovskite layers $(A_{m-1}B_mO_{3m+1})^{2-}$. The general formula for the compound is $(Bi_2O_2)^{2+}$ $(A_{m-1}B_mO_{3m+1})^{2-}$, where m is the number of perovskite unit sandwiched between bismuth oxide layers, and A is a mono, di or trivalent cation, and B is a tri, tetra, penta or hexavalent cation. In the $(A_{m-1}B_mO_{3m+1})^{2-}$ unit cell, B-site ions are enclosed by oxygen octahedral and A-site ions occupy spaces in the framework of BO_6 octahedral.

In the family of layered structured materials, the best-known compound is $Bi_4Ti_3O_{12}$. $Bi_4Ti_3O_{12}$ can be represented by the general formula (Bi_2O_2) $[(A_{m-1}B_mO_{3m+1})]$ with m = 3, consists of $(Bi_2O_2)^{2+}$ sheets alternating with $(Bi_2Ti_3O_{10})^{2-}$ perovskite like layers [27] as shown in Figure 3.

Materials Research Foundations **112** (2021) 311-335 https://doi.org/10.21741/9781644901595-9

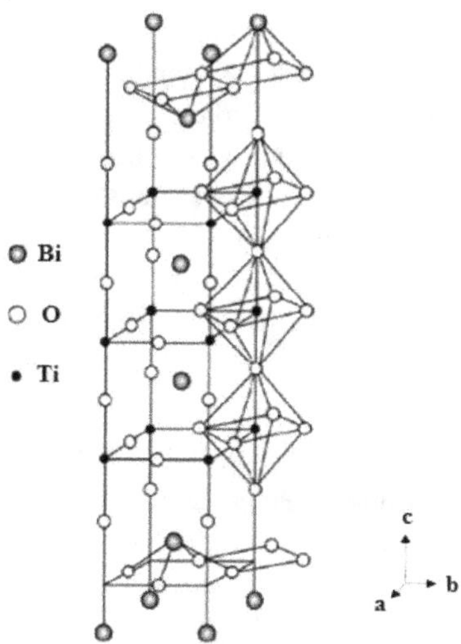

Figure 3: $Bi_4Ti_3O_{12}$ Crystal Structure [27].

In Figure 3, the Ti ions are enclosed by oxygen octahedral, whereas, Bi-ions occupy the spaces in the framework of TiO_6 octahedral. Along the c-axis, the Perovskite units $(Bi_2Ti_3O_{10})^{2-}$ are enclosed with $(Bi_2O_2)^{2+}$ layers. Subbarao has first discovered ferroelectricity in $Bi_4Ti_3O_{12}$ [29]. The $Bi_4Ti_3O_{12}$ has attracted great interest due to its applications in optical devices and ferroelectric memories [25]. The $Bi_4Ti_3O_{12}$ exhibit orthorhombic crystal structure and has Curie temperature (T_c) around 675 °C [30-32].

5. Origin of multiferroicity in $Bi_4Ti_3O_{12}$

Multiferroic materials are of great technological importance, since ferroelectricity and ferromagnetism co-exist in same phase at room temperature. Due to the existence of coupling between the two orders, it will be possible to write the data bit with an electric field and read it with a magnetic field and vice-versa. Various research groups are working on $Bi_4Ti_3O_{12}$, to bring the two ferroic orders in single phase material along with coupling between them at room temperature.

In the field of ferroelectric oxides, study of the Aurivillius family multiferroic materials is a challenging area for diverse applications. The first report on the possibility of

Materials Research Foundations **112** (2021) 311-335 https://doi.org/10.21741/9781644901595-9

multiferroic properties was published by Lu *et al.* [33], who have investigated that partly substituting Ti ions with Fe ions leads to weak ferromagnetism at room temperature. In the year 2010 Chen *et al.* [34] reported the room temperature magnetoelectric coupling in $Bi_4(Ti_1Fe_2)O_{12-\delta}$ ceramics. They use conventional solid state reaction method to synthesize the compound. In the same year Chen *et al.* [35] has again reported the enhancement in multiferroic properties $Bi_4(Ti_1Fe_2)O_{12-\delta}$ (x = 0, 1, 1.5, 2) ceramics. They reported maximum remnant polarization ($2P_r$ = 4.1 $\mu C/cm^2$) and maximum remnant magnetization (M_r = $3x10^{-3} emu/g$). Due to high losses and conduction, the materials exhibit high losses and results in poor ferroelectric properties.

It has been reported that ferroelectric properties of $Bi_4Ti_3O_{12}$ have been improved by substitutions at either A-site or at the B-site or at the combination of both sites. It was reported by Park *et al.* that La-substituted $Bi_4Ti_3O_{12}$ ($Bi_{3.25}La_{0.75}Ti_3O_{12}$) shows large remnant polarization, low synthesis temperature and good fatigue endurance [36]. Kim *et al.* has reported that the substitution of Nd^{3+} ions in $Bi_4Ti_3O_{12}$ thin films was more effective in improving the ferroelectric properties than La-substitution [37]. Mao *et al.* has also reported that substitution of Nd^{3+} ions in $Bi_4Ti_3O_{12}$, ($Bi_{4-x}Nd_xTi_3O_{12}$, x = 0 to x = 0.09) improved the ferroelectric and dielectric properties by suppressing the oxygen vacancies [38]. Tomar *et al.* reported that the substitution of Nd^{3+} and Sm^{3+} in $Bi_4Ti_3O_{12}$ as $Bi_{4-x}M_xTi_3O_{12}$, (M = Nd, Sm) has improved the ferroelectric properties by reducing the leakage current [39]. Shigyo *et al.* has used Sm^{3+} ion in $Bi_4Ti_3O_{12}$ and improved the dielectric and ferroelectric properties by suppressing the concentration of oxygen vacancies [40]. Noguchi *et al.* has uses vanadium ions in $Bi_4Ti_3O_{12}$ and achieve remnant polarization ($2P_r$) of over $40\mu C/cm^2$ without compensating other physical properties [41].

From the above discussions, it has been concluded that substitution of La, Nd, Sm, Gd, Dy, and V were improving the ferroelectric and magnetic properties of $Bi_4Ti_3O_{12}$ by suppressing the concentration of oxygen vacancies and reducing the conduction losses. In this context Paul et. al. in 2014 studied the multiferroic and magnetoelectric properties of $Bi_{4-x}Sm_xTi_{3-x}Fe_xO_{12\pm\delta}$, $Bi_{4-x}Sm_xTi_{3-x}Co_xO_{12\pm\delta}$ and $Bi_{4-x}Sm_xTi_{3-x}Ni_xO_{12\pm\delta}$ ceramics [42-44]. They have found that the substitution of Sm at Bi-site and substitution of Fe, Co, and Ni at Ti-site improved the ferroelectric, ferromagnetic as well as magnetoelectric coupling of the ceramic compounds. The authors have used solid state reaction method to fabricate the compounds and proposed the F-center exchange mechanism and super-exchange interaction mediated through bridging oxygen to explain the source of ferromagnetism in ferroelectric $Bi_4Ti_3O_{12}$. Up to these findings, only single dopant at Ti-site has been investigated to enhance the multiferroic and magnetoelectric properties of $Bi_4Ti_3O_{12}$. On the similar framework, multi dopants have been used subsequently to improve the multiferroic properties of Aurivillius based compounds.

Extending our discussion on multi dopants systems, Jun Xiao et al. in 2015 has worked on four layered structured ferroelectrics on Ni-doped $Bi_4NdTi_3Fe_{1-x}Ni_xO_{15}$ (x=0.1, 0.3, 0.5 and 0.7) ceramics. The ferroelectric and magnetic properties together the optimum room temperature multiferroic property was acquired for x=0.3 composition [45]. In the same year Ruixia Ti et al. has worked on $Bi_{3.25}La_{0.75}(Ti_{3-x}Fe_{x/2}Co_{x/2})O_{12}$ ($0 \leq x \leq 0.5$) polycrystalline ceramics synthesized by the conventional solid state reaction method. The ceramics with x = 0.25 composition possess the best properties, including phase purity, large ferroelectric polarization and remnant magnetization compared with other samples [46]. In the year 2017, the D. L. Zhang et al. reported the structure evolution, magnetic and ferroelectric properties in Co-doped four and three-layered intergrowth Aurivillius compounds $Bi_4NdTi_3Fe_{1-x}Co_xO_{15}$-$Bi_3NdTi_2Fe_{1-x}Co_xO_{12-\delta}$ [47].

In continuation of this work, Yuying Wu et al. in 2017 and Ruixia Tia et al. in 2019 have synthesized $Bi_{4.25}Gd_{0.75}Ti_3Fe_{0.5}Co_{0.5}O_{15}$ and $Bi_4(Ti_{3-x}Fe_{x/2}Co_{x/2})O_{12}$ ($0 \leq x \leq 0.4$) polycrystalline ceramics. The low temperature ferromagnetism is successfully induced by Fe/Co co-doping. Particularly, the x=0.3 ceramics possess the largest remnant magnetization (2Mr) of about 0.272 emu/g at 100 K [48, 49].

From the existing literature on multiferroicity in $Bi_4Ti_3O_{12}$, it has been concluded that the research on Aurivillius based structured ferroelectric compounds is the burning topic amongst the scientific community all around the world. The search has been going on to synthesize new class of materials and to enhance the multiferroic as well as magnetoelectric coupling of the existing Aurivillius based compounds for the futuristic applications.

Now we are concentrating our discussions on some of the characteristic properties of $Bi_4Ti_3O_{12}$ based Aurivillius compounds investigated by different researchers, so that readers have some basic understanding, that how the substitutions at A- and B-sites affect the structural, multiferroic and magnetoelectric properties of $Bi_4Ti_3O_{12}$ compound.

6. Structural studies of $Bi_4Ti_3O_{12}$

The pure and substituted samples of $Bi_4Ti_3O_{12}$ multiferroic ceramics were mostly synthesized by solid state reaction method by different authors. Generally, high purity oxide precursor materials have been used to synthesize $Bi_4Ti_3O_{12}$. The sintering temperature were varied from 850 °C to 1100 °C depending upon the dopant used. Chen *et al.* [34] have first time synthesized the $Bi_4(Ti_1Fe_2)O_{12-\delta}$ multiferroic ceramics using solid state reaction method. They have found single phase material with no impurity peaks. The lattice parameters and the unit cell volume were found to increases with substitution, which may be attributed to the larger radius of Fe^{3+} ions (0.0645 nm) in comparison to Ti^{4+} ions (0.0605 nm). J. Paul et al. have substitute the Sm^{3+} ions at Bi-site and Fe/Co/Ni ions at Ti-site in $Bi_4Ti_3O_{12}$. They found that, in case of Fe substitution, Fe_2O_3 replaces Ti-sites successfully without forming any impurity phase up to $x \leq 0.3$ composition, whereas, for $x > 0.3$, the structure of the system changes from orthorhombic to tetragonal [42]. On the other hand, in the case of Co and Ni doping, the optimum point of Co_3O_4 and NiO in $Bi_4Ti_3O_{12}$ is $x = 0.07$. For $x > 0.07$ the Co_3O_4 and NiO will start forming the secondary phase in the system [43, 44]. The X-ray diffraction pattern of $Bi_{4-x}Sm_xTi_{3-x}Co_xO_{12}$ has been shown in Figure 4. Upto x=0.07, the pure phase is obtained while at $x > 0.07$, an impurity peak at around $2\theta = 27.89°$ (indicated by *) has been obtained, which can be attributed to the formation of secondary phase and is identified as bismuth cobalt oxide ($Bi_{25}CoO_{40}$).

Similar results have been observed in X-ray diffraction studies by Ruixia Ti et al. [46], where the authors observed the impurity peaks in Fe/Co co-doped $B_{3.25}La_{0.75}Ti_3O_{12}$ ceramic system. They have observed weak impurity peaks related to $Bi_2Ti_2O_7$ (for x=0.01) and Bi_2O_3 (for x=0.05). These impurity peaks were attributed to the evaporation of Bi during the high temperature sintering and non-stoichiometry induced by doping. The orthorhombicity defined by the formula, 2(a-b)/(a+b), where a and b are the lattice parameters was found to decrease with increasing substitution. Similar results have also been obtained for other bismuth layered structured compounds [47, 48].

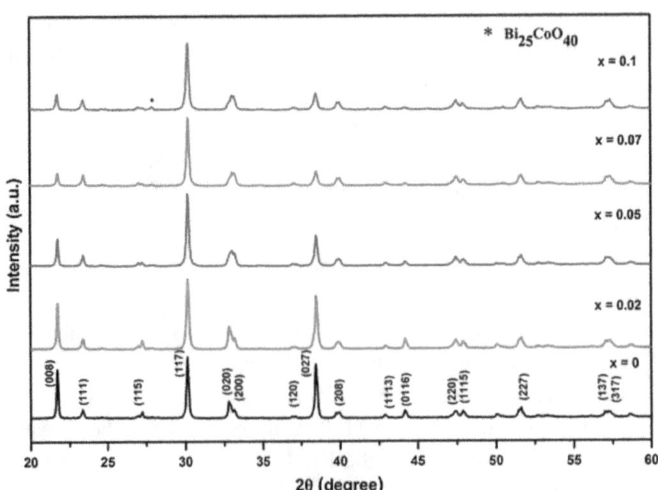

Figure 4: X-ray diffraction pattern of $Bi_{4-x}Sm_xTi_{3-x}Co_xO_{12}$ (Reprinted from reference [43] with kind permission from Springer Nature).

Materials Research Foundations **112** (2021) 311-335 https://doi.org/10.21741/9781644901595-9

From the findings of X-ray diffractions studies, it has been concluded that the substituent used has altered the structure of the parent Bi-layered compounds. The occurrence of the impurity peaks will be dependent on the ionic radii of the dopant at A- and B-sites, the composition, the reactivity of the dopant with the other precursor materials and the sintering temperature also.

To study the structure of the synthesized compounds Raman spectroscopy is a very useful technique and frequently used to investigate the structure of $Bi_4Ti_3O_{12}$ and related compounds. It has been published in different reports that orthorhombic $Bi_4Ti_3O_{12}$ has 24 Raman active modes [50, 51]. The Raman spectra of $Bi_{4-x-}Sm_xTi_{3-x}Co_xO_{12}$ has been shown in Figure 5. The low frequency modes i.e., modes observed below 200 cm^{-1} are ascribed to the vibrations of Bi^{3+} ions in the perovskite layer, while the high frequency modes above 200 cm^{-1} are attributed to the bending and stretching of TiO_6 octahedra.

In Figure 5, the Raman modes were observed at around 116, 144, 224, 268, 328, 357, 535, 564, 610 and 850 cm^{-1}. It has been found that mode at 224 cm^{-1} is getting suppressed with increasing substitution of Co. This can be ascribed to the decrease in distortion of TiO_6 octahedra and hence the decrease in orthorhombicity [52]. The modes observed at 116 and 144 cm^{-1} reflects the vibration of Bi^{3+} ions in layer structured pervoskite. The more fascinating result is the Raman mode at about 716 cm^{-1}, which becomes more prominent with increasing Co content. The occurrence of new Raman mode is ascribed to the vibration of Co-O bond, due to the substitution of Ti by Co ions whose intensity increases gradually with increase in Co substitution. This confirms the substitution of Co ions in pervoskite unit cell.

Similar results of Raman spectroscopy have been observed for Fe/Co co-doped $B_{3.25}La_{0.75}Ti_3O_{12}$ ceramics by Ruixia Ti et al. [46].

It has been reported that after co-doping of Fe and Co, the relative intensity of common modes decreases gradually. The bands observed at 269, 557, and 851 cm^{-1} becomes broader and could be attributed to the lattice distortion and stress induced by doping. A new vibration mode at 732 cm^{-1} comes into existence after doping, whose intensity increases with doping. The authors ascribed this new mode to the FeO_6 and CoO_6 octahedral.

Finally, from the structural studies, it has been concluded that the substitution at A-and B-site alter the crystal structure of $Bi_4Ti_3O_{12}$. The structure transforms from orthorhombic to tetragonal at some optimum compositions and the orthorhombicity decreases with substitution due to relaxation of orthorhombic distortions. It was confirmed from Raman studies that the intensity of common Raman active modes decreases, whereas new modes come into picture with substitution.

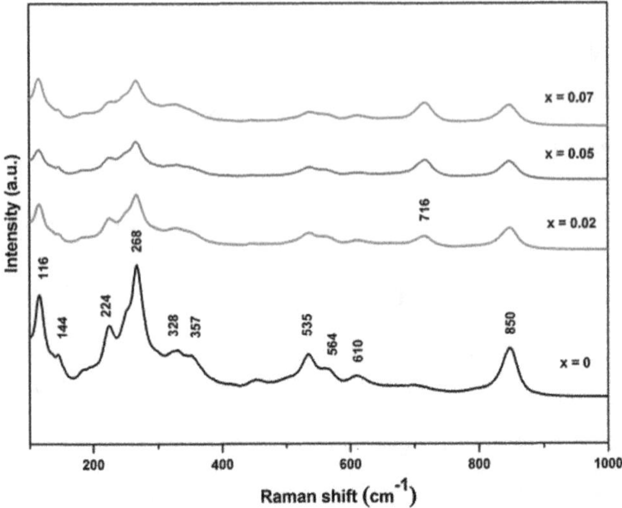

Figure 5: Raman Spectra of $Bi_{4-x}Sm_xTi_{3-x}Co_xO_{12}$ (Reprinted from reference [43] with kind permission from Springer Nature).

7. Ferroelectric studies of $Bi_4Ti_3O_{12}$

Ferroelectrics are technologically important materials, for applications in non-volatile ferroelectric random-access memories (Fe-RAM). One of the characteristic properties of a multiferroic material is that it shows saturated polarization-electric field (P-E) hysteresis curve. $Bi_4Ti_3O_{12}$ Aurivillius-type compounds are the promising candidates for such applications. These materials offer several advantages such as fatigue-free, lead free, low operating voltage and high curie temperature. The fatigue problem of the $Bi_4Ti_3O_{12}$ compound was resolved by the addition of different dopants (La, Nd, Sm, V etc.) at Bi^{3+} site, which supresses the oxygen vacancies and improve the ferroelectric properties of the compound.

Figure 6 shows the room temperature P-E hysteresis curve of $Bi_{4-x}Sm_xTi_{3-x}Co_xO_{12-\delta}$ ceramics. It was observed that the shape of ferroelectric loop improves with increasing content of Sm^{3+} and Co^{3+} ions, which indicate the reduction in conduction losses. With increase in substitution, there is significant enhancement in remnant polarization ($2P_r$) and its value increases from 8.66 $\mu C/cm^2$ (for x=0) to 15.82 $\mu C/cm^2$ (for x=0.07). This enhancement was attributed to the decrease in orthorhombicity which reduces the defects related to segregation of vacancies and enhances the movement of domain walls [53].

Materials Research Foundations **112** (2021) 311-335 https://doi.org/10.21741/9781644901595-9

Similar results have also been observed for Fe, Ni and other co-doping systems at different concentrations [42, 44, 46]. One of the most important factors, that affect the ferroelectric properties of $Bi_4Ti_3O_{12}$, is how the grain size vary with dopant concentration. It was proposed that if the grain size increases with substitution, the volume fraction of grain boundaries reduces. This decreases the probability of trapping the space charge at the grain boundaries. As a result, the pinning of neighbouring domains gets reduced, which makes domain reorientation easier and enhances the remnant polarization [54].

It was concluded from ferroelectric studies that substitution of different ions at Bi-site has profound effect on the ferroelectric properties of $Bi_4Ti_3O_{12}$.

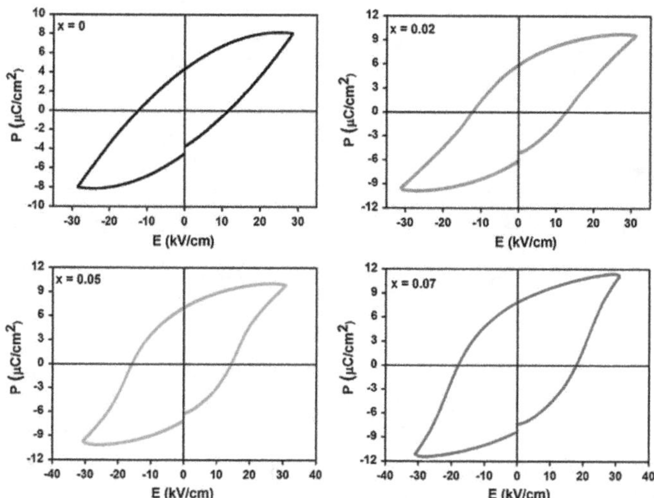

Figure 6: Ferroelectric hysteresis loops of $Bi_{4-x}Sm_xTi_{3-x}Co_xO_{12-\delta}$ (x = 0, 0.02, 0.05, 0.07) (Reprinted from reference [43] with kind permission from Springer Nature).

8. Multiferroic studies of $Bi_4Ti_3O_{12}$

The co-existence of ferroelectricity and ferromagnetism are the pre-requisite of the multiferroic material. Multiferroic are of great technological potential, since ferroelectricity and ferromagnetism co-exist at room temperature in same phase. In $Bi_4Ti_3O_{12}$ ferromagnetism has been introduced by incorporating the partially filled d-orbitals elements such as Fe, Ni, Co, Mn, Gd, Cr etc. or their combination at the Ti^{4+} site.

Figure 7 shows the magnetization curve for $Bi_{4-x}Sm_xTi_{3-x}Co_xO_{12\pm\delta}$ (x = 0, 0.02, 0.05, 0.07) ceramics at room temperature.

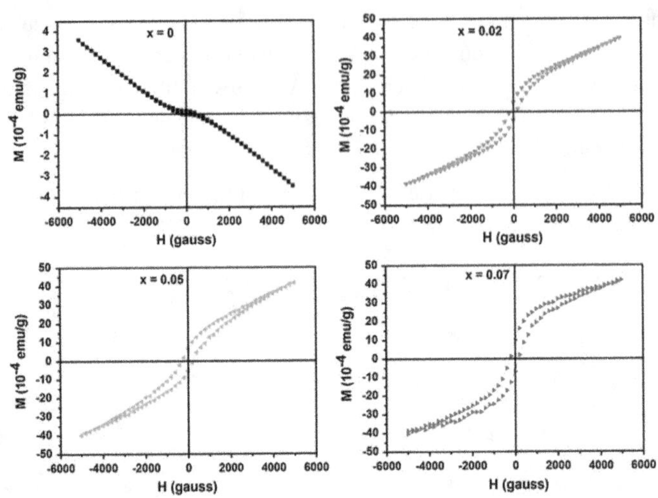

Figure 7: Room temperature M-H curve for $Bi_{4-x}Sm_xTi_{3-x}Co_xO_{12-\delta}$ (x = 0, 0.02, 0.05, 0.07) ceramics (Reprinted from reference [43] with kind permission from Springer Nature)

It was observed from Figure 7 that for pure $Bi_4Ti_3O_{12}$, an anti S-type nature of loop has been observed, which confirm the diamagnetic nature of $Bi_4Ti_3O_{12}$. Whereas, with substitution of Sm and Co, the loop gradually changes into a S-type hysteresis loop, indicating the origin of weak ferromagnetism. The remnant magnetization was found to increase with increasing content of Co ions.

The authors have proposed different mechanism to explain the origin of weak ferromagnetism:

i. In oxide materials the oxygen vacancies were created naturally to compensate the charge neutrality. In $Bi_4Ti_3O_{12}$, when Ti^{4+} site is substituted by Co^{3+} ions, there is a formation of Co^{3+} - □ - Co^{3+} network structure where (-□- denotes the oxygen vacancy). There is creation of F-centres, when an electron having a down spin trapped in between these vacancies [35, 55]. Subsequently, Co^{3+} have an unoccupied $3d^6$ spin orbital, so it might be possible that F-centre exchange mechanism is suitable in Co^{3+} - □ - Co^{3+} network [56]. The direct ferromagnetic coupling is possible through exchange interactions between the two neighbouring Co^{3+} ions via F-centre.

ii. In another explanation, it was proposed that Co^{3+}-O-Co^{3+} coupling might occur, which also leads to ferromagnetic state.

iii. The valence state of Co ions also affects the existence of ferromagnetic state. Co ion may exist in +2 valence state also. This also favour the creation of Co^{3+}-O-Co^{2+} network structure. Therefore, the existence of double exchange interactions also favours the ferromagnetic state in doped $Bi_4Ti_3O_{12}$ ceramics [56].

Similar explanations have been given by Ruixia Ti et al. [46] in Fe/Co co-doped $B_{3.25}La_{0.75}Ti_3O_{12}$ ceramics. It was reported by Ruixia Ti et al that, x=0.1 sample has nonferromagnetic properties, whereas ferromagnetic nature comes into existence for all other compositions. It was observed for x=0.25 sample, maximum remnant magnetization (M_r) and smallest saturation magnetization (M_s), or we can say that largest remnant magnetization ratio M_r/M_s was observed, which shows largest ferromagnetism in the considered sample.

The authors have proposed similar explanations for the observed ferromagnetism such as:

i. Ferromagnetic interaction between bound magnetic polarons, and

ii. Ferromagnetic double exchange interactions between different valence states of Fe and Co ions.

From the above discussions, it has been concluded that, it is possible to introduce the magnetic properties in ferroelectric $Bi_4Ti_3O_{12}$ and hence the multiferroism. Different mechanisms have been proposed to explain the origin of ferromagnetism in $Bi_4Ti_3O_{12}$.

9. Magnetoelectric coupling in $Bi_4Ti_3O_{12}$

The magneto electric effect provides an additional degree of freedom for device fabrication. To determine the order of coupling between ferroelectric and ferromagnetic orders, magnetoelectric voltage is measured across the sample under applied dc magnetic field. The ME coupling co-efficient ($\alpha = dE/dH$), is calculated from the output voltage [57]. The variation of α as a function of applied dc magnetic field (H_{dc}) at room temperature is shown in Figure 8 for $Bi_{4-x}Sm_xTi_{3-x}Co_xO_{12-\delta}$ (x = 0.02, 0.05 & 0.07).

From Figure 8, it has been observed that the coupling coefficient has nonlinear behaviour with the applied dc magnetic field. The α is also observed to increase with increase in substitution. These results confirm the coupling between two ferroic orders. The coupling between two different ferroic orders is attributed due to the interaction between sub-

Materials Research Foundations **112** (2021) 311-335 https://doi.org/10.21741/9781644901595-9

lattices of electric and magnetic phases, through the transmission of stress and strain from one sub-lattice to another sub-lattice [58]. In the presence of magnetic field, the strain-induced magnetic sub-lattice induces a stress on the electrical sub-lattice which is realized as ME output.

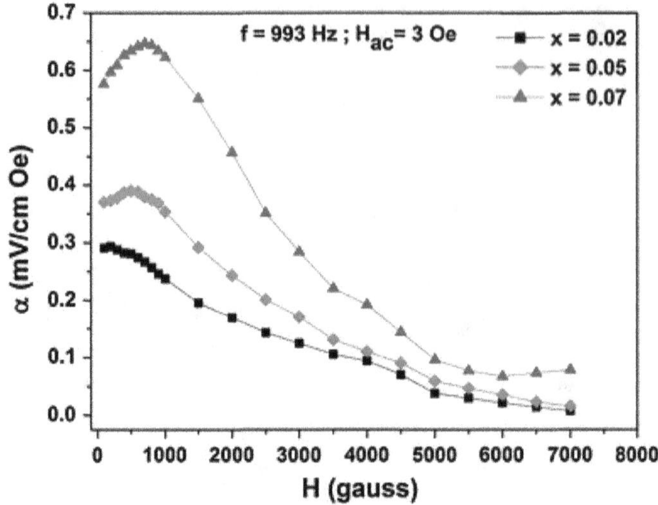

Figure 8: The variation of Magnetoelectric coefficient as a function of dc magnetic field (Reprinted from reference [43] with kind permission from Springer Nature)

Similar results have been observed by different authors, in Fe, Ni, Fe/Co co-doped $Bi_4Ti_3O_{12}$ compounds, which confirm the coupling between ferroelectric and ferromagnetic orders.

Finally, it would be concluded here that after substituting and optimizing the composition of substituents at A-site (Bi) and the B-site (Ti) in $Bi_4Ti_3O_{12}$, one can obtain the new class of multiferroic materials and the corresponding magnetoelectric coupling even at room temperature.

Table 1 gives the summary of different Aurivillius based compounds synthesized for possible applications as multiferroic materials.

Materials Research Foundations **112** (2021) 311-335 https://doi.org/10.21741/9781644901595-9

Table 1: Summary of Aurivillius based compound synthesized for multiferroic applications.

S.No.	Type of Aurivillius Compound	Dopant Used	Year	Reference
1.	$RBi_4FeTi_3O_{15}$	R=Nd, Sm, Gd, and Dy	2003	[58]
3.	$Bi_4Ti_3O_{12}$	Partially substitution of Fe at Ti site	2008	[33]
4.	$Bi_4\ (Ti_1Fe_2)\ O_{12-\delta}$	Substituting Fe at Ti site	2010	[34]
5.	$Bi_4\ (Ti_1Fe_2)\ O_{12-\delta}$	Substituting Fe at Ti site	2010	[35]
6.	$Bi_{4-x}Sm_xTi_{3-x}Ni_xO_{12\pm\delta}$	Substitution of Sm at Bi site and Ni at Ti site	2014	[44]
7.	$Bi_{4-x}Sm_xTi_{3-x}Co_xO_{12\pm\delta}$	Substitution of Sm at Bi site and Co at Ti site	2014	[43]
8.	$Bi_{4-x}Sm_xTi_{3-x}Fe_xO_{12\pm\delta}$	Substitution of Sm at Bi site and Fe at Ti site	2015	[42]
9.	$Bi_4NdTi_3Fe_{1-x}Ni_xO_{15}$	Substitution of Nd at Bi site and Ni/Fe at Ti site	2015	[45]
10.	$Bi_{3.25}La_{0.75}(Ti_{3-x}Fe_{x/2}Co_{x/2})O_{12}$	Substitution of La at Bi site and Fe/Co at Ti site	2015	[46]
11.	$Bi_4LaTi_3FeO_{15}$	Substitution of La at Bi site and Fe at Ti site	2015	[60]
12.	$Bi_4NdTi_3Fe_{1-x}Co_xO_{15}$-$Bi_3NdTi_2Fe_{1-x}Co_xO_{12-\delta}$	Nd at Bi site and Fe/Co at Ti site	2017	[47]
13.	$Bi_{4.25}Gd_{0.75}Ti_3Fe_{0.5}Co_{0.5}O_{15}$	Gd at Bi site and Fe/Co at Ti site	2017	[48]
14.	$Bi_4(Ti_{3-x}Fe_{x/2}Co_{x/2})O_{12}$ $(0\leq x\leq 0.4)$	Fe/Co at Ti site	2019	[49]

10. Future prospective of Aurivillius based multiferroic ceramics

Aurivillius based layered structured compounds have wide applications in ferroelectric random-access memories, high temperature piezoelectricity, photocatalytic applications and many more. From the technological point of view, the fabrication of Aurivillius based single-phase multiferroic materials, having applications at room temperature, is a complicated task. Various efforts are being made in this direction. The outcome of various research has confirmed that newly synthesized systems exhibit both the ferroelectric and ferromagnetic orders simultaneously, and these orders are coupled to each other at room temperature. Moreover, the ME coupling in Aurivillius based ceramic systems is small, and more efforts are required to enhance the ME coupling at room temperature, so that further applications can be extracted from the material. The room temperature single-phase multiferroic materials may definitely be used for the fabrication of data storage devices, multiple-state memory elements and sensors.

Moreover, in case of Aurivillius compounds, the requirement of high coercive field can make tough the required poling which is a disadvantage for the potential applications for these materials. The fabrication of thin films can overwhelm this problem, and enhances the applicability for Bi-based Aurivillius compounds. By reducing the thickness of the material, less voltage is required to apply high electric field, which enhances its utility in microelectronic devices. The ferroelectric properties in Aurivillius based thin films are studied well in the literature for optical applications. The growth of $Bi_4Ti_3O_{12}$ thin film depends on various parameters such as microstructure, orientation, and composition of the grown thin films. The substrate also plays an important role for optimizing the properties of the thin films. In the field of multiferroic thin films of Aurivillius based compounds, there are very less reports in the literature. The growth of Aurivillius based multiferroic thin films and their mechanism involved, is a burning topic amongst the scientific community. Different thin film techniques such as sol-gel spin coating, pulse laser deposition, and Magnetron sputtering can be used and optimized for the growth of $Bi_4Ti_3O_{12}$ based multiferroic thin films for futuristic applications.

11. Comparison of multiferroic and ferrite materials

Multiferroic materials are those, that exhibit two or more than two orders, including electric, magnetic, and elastic. Electric order originates form the ordering of electric dipoles, and magnetic order results due to the exchange interactions between the magnetic dipoles. Multiferroic materials have many potential applications in multifunctional devices such as multiple-state memory elements, transducers, spintronics, sensors, and tetrahertz radiations Bi-based Aurivillius-type oxides system having layered perovskite structure, attracted increasing research interest due to their wide applicability.

The existence of simultaneous electric and magnetic ordering at room temperature, makes this compound a good candidate for multiferroic applications. This Aurivillius-type oxides family can accommodate different elements at A- and B-sites and exhibit a rich spectrum of properties which enables the fabrication of new functionalized devices.

On the other hand, ferrites are oxides materials with outstanding magnetic properties. They are composed of iron oxide as main constituent and metal oxides. On the basis of crystal structure, ferrites are classified into: (i) Spinel ferrite, (ii) Garnet, (iii) Ortho-ferrite and, (iv) Hexagonal ferrites. They have broad range of industrial applications as high and low frequency transformer cores, electromagnetic interference (EMI) suppressions, antenna rods, core of coils in microwave frequency devices, computer memory core elements etc. These broad ranges of applications are owed to their low eddy currents, high resistivity, high saturation magnetization, and high Curie temperature. Moreover, ferrite nanoparticles have extraordinary chemical and physical properties, which has extended their scope in magnetically driven drug delivery system, MRI, and hyperthermia applications.

The main issue in single phase multiferroic materials is their low magnetoelectric coupling. To overcome this issue, lot of research has been carried out on multiphase multiferroic composites. The results shows that multiphase multiferroic composites have large value of saturation magnetization and polarization in comparison to single phase multiferroic systems. Composite's nanostructure is a good approach in this direction. Some of the composite system studied are $NiFe_2O_4$-$BaTiO_3$, (Ni, Zn) Fe_2O_4-$BaTiO_3$, $CoFe_2O_4$-$BaTiO_3$, (Ba, Sr) TiO_3-(Ni, Zn) Fe_2O_4 etc. [60-64]. Large number of applications can be extracted from the composite systems consisting of ferrite as one phase and ferroelectric as second phase. We can fabricate multiferroic composite using different single and mixed ferrites for magnetic properties and doped Bi-based Aurivillius compounds for ferroelectric properties. While designing multiferroic composites, parameters such as difference in thermal expansion, mismatch in crystal lattices and grain boundary has to be taken care.

Conclusions

In this chapter, authors have tried to give an overview of multiferroic materials their types and magnetoelectric coupling. Then, authors have included the work done so far on Aurivillius based multiferroic material ($Bi_4Ti_3O_{12}$) and discussions on its properties. At the end authors have incorporate comparison between ferrites and multiferroic materials and their future prospective.

Conflict of Interest

The authors declare that they have no conflict of interest.

Acknowledgment:

The authors are grateful to the research and scientific community who have made relentless efforts to bring novel data in the fields and topics of research and development discussed in this review. This work did not receive any funding.

References

[1] H. Schmid, Multi-ferroic magnetoelectrics, Ferroelectrics,162 (1994) 317-338. https://doi.org/10.1080/00150199408245120

[2] Y. Tokura, and S. Seki, Multiferroics with spiral spin orders, Adv. Mater., 22 (2010) 1554-1565. https://doi.org/10.1002/adma.200901961

[3] T. Kimura, Spiral magnets as magnetoelectrics, Annu. Rev. Mater. Res., 37 (2007) 387-413. https://doi.org/10.1146/annurev.matsci.37.052506.084259

[4] J. Vanden Brink, and D. Khomskii, Multiferroicity due to charge ordering, J. Phys.: Condens. Matter, 20 (2008) 434217. https://doi.org/10.1088/0953-8984/20/43/434217

[5] P. Curie, "Sur la symetriedans les phenomenes physiques, symetrie d'un champ electrique et d'un champ magnetique, J. Phys. Theor. Appl., 3(1) (1894) 393-415. https://doi.org/10.1051/jphystap:018940030039300

[6] P. G. Radaelli, L. C. Chapon, A. Daoud-Aladine, C. Vecchini, P. J. Brown, T. Chatterji, S. Park, and S. W. Cheong, Electric field switching of antiferromagnetic domains in YMn_2O_5: A probe of the multiferroic mechanism, Phys. Rev. Lett., 101 (2008) 067205. https://doi.org/10.1103/PhysRevLett.101.067205

[7] M. Fiebig, Revival of the magnetoelectric effect, J. Phys. D: Appl. Phys., 38 (2005) R123. https://doi.org/10.1088/0022-3727/38/8/R01

[8] W. Eerenstein, N. D. Mathur, and J. F. Scott, Multiferroic and magnetoelectric materials, Nature, 442 (2006) 759-765. https://doi.org/10.1038/nature05023

[9] D. I. Khomskii, Multiferroics: Different ways to combine magnetism and ferroelectricity, Journal of Magnetism and Magnetic Materials, 306 (2006) 1-8. https://doi.org/10.1016/j.jmmm.2006.01.238

[10] S. W. Cheong, and M. Mostovoy, Multiferroics: a magnetic twist for ferroelectricity, Nature Materials, 6 (2007) 13-20. https://doi.org/10.1038/nmat1804

[11] J. P. Velev, S. S. Jaswal, and E. Y. Tsymbd, Multiferroic and magnetoelectric materials and interfaces, Philosophical Transactions of the Royal Society A, Mathematical, Physical and Engineering Sciences, 369 (2011) 3069. https://doi.org/10.1098/rsta.2010.0344

[12] D. Khomskii, Classifying multiferroics: Mechanisms and effects, Physics, 2 (2009) 20. https://doi.org/10.1103/Physics.2.20

[13] T. Kimura, T. Goto, H. Shintani, K. Ishizaka, T. Arima, and Y. Tokura, Magnetic control of ferroelectric polarization, Nature 426 (2003) 55. https://doi.org/10.1038/nature02018

[14] T. Lottermoser, T. Lonkai, U. Amann, D. Honlwein, J. Ihringer, and M. Fiebig, Magnetic phase control by an electric field, Nature, 430 (2004) 541-544. https://doi.org/10.1038/nature02728

[15] W. Eerenstein, N. D. Mathur, and J. F. Scott, Multiferroic and magnetoelectric materials, Nature, 442 (2006) 759-765. https://doi.org/10.1038/nature05023

[16] M. Gajek, M. Bibes, S. Fusil, K. Bouzenouane, J. Fontcuberta, A. Barthelemy, and A. Fort, Tunnel junction with multiferroic barriers, Nat. Mater., 6 (2007) 296. https://doi.org/10.1038/nmat1860

[17] W. C. Rontgen, Uber die durch Bewegung eines in homogenous electrischen Felde befindlichen Dielectricums hervorgerufene electrodynamische Kraft, Ann. Phys., 35, (1888) 264-270. https://doi.org/10.1002/andp.18882711003

[18] H. A. Wilson, Phil. Trans. R. Soc. A, 204 (1905) 129-136.

[19] I. Dzyaloshinskii, On the magnetoelectric effect in antiferromagnets, Sov. Phys. JETP, 10 (1959) 628.

[20] D. N. Astrov, The magnetoelectric effect in antiferromagnetics, Sov. Phys. JETP, 11 (1960) 708.

[21] D. N. Astrov, Magnetoelectric effect in chromium oxide, Sov. Phys., 13, (1961) 729-733.

[22] W. J. Merz, Domain formation and domain wall motions in ferroelectric $BaTiO_3$ single crystals, Phys. Rev., 95 (1954) 690. https://doi.org/10.1103/PhysRev.95.690

[23] R. D. King-Smith, and D. Vanderbilt, First-principles investigation of ferroelectricity in perovskite compounds, Phys. Rev. B, 49 (1994) 5828. https://doi.org/10.1103/PhysRevB.49.5828

[24] G. Burns, and B. A. Scott, Lattice modes in ferroelectric Perovskite: $PbTiO_3$, Phys. Rev. B, 7 (1973) 3088. https://doi.org/10.1103/PhysRevB.7.3088

[25] J. M. Herbert, Electrocomponent Science Monographs, Gordon and Breach, New York, 1982.

[26] B. Aurivillius, Mixed bismuth oxides with layer lattices: The structure type of $CaNb_2Bi_2O_9$, Arkiv for kemi., 1 (1949) 463-480.

[27] B. Aurivillius, Mixed bismuth oxides with layer lattices, structure of $Bi_4Ti_3O_{12}$, Arkivforkemi1, 58 (1949) 499-512.

[28] B. Aurivillius, Mixed oxides with layer lattices, structure of $BaBi_4Ti_4O_{15}$, Arkiv for kemi2, 37 (1950) 519-527.

[29] E. C. Subbarao, Ferroelectricity in $Bi_4Ti_3O_{12}$ and its solid solutions, Phys. Rev., 122 (1961) 804-807. https://doi.org/10.1103/PhysRev.122.804

[30] T. Takenaka, K. Sakata, Grain orientation effects on electrical properties of bismuth layer-structured ferroelectric solid solution, ceramics of ferroelectric bismuth compound with layer structure, Jpn. J. Appl. Phys., 5513 (1984) 1092-1099. https://doi.org/10.1063/1.333198

[31] Q. Zhou, and B. J. Kennedy, Structural studies of the ferroelectric phase transition in $Bi_4Ti_3O_{12}$, Chem. Mater., 15 (2003) 5025-5028. https://doi.org/10.1021/cm034580l

[32] B. D. Stojanovic, C. O. Paiva-Santos, M. Cilense, C. Jovalekic, and Z. Z. Lazarevic, Strictly study of $Bi_4Ti_3O_{12}$ produced via mechano-chemically assisted synthesis, Mater. Res. Bull., 43 (2008) 1743-1753. https://doi.org/10.1016/j.materresbull.2007.07.007

[33] J. Lu, L. J. Qiao, and W. Y. Chu, Comparison of room temperature multiferroics in $Bi_4Fe_2TiO_{12}$ film and bulk, J. Univ. Sci. Technol. Beijing, 15 (2008) 782. https://doi.org/10.1016/S1005-8850(08)60287-X

[34] X. Q. Chen, F. J. Yang, W. Q. Cao, D. Y. Wang, and K. Chen, Room temperature magnetoelectric coupling in Bi_4 (Ti_1Fe_2) $O_{12-\delta}$ system, J. Phys. D: Appl. Phys., 43 (2010) 065001. https://doi.org/10.1088/0022-3727/43/6/065001

[35] X. Q. Chen, F. J. Yang, W. Q. Cao, H. Wang, C. P. Yang, D. Y. Wang, and K. Chen, Enhanced multiferroic characteristics in Fe-doped $Bi_4Ti_3O_{12}$ ceramics, Solid State Commun., 150 (2010) 1221-1224. https://doi.org/10.1016/j.ssc.2010.04.002

[36] B. H. Park, B. S. Kang, S. D. Bu, T. W. Noh, J. Lee, W. Jo, Lanthanum-substituted bismuth titanate for use in non-volatile memories, Nature, 401 (1999) 682-684. https://doi.org/10.1038/44352

[37] J. S. Kim, and S. S. Kim, Ferroelectric properties of Nd-substituted bismuth titanate thin films processed at low temperature, Appl. Phys. A: Materials science & amp; processing, 81 (2005) 1427-1430. https://doi.org/10.1007/s00339-004-3190-0

[38] X. Y. Mao, F. W. Mao, and X. B. Chen, Ferroelectric and dielectric properties of $Bi_{4-x}Nd_xTi_3O_{12}$ ceramics, Integrated Ferroelectrics: An international Journal, 79 (2011) 155-161. https://doi.org/10.1080/10584580600659365

[39] M. S. Tomar, R. E. Melgarejo, and S. P. Singh, Leakage current and ferroelectric memory in Nd and Sm substituted $Bi_4Ti_3O_{12}$ films, Journal of Microelectronics, 36 (2005) 574-577. https://doi.org/10.1016/j.mejo.2005.02.088

[40] T. Shigyo, H. Kiyono, J. Nakano, H. Itoh, and J. Takahashi, Synthesis and dielectric-magnetic properties of rare-earth (La, Nd, Sm)-modified $Bi_4Ti_3O_{12}$, Jpn. Soc. Appl. Phys., 47 (2008) 7617-7622. https://doi.org/10.1143/JJAP.47.7617

[41] Y. Noguchi, and M. Miyayama, Large remanent polarization of vanadium-doped $Bi_4Ti_3O_{12}$, Appl. Phys. Lett., 78 (2001) 13. https://doi.org/10.1063/1.1357215

[42] J. Paul, S. Bhardwaj, K. K. Sharma, R. K. Kotnala, R. Kumar, Room temperature multiferroic behaviour and magnetoelectric coupling in Sm/Fe modified $Bi_4Ti_3\ O_{12}$ ceramics synthesized by solid state reaction method, J. Alloys Compd., 634 (2015) 58-64. https://doi.org/10.1016/j.jallcom.2015.01.259

[43] J. Paul, S. Bhardwaj, K. K. Sharma, R. K. Kotnala, R. Kumar, Room-temperature multiferroic properties and magnetoelectric coupling in $Bi_{4-x}Sm_xTi_{3-x}Co_xO_{12-\delta}$ ceramics, J. Mater. Sci., 49 (2014) 6056-6066. https://doi.org/10.1007/s10853-014-8328-7

[44] J. Paul, S. Bhardwaj, K. K. Sharma, R.K. Kotnala, R. Kumar, Room temperature multiferroic properties and magnetoelectric coupling in Sm and Ni substituted $Bi_{4-x}Sm_xTi_{3-x}Ni_xO_{12-\delta}$ (x = 0, 0.02, 0.05, 0.07) ceramics, J. Appl. Phys., 115 (2014) 204909. https://doi.org/10.1063/1.4880159

[45] J. Xiao, H. Zhang, Y. Xue, Z. Lu, X. Chen, P. Su, F. Yang, X. Zeng, The influence of Ni-doping concentration on multiferroic behaviors in $Bi_4NdTi_3FeO_{15}$ ceramics, Ceram. Int., 41 (2015) 1087-1092. https://doi.org/10.1016/j.ceramint.2014.09.033

[46] R. Ti, X. Lu, J. He, F. Huang, H. Wu, F. Mei, M. Zhou, Y. Li, T. Xu and J. Zhu, Multiferroic properties and magnetoelectric coupling in Fe/Co co-doped

$Bi_{3.25}La_{0.75}Ti_3O_{12}$ ceramics, J. Mater. Chem. C, 3 (2015) 11868-11873. https://doi.org/10.1039/C5TC02399H

[47] D. L. Zhang, W. C. Huang, Z. W. Chen, W. B. Zhao, L. Feng, M. Li, Y. W. Yin, S. N. Dong and X. G. Li, Structure Evolution and Multiferroic Properties in Cobalt Doped $Bi_4NdTi_3Fe_{1-x}Co_xO_{15}$-$Bi_3NdTi_2Fe_{1-x}Co_xO_{12-\delta}$ Intergrowth Aurivillius Compounds, Sci. Rep., 7 (2017) 43540. https://doi.org/10.1038/srep43540

[48] Y. Wu, T. Yao, Y. Lu, B. Zou, Magnetic, dielectric, and magnetodielectric properties of Bi-layered perovskite $Bi_{4.25}Gd_{0.75}Fe_{0.5}Co_{0.5}Ti_3O_{15}$, J. Mater. Sci., 52 (2017) 7360-7368. https://doi.org/10.1007/s10853-017-0971-3

[49] R. Tia, C. Wanga, H. Wua, Y. Xua, C. Zhangb, Study on the structural and magnetic properties of Fe/Co co-doped $Bi_4Ti_3O_{12}$ ceramics, Ceram. Int., 45 (2019) 7480-7487. https://doi.org/10.1016/j.ceramint.2019.01.040

[50] S. Kojima, and S. Shimada, Soft mode spectroscopy of bismuth titanate single crystals, Physica B Condens. Matter., 219-220 (1996) 617-619. https://doi.org/10.1016/0921-4526(95)00830-6

[51] S. Kojima, Raman spectroscopy of bismuth layer structured ferroelectrics, Ferroelectrics, 239 (2000) 55-62. https://doi.org/10.1080/00150190008213305

[52] J. Zhu, X. B. Chen, Z. P. Zhang, and J. C. Shen, Raman and X-ray photoelectron scattering study of lanthnum-doped strontium bismuth titanate, Acta Mater., 53 (2005) 3155-3162. https://doi.org/10.1016/j.actamat.2005.03.020

[53] Q. Y. Jiang, E. C. Subbarao, and L. E. Cross, Effect of composition and temperature on electric fatigue of La-doped lead zirconate titanate ceramics, J. Appl. Phys., 75 (1994) 7433. https://doi.org/10.1063/1.356637

[54]. W. Y. Yi, W. X. Hui, and L. L. Tu, Ferroelectric and dielectric properties of La/Mn co-doped $Bi_4Ti_3O_{12}$ ceramics, Chin. Phys. B, 19 (2010) 037701. https://doi.org/10.1088/1674-1056/19/3/037701

[55]. J. M. D. Coey, A. P. Douvalis, C. B. Fitzgerald, and M. Venkatesan, Ferromagnetism in Fe-doped SnO_2 thin films, Appl. Phys. Lett., 84 (2004) 1332-1334. https://doi.org/10.1063/1.1650041

[56] X. Chen, C. Wei, J. Xiao, Y. Yue, X. Zeng, F. Yang, P. Li, and Y. He, Room temperature multiferroic properties and magneto-capacitance effect of modified ferroelectric $Bi_4Ti_3O_{12}$ ceramics, J Phys D: Appl Phys., 46 (2013) 425001. https://doi.org/10.1088/0022-3727/46/42/425001

[57] A. Singh, A. Gupta, and R. Chatterjee, Enhanced magnetoelectric coefficient in the modified $BiFeO_3$-$PbTiO_3$ system with large La substitution, Appl. Phys. Lett., 93 (2008) 022902. https://doi.org/10.1063/1.2945638

[58] A. Srinivas, D. W. Kim, K. S. Hong, and S. V. Suryanarayan, Observation of ferroelectromagnetic nature in rare-earth-substituted bismuth iron titanate, Appl. Phys. Lett., 83 (2003) 2217-2219. https://doi.org/10.1063/1.1610255

[59] R. Ti, F. Huang, W. Zhu, J. He, T. Xu, C. Yue, J. Zhao, X. Lu, J. Zhu, Multiferroic and dielectric properties of $Bi_4LaTi_3FeO_{15}$ ceramics, Ceram. Int., 41 (2015) S453-S457. https://doi.org/10.1016/j.ceramint.2015.03.157

[60] Y. Liu, Y. Wu, D. Li, Y. Zhang, J. Zhang, A study of structural, ferroelectric, ferromagnetic, dielectric properties of $NiFe_2O_4$–$BaTiO_3$ multiferroic composites, J. Mater Sci: Mater Electron 24 (2014) 1900-1904. https://doi.org/10.1007/s10854-012-1032-y

[61] C. Harnagea, L. Mitoseriu, V. Buscaglia, I. Pallecchi, P. Nanni, Magnetic and Ferroelectric Domain Structures in $BaTiO_3$–$(Ni_{0.5}Zn_{0.5})Fe_2O_4$ Multiferroic ceramics, J. Eur. Ceram. Soc. 27 (2007) 3947-3950. https://doi.org/10.1016/j.jeurceramsoc.2007.02.072

[62] S. S. Chougule, B. K. Chougule, Response of dielectric behaviour on ferroelectric rich $(y)Ni_{0.8}Zn_{0.2}Fe_2O_4 + (1 - y)$ PZT ME composites, Mater. Chem. Phys. 108 (2008) 408-412. https://doi.org/10.1016/j.matchemphys.2007.10.016

[63] S. G. Lu, Z. K. Xu, Y. P. Wang, S. S. Guo, Effect of $CoFe_2O_4$ content on the dielectric and magnetoelectric properties in $Pb(ZrTi)O_3/CoFe_2O_4$ composite, J. Electro. Ceram. 21 (2008) 398-400. https://doi.org/10.1007/s10832-007-9217-0

[64] K. K. Patankar, S. A. Kanade, D. S. Padalkar, B. K. Chougule, Complex impedance analyses and magnetoelectric effect in ferrite–ferroelectric composite ceramics, Phys. Lett. A 361 (2007) 472-477. https://doi.org/10.1016/j.physleta.2006.05.016

Materials Research Foundations **112** (2021) 336- https://doi.org/10.21741/9781644901595-10

Chapter 10

Hexagonal Ferrites, Synthesis, Properties and Their Applications

Pooja Dhiman[1,2,*], Rohit Jasrotia[1], Dipanshi Goyal[1], Genene Tessema Mola[3]

[1]School of Physics & Materials Science, Shoolini University of Biotechnology and Management Sciences, Bajhol, Solan (H.P.) 173229, India

[2]International Research Centre of Nanotechnology for Himalayan Sustainability (IRCNHS), Shoolini University, India

[3]School of Chemistry & Physics, University of KwaZulu-Natal, Pietermaritzburg Campus, Private Bag X01, Scottsville, 3209, South Africa

* dhimanpooja85@gmail.com

Abstract

We intend to report on possible fabrication routes for all types of hexagonal ferrites which are known for their wide area of use and applications. Hexagonal ferrites have now become an intense topic of research as they are the part of most of magnetic recording and data storage applications globally. Hexagonal or popularly known as 'Heaxa-ferrites' are known for their utilization in permanent magnets and their utilization in electrical devices being operated at high frequencies especially at GHz frequencies. We have presented in this chapter all main six types of hexagonal ferrites i.e. M Type, Z-Type, Y-type, W-type, X-Type and U-type hexa-ferrites. Hexaferrites belong to ferromagnetic class of magnetic materials and their properties are purely dependent on intrinsic structure of ferrites. In this chapter, we aim to discuss more on M-type of hexa-ferrites, their properties and their applications. Also, recent advances on M-type ferrites are also a part of this chapter.

Keywords

Hexaferrites, M-Type, Anisotropy, High Frequency Applications

Materials Research Foundations **112** (2021) 336- https://doi.org/10.21741/9781644901595-10

Contents

Materials Research Foundations **112** (2021) 336- https://doi.org/10.21741/9781644901595-10

1. Introduction

Ferrites belong to the class of most widely used magnetic materials with diverse application field. The applications covers a broad spectrum of today's technological devices whether electronic, storage devices, antennas and much more for high frequency applications. Ferrites are mainly classified into spinel, garnet, hexagonal, and perovskite types. Each type of these ferrites are unique in term of properties and corresponding applications [1, 2]. With the discovery of hexagonal ferrites in 1950's, there has been a continuous completion in exploring more exotic properties of hexa-ferrites. Most of the research works on hexa-ferrites are concentrated on their complex structure, structure modified properties and applications of theses ferrites [3-6]. Growing interest in the field of hexagonal ferrites can be analysed viewing the number of papers published on hexagonal ferrites during 2010-2020.

Figure 1 shows the publications data gathered from the Scopus database. We can see the tremendous increase in paper published on hexagonal ferrites. Hexaferrites finds their space in today's electronic devices, mobile as well wireless communication at GHz frequencies, EMI shielding, multiferroics, microwave absorbers, antenna applications etc. [7,8]. Hexagonal ferrites have certain advantages over their spinel counterparts, in term of higher ferromagnetic resonance frequency due to which they become suitable for being utilize in microwave regime to the preferred plane of magnetization [9].

Hexaferrites are mainly classified into six categories named as M-type (BaM or $BaFe_{12}O_{19}$, SrM or $SrFe_{12}O_{19}$), Z-type (Ba–Sr, Co hexagonal ferrite), Y-type ($Ba_2Me_2Fe_{12}O_{22}$), W-type ($BaMe_2Fe_{16}O_{27}$), X-type ($Ba_2Me_2Fe_{28}O_{46}$), U-type ($Ba_4Me_2Fe_{36}O_{60}$).

Out of possible types of hexaferrites, BaM constitutes the most of the commercially available magnetic material for diverse technological applications. In this chapter, we have focused on technological importance of hexaferrites, their classifications and most important properties and applications of hexaferrites.

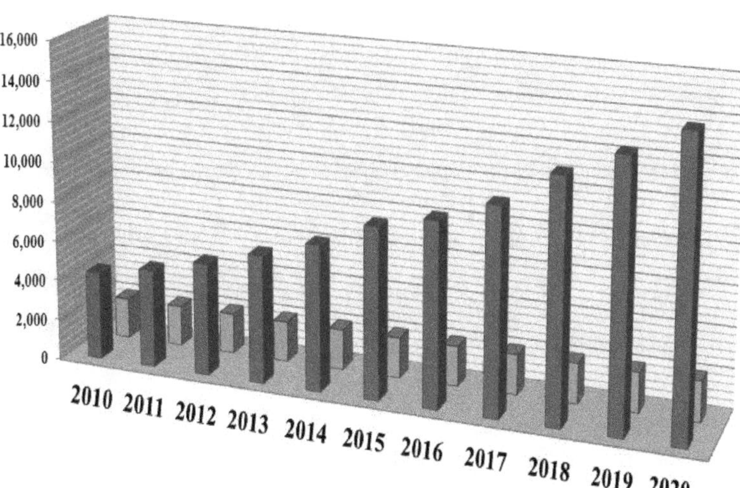

Figure 1: Ten Years data on hexaferrites publications from 2010-2020 (searched using keyword 'hexaferrites' on Scopus).

2. Classification and structure of hexaferrites

All magnetic oxides containing iron as a primary metallic element are referred to as ferrites. Due to their ferromagnetic and insulating characteristics at room temperature, they have made a significant contribution to various technological applications [10, 11]. According to chemical content and crystal structure, hexagonal ferrites are classified into six types: M ($AFe_{12}O_{19}$), W ($AMe_2Fe_{16}O_{27}$), X ($A_2Me_2Fe_{28}O_{46}$), Y ($A_2Me_2Fe_{12}O_{22}$), Z ($A_3Me_2Fe_{24}O_{41}$), and U ($A_4Me_2Fe_{36}O_{60}$); where A = Ba^{2+}, Sr^{2+}, La^{2+}, Pb^{2+} and Me = a bivalent transition metal [12, 13]. Hexaferrites are categorized into six primary categories based on their chemical formulae, as shown in Table 1. Figure 2 shows the schematic crystal structures of each hexaferrite. The M-type (magnetoplumbite) is the very well-known hexaferrite, whose structure (Figure 2) can be built up from S blocks interspersed by R blocks and symbolically expressed as RSR*S*. (The *equivalent block has been rotated 180 degrees around the hexagonal c-axis). The construction of the Y-type seen in Figure 2(d) can be thought of as an alternating stacking of the S and T blocks along the c-axis TST'S'T"S" where the apostrophe indicates that the accompanying block is rotated 120 degrees around this axis. Ferrimagnetism is one of the most essential features of hexaferrite from a technical perspective. On the other hand, not all hexaferrites have a ferrimagnetic ordered state as the ground state. Some hexaferrites, such as M-type Ba (Fe,Sc) $_{12}O_{19}$ [14], W-type $BaNi_2Sc_2Fe_{16}O_{27}$ [15], and Y-type $(Ba,Sr)_2Zn_2Fe_{12}O_{22}$, are reported to explain a non-collinear magnetic order [16, 17].

Table 1: Hexaferrite types with their chemical formula, composition, unit cell, space group and stacking order.

Ferrite	Chemical Formula	Space Group	C(Å)	Unit Cell	Stacking order
M	$(Ba,Sr)Fe_{12}O_{19}$	*P6₃/mmc*	~22	2M	RSR*S*
W	$(Ba,Sr)Me_2Fe_{16}O_{27}$	*P6₃/mmc*	~33	2W	$RS_2R^*S^*_2$
X	$(Ba,Sr)_2Me_2Fe_{28}O_{46}$	$R\bar{3}m$	~84	3X	$(RSR^*S^*_2)_3$
Y	$(Ba,Sr)_2Me_2Fe_{12}O_{22}$	$R\bar{3}m$	~43	3Y	TST'S'T"S"
Z	$(Ba,Sr)3Me_2Fe_{24}O_{41}$	*P6₃/mmc*	~52	2Z	RSTSR*S*T*S*
U	$(Ba,Sr)_4Me_2Fe_{36}O_{60}$	$R\bar{3}m$	~113	U	$(RSR^*S^*TS^*)_3$

S = Fe_6O_8 $^{2+}$ (spinel), R = $BaFe_6O_{11}$ $^{2-}$ (hexagonal), and T = $Ba_2Fe_8O_{14}$ are the sub-units for stacking order (hexagonal). The asterix (*) denotes a 180° rotation of the relevant sub-unit around the hexagonal axis.

2.2 W-type hexaferrite

W-type hexaferrite is constructed of a superposition of S blocks and R blocks along the hexagonal c-axis with the structure $RS_2R^*S^*_2$, where R is a three oxygen layer block with composition $BaFe_6O_{11}$, S (Spinel block) is a two oxygen layer block with composition Fe_6O_8, and '*' indicates that the respective block has been turned 180 degrees around the hexagonal axis. The R block still possesses a mirror plane, and the unit cell is built up of two molecular W units, yielding $RS_2R^*S^*_2$. The O^{2-} ions are found in close-packed layers, one Ba^{2+} (or Sr^{2+}) ion is located in R block replacing an O^{2-} ion, and the Fe^{3+} ions are located in five separate interstitial sublattices. The cell length of Fe_2W is 32.84 Å [12], and the unit cell is made up of two molecules of W units. It belongs to the *P6₃/mmc* space group.

2.3 X- type hexaferrite

The crystal structure of X-type hexaferrite is made up of one M and one W unit, and it can be seen as a stacking of R and S blocks along the hexagonal c-axis in the order $(RSR^*S^*_2)_3$ with the blocks of the W segment rotated through 180 degrees relative to those of the M section. The unit cell is made up of four different layers of M and W type structures. Three chemical formula units (Z = 3) are contained in the elementary cell (a = 5.88 Å, c = 84.11 Å), and space group is $R\overline{3}m$ [12, 20, 21].

2.4 Y-type hexaferrite

The crystalline structure of these ferrites is made up of one T block and one S block. So because the T block lacks a mirror plane, the overlap of hexagonal and cubic close-packed layers requires three T blocks, with the relative placements of the barium atoms repeating every three T blocks. The unit cell formula is 3 (ST), and it has hexagonal symmetry, with 18 oxygen layers and a repeat distance of only six oxygen layers, and a length of 43.56 Å on the c-axis. The crystal structure of Y-type hexaferrite is shown in Figure 2(d). This compound's fundamental structure belongs to the space group $R\overline{3}m$ and rhombohedral crystal structure, which prevents spontaneous polarisation [6].

2.5 Z-type hexaferrite

M and Y type hexaferrites make up Z-type hexaferrite, making it ST + SR, with a mirror plane in the R block and 11 layers of oxygen layer repeat distance. With a c-axis length of 52.30 Å [22], the crystal structure has 33 close-packed layers stacked along the hexagonal c-axis, giving order RSTSR*S*T*S*. A Z-type unit cell has 140 atoms; two molecular units combine to produce a single Z-type ferrite structure, which belongs to the $P6_3/mmc$ space group [6].

2.6 U-type hexaferrite

Block structure of U-type is $(RSR^*S^*TS^*)_3$ with the space group $R\overline{3}m$, and they are a superposition of 2M and Y-blocks along the c-axis. HRTEM pictures revealed the MMY stacking sequence of U ferrite, as well as stacking defects where the crystal structure was damaged by the addition of extra Y blocks [23]. The U unit cell, which is made up of three molecular units with a = 5.88 Å and c = 38.16 Å [6]. $Sr_4Co_2Fe_{36}O_{60}$ has lattice parameters of a = 5.86 Å and c = 112.3 Å [12].

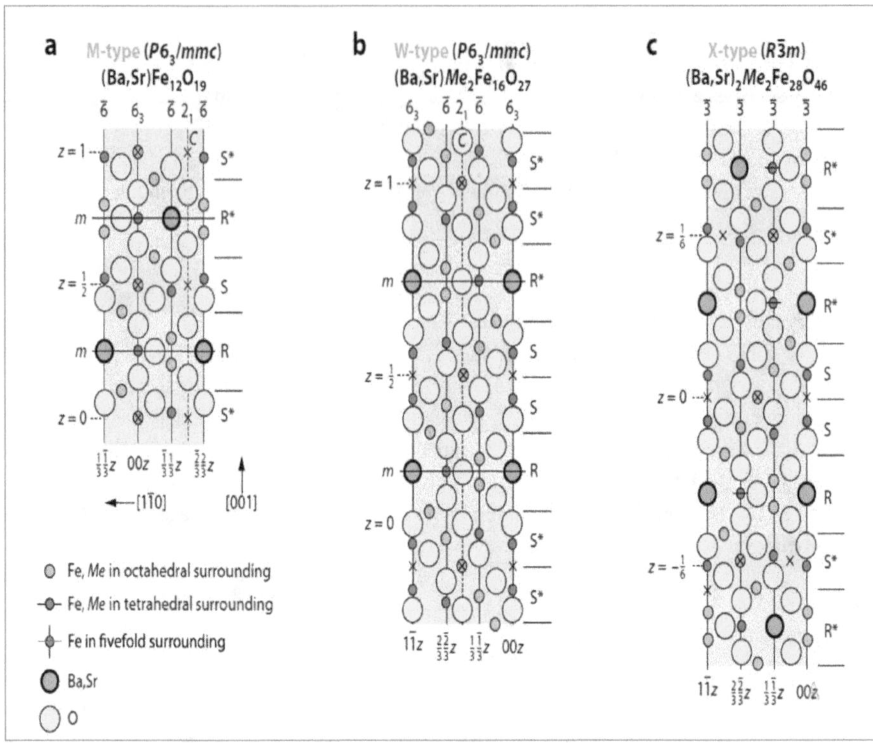

Figure 2: Schematic (a) crystal and (b-d) proposed magnetic structures of magnetoelectric Y-type hexaferrites. In the Y-type hexaferrites, (b) the proper-screw (or longitudinal-conical) magnetic structure transforms into (c) the transverse-conical one, which allows finite polarization by applying a magnetic field along the c plane. By applying higher magnetic fields (Reprinted with the permission from ref. [24]).

Materials Research Foundations **112** (2021) 336- https://doi.org/10.21741/9781644901595-10

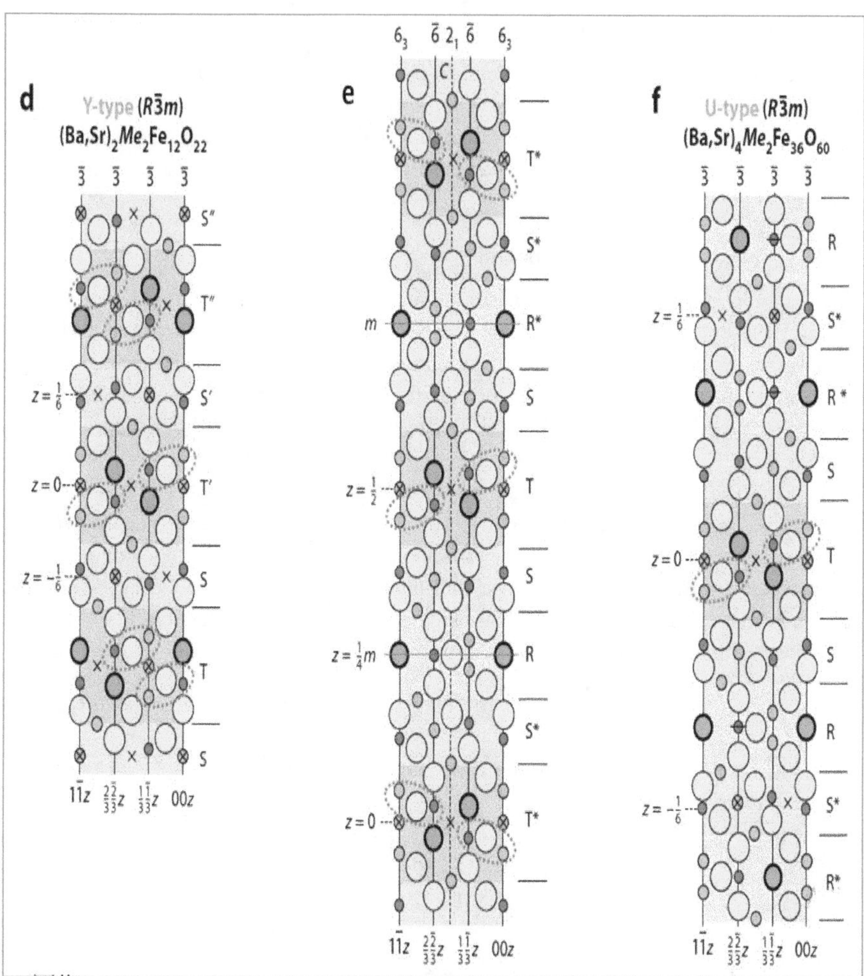

Figure 2: (d-f) the collinearly ferrimagnetic ordered state appears. The ratio of Sr to Ba has a significant impact on the (Fe, Me)-O-(Fe, Me) bond angles surrounded by dashed blue ellipsoids (Reprinted with the permission from ref.[24]).

3. Synthesis routes for hexaferrites

Hexagonal ferrites are highly difficult to prepare and involve much complicated steps in synthesis routes. Most of methods when utilized yield some by-products along with the required phases. A number of the researchers have opted various synthesis route like sol-gel [25], co-precipitation method [26], solution combustion method [27], hydrothermal method [28].

3.1 Traditional ceramic method

This process entails combining raw components by either grinding or melting them together, or just applying heat to a mixture of raw components. In the absence of solvents, it can happen with both wet and dry procedures. Wet mixing methods use an aqueous suspension, agitator mill, or vibration drum. This process is quite effective, but it demands a lot of energy for dewatering and drying. Dry mixing is done in a drum or ball mill, which grinds and mixes the components. Yang et al. [29] fabricated M-type Mn, Zn doped SrCaLa hexaferrite by traditional ceramic method. In the typical procedure, $SrCO_3$, $CaCO_3$, La_2O_3, ZnO, Mn_3O_4 and Fe_2O_3 were taken as raw materials. The needed weights of various starting ingredients for different stoichiometric ratio of compositions were weighted in estimate quantities and ball-mixed in a ball mill for 8 hours with distilled water. In an electric furnace, the combined powders were presintered at 1240 °C in air for 2 hours. Table 2 shows the effect of dopant on M-type SrCaLa hexaferrite.

Table 2: Magnetic parameters of different substituted SrCaLa (Reprinted with permission from ref. [29]).

MnZn content (x)	Ms (emu/g)	Mr (emu/g)	Hc (Oe)	Mr/Ms
0.0	69.15	25.32	957.0	0.366
0.1	70.52	14.32	417.6	0.203
0.2	69.10	12.97	399.2	0.188
0.3	65.69	16.11	967.5	0.356
0.4	62.43	22.25	967.5	0.356
0.5	59.62	23.45	1293.5	0.393

Figure 3: FE-SEM micrographs of the M-type $Ca_{0.4}Sr_{0.6-x}Nd_xFe_{12.0-x}(Nb_{0.5}Zn_{0.5})_xO_{19}$ hexaferrites with Nd-NbZn content (x) of (a) x=0.00, (b) x=0.08, (c) x=0.16, (d) x=0.24, and (e) x=0.32 [30] (Copyright permission with the licence number: 5087660456363).

Modifications to the traditional ceramic technique yielded $PbFe_{12}O_{19}$ hexaferrite powders with strong coercivity. At this temperature (presintered at 900°C/2 h and sintered at 1000°C/2 h), the highest parameters of saturation magnetization and coercive field were 62.0 emu/g and 4.0 kOe, respectively [31]. The traditional ceramic technique was used to make Nd-NbZn co-substituted M-type Ca-Sr hexaferrites with nominal compositions of $Ca_{0.4}Sr_{0.6-x}Nd_xFe_{12.0-x}(Nb_{0.5}Zn_{0.5})_xO_{19}$ (x = 0.00-0.32) [30]. The grains have platelet-like morphologies, according to FE-SEM micrographs (Figure 3). As can be seen, following Nd-NbZn co-substitution, the average grain size does not change significantly, and the grain size is clearly greater than the crystallite size(41.4-56.8nm) [30].

3.2 Co-precipitation method

Co-precipitation method is a very old method and being used for preparation of ferrites for decades. Co-precipitation method involves salts like chlorides or nitrates and a base as raw materials for co-precipitation of desired phase. A number of reports are available on M-type ferrites synthesized through co-precipitation method. However, this method is much more sensitive to the experimental conditions involved in synthesis route. In general, it has been noticed that for the preparation of BaM ferrites, this method requires Fe deficient environment and non-stoichiometric mixture of salts [32]. In a recent work, Hg doped BaM ferrite were synthesized using co-precipitation method [33]. In the typical procedure, barium chloride, ferric chloride, and sodium hydroxide were taken as raw materials. Chlorides taken in stoichiometric ratio were dissolved in distilled water and magnetically stirred at 60 ˚C. pH of the solution was maintained at 12 with the drop wise addition of alkaline NaOH solution. After two hours of stirring, precipitates were collected with the help of magnet and washed a number of times till the pH of the filtrate again reached 7. After drying, the particles are sintered at variable sintering temperature. The formation of BaM takes place but with an additional phase of Fe_2O_3 at some sintering temperatures of 700, 800, 900, 1000˚C. Results shows that sintering temperature must be optimized for the single phased BaM samples (Table 3). The magnetic parameters like saturation magnetization, remnant magnetization and coercive field increased with the Hg content, however intrinsic coercivity field decreased.

Table 3: Barium hexaferrite (synthesized at 700, 800, 900, and 1000 ˚C) and structural parameters(Reprinted with permission from ref. [33]).

Sample	$BaFe_{12}O_{19}$ (%)	Fe_2O_3 (%)	a (Ã…)	c (Ã…)	L (nm)
B-700	69.7	31.3	5.87	23.239	32.88
B-800	82.2	17.8	5.875	23.206	34.76
B-900	91	9	5.809	23.234	35.45
B-1000	100	0	5.893	23.132	36.08

3.3 Sol-gel route

Sol-gel method is known for mixing of particles of inorganic or metallo-organic precursors on the colloidal scale. Sol gel method involves mostly mixing of metal nitrates in desired solvent followed by continuous stirring of the solution. In general, citric acid solution is added to the above solution where citric acid works as chelating agent and helps in the formation of complex structure [34]. pH of the solution is maintained by dropwise addition of ammonia solution. Later on with heating, gel formation takes place. This gel on annealing at optimized temperature yields the desired material [35]. However, the experimental conditions like heating conditions, pH, ratio of chelating agent plays a vital role in morphology, shape and properties of ferrites prepared through sol-gel route. Pullar et al. devised an unique aqueous sol–gel approach for the synthesis of a variety of hexagonal ferrites from stoichiometric precursors, by adding barium, strontium, and cobalt salts to an acidpeptised iron(III)hydroxide (FeOOH) sol. This was utilised to make BaM [36], SrM [37], Co_2Y [38], Co_2W [39], Co_2Z [40], and Co_2X [40] ferrites, which were then blow-spun into continuous filaments. In a recent work, BaM nano-hexaferrites are synthesized by this route at various experimental conditions. Pictorial presentation of gel conversion into nanoferrites ash (tree like) is shown in Figure 4. Particles morphology, shape and size change changes with the change in annealing temperature which shows the sensitivity of the method (Figure 5). The sol–gel auto-combustion process was used to make M-type hexaferrites with Tb^{3+} substitution and a nominal composition of $Ba_{0.5}Sr_{0.5x}Tb_xAlFe_{11}O_{19}$ (x = 0.0, 0.05, 0.1, 0.15, 0.2, and 0.25) reported by Ali et al. The M-H loops were used to calculate magnetic parameters such as retentivity (M_r), saturation magnetization (M_s), coercivity (H_c), and squareness ratio (M_r/M_s). The saturation magnetization (M_s) and retentivity (M_r) reduced from 48.9 to 26.9 and 36.8 to 18.1emu/g, respectively, possibly due to spin canting and hence reduced super-exchange interactions. Coercivity is being improved (H_c) Higher magneto-crystalline anisotropy, which is attributed to the Fe^{2+} ions situated on a 2a site, could explain the increase in coercivity (H_c) from 1825 to 4440 G[41]. By using the sol–gel technique at 850 °C, a series of geometrically pure Co–Zr doped ferrites of the composition $BaCo_xZr_xFe_{(12-2x)}O_{19}$ were effectively produced. On substitution, M_S decreases somewhat from 63.63 to 56.94 emu/g, whereas H_C decreases dramatically from 5428 to 630 Oe, according to the magnetization data. Substituted compounds have a high M_S and a low H_C, making them ideal for data recording[42].

Figure 4: Conversion of gel into nano-hexaferrites [43] (Copyright permission with the licence number: 5087650493817).

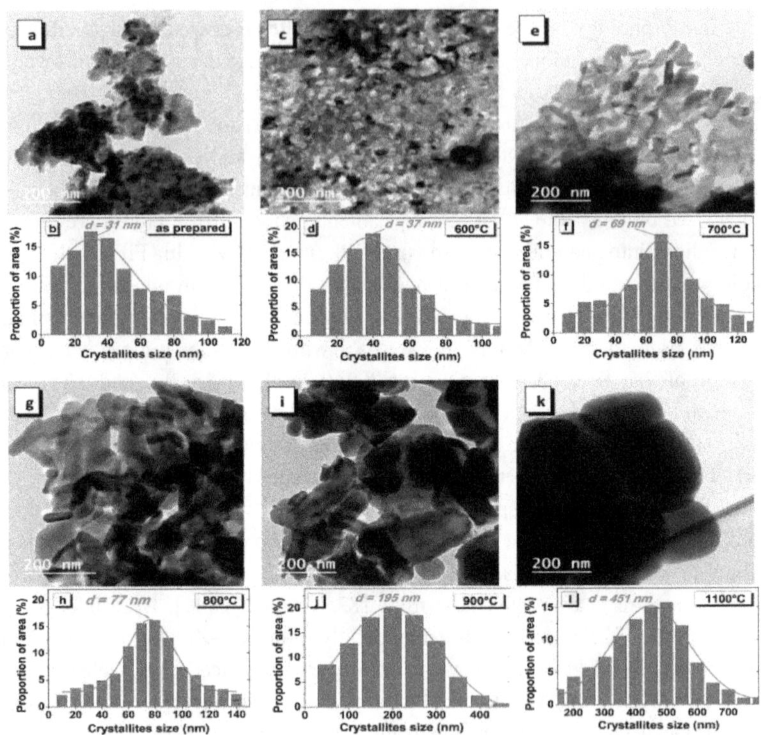

Figure 5: (a-l) TEM images and particle size histogram for BaM nanohexaferrites annealed at 600, 700, 800, 900, 1100 ℃[43] (Copyright permission with the licence number: 5087650493817).

3.4 Solution combustion method

Solution combustion method, a facile, economic and easy method which generally is used for preparation of oxide nano-particles [44]. In this method, metal nitrates are dissolved in suitable solvent with continuous stirring and heating arrangement. The fuel or reducing agent is then added for the open air combustion of the mixture at some moderate heating temperature [45, 46]. The solution combustion approach was used to make nanocrystalline M-type Al^{3+} substituted barium hexaferrite samples with the general formula $BaFe_{12x}Al_xO_{19}$ (where x = 40.00, 0.25, 0.50, 0.75, 1.00). The precursors were synthesized with stoichiometric quantities of Ba^{2+}, Fe^{3+}, and Al^{3+} nitrate solutions chelated citric with. The ratio of barium nitrate to citric acid was 1:2, and the pH of the solution was maintained at 8. With a rise in Al^{3+} substitution x from x=0.0 to 1.0, the

Materials Research Foundations **112** (2021) 336- https://doi.org/10.21741/9781644901595-10

saturation magnetization (Ms) and magneton number (n_B) decrease from 38.567 to 21.732 emu/g and from 7.6752 to 4.2126μ_B, respectively[47].

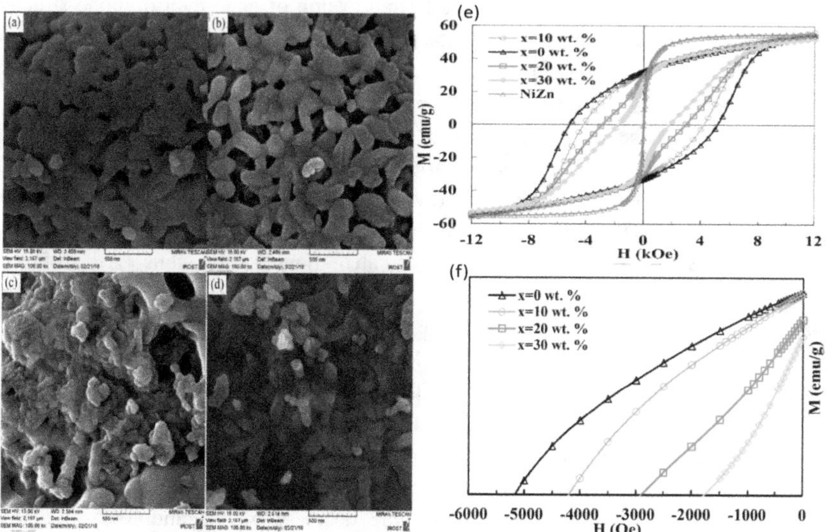

Figure 6: SEM micrographs of the $SrFe_{12}O_{19-x}$ $Ni_{0.6}Zn_{0.4}Fe_2O_4$ composites; (a) $SrFe_{12}O_{19}$, (b) x=10 wt. %, (c) x=20 wt. %, (d) x=30 wt. % (e) Magnetization curves of the $SrFe_{12}O_{19-x}Ni_{0.6}Zn_{0.4}Fe_2O_4$ composites and (f) Demagnetizing curves of the magnetic nanocomposites with different soft phase content[48](Copyright permission with the license number: 5087630182674).

Abasht et al. prepared $Ca_{1-x}La_xFe_{12}O_{19}$ (x = 0 and 0.4) by solution combustion method and calcium hexaferrite particles that were calcined at 1200 °C for 1 hour produced low levels of coercive field (23.52 kA m^{-1}) and maximum magnetization (58.75 Am^2 kg^{-1})[49]. $SrFe_{12}O_{19-x}$ $Ni_{0.6}Zn_{0.4}Fe_2O_4$ particles were synthesized using a one-pot solution combustion technique. The spherical and platelet particles in the composites exhibit two unique morphologies, which are connected to spinel and hexagonal ferrites, respectively (Figure 6(a-d)). The exchange coupling, as evidenced by its single hysteresis loop, was responsible for the increase in saturation magnetization from 54 to 56 emu/g for x= 10%. With the addition of more NiZn ferrite phase, a quirk in the hysteresis loop was discovered, and Ms was reduced to 52emu/g. Furthermore, when the NiZn ferrite concentration increased, the coercivity reduced from 5143 to 1778 Oe (Figure 6 (e)) [48].

3.5 Hydrothermal method

The hydrothermal approach has piqued the interest of researchers and scientists during the last 15 years, and it is beneficial in the fabrication of nano-hybrid, mono-structural, and nano-composite materials [50,51,52]. A mixture of metal salts and a base is autoclaved under pressure to produce the product in hydrothermal synthesis. The final product is frequently a mixed phase containing unreacted precursors and occasionally α-Fe_2O_3, which can be eliminated by dilute HCl washing. One of the most appealing aspects of hydrothermal process is the variety of ways it may be coupled with some other techniques to yield benefits such as improved reaction kinetics or the possibility to create novel materials [53]. Hydrothermal method is used to successfully produce M-type hexaferrites $BaMeFe_{11}O_{19}$ replaced with trivalent nonmagnetic ions (Al^{3+}, Bi^{3+}) and magnetic ions (Cr^{3+}, Mn^{3+}) reported by Ghzaiel et al. [54]. In comparison to pure $BaMeFe_{11}O_{19}$, the magnetic characteristics of M-type hexaferrite i.e. Ms, Mr, and Hc increases with the magnetic ion (Cr^{3+}, Mn^{3+}) substitution. On the other hand these magnetic characteristics decrease with the non-magnetic trivalent ions (Al^{3+}, Bi^{3+}) substitution. Table 4 shows that the produced samples have ferromagnetic properties, which are influenced by the magnetic behaviour of trivalent ion replacements and their occupied location in the five distinct Fe^{3+} lattice positions. Non-magnetic substitution by Al^{3+} and Bi^{3+} ions results in a considerable decrease of Ms and Mr. As a result, hydrothermal synthesising affects the substitution of Fe^{3+} by these trivalent ions at various places.

Table 4: Magnetic parameters of different substituted Barium hexaferrites (Reprinted with permission from ref. [54]).

Composition	µHc (T)	Ms (emu/g	Mr (emu/g)	Mr/Ms
$BaFe_{12}O_{19}$	0.241	61.84	33.95	0.567
A1900	0.335	51.89	28.86	0.535
A1950	0.329	52.86	30.05	0.538
Al1000	0.339	54.98	32.25	0.556
Bi900	0.488	46.27	24.54	0.531
Bi950	0.481	47.72	25.43	0.530
Bi1000	0.376	49.50	25.84	0.522
Cr900	0.449	36.47	19.41	0.532
Cr950	0.345	55.61	29.22	0.525
Cr1000	0.230	60.33	35.68	0.581
Mn900	0.434	50.16	27.96	0.546
Mn950	0.431	51.68	28.45	0.551
Mn1000	0.426	61.10	33.26	0.550

Figure 7: shows the SEM images of specimens hydrothermally synthesized with different atomic ratios of Fe/Sr ($R_{F/S}$). (a)–(g) $R_{F/S}$=2–8) [55] (Copyright with the permission of license number: 5087680143574).

Banihashemi et al. [56] fabricated cerium doped strontium hexaferrite by hydrothermal technique. FESEM revealed that the prepared nanoparticles are agglomerated in irregular platelets. The ferrimagnetic behaviour was confirmed by magnetic hysteresis loop experiments at room temperature, and the saturation magnetization decreased as the Ce content increased. The hydrothermal process was used to make hexagonal M type $SrFe_{12}O_{19}$ ferrites. SEM micrographs of SrM materials with various RF/S are shown in Figure 7. The image contains a large number of hexagonal plate-like formations, which aids in proving the synthesis of hexagonal SrM ferrites [55]. It's also been discovered that sintering specimens at high temperatures improves their magnetic characteristics.

3.6 Molten salt method

A solution of metal chlorides is coprecipitated by NaOH, Na_2CO_3, or $NaHCO_3$, and then the salts NaCl and/or KCl are added to this mixture, which is then dried and heated to a flux between 600 and 1100 °C, yielding a mixture of ferrite material and alkaline metal salts on cooling [57]. The original procedure was simpler, requiring only a mixture of conventional ceramic precursors $BaCO_3$ and Fe_2O_3 warmed in a NaCl–KCl flux [57], but the co-precipitation stage produces a superior product. Although the sodium and potassium salts can be completely recovered from ferrite by washing it with water or weak acid, the ferrite product will inevitably be contaminated with the alkaline metals. BaM hexaferrite was fabricated by molten salt method and they show well crystallized powders with high anisometric morphology. The saturation magnetization of barium hexaferrite is 90emu/g at 10k and 59 emu/g at 300k[58]. Li et al.[59] prepared BaM hexaferrite by MSM as shown in figure 8.

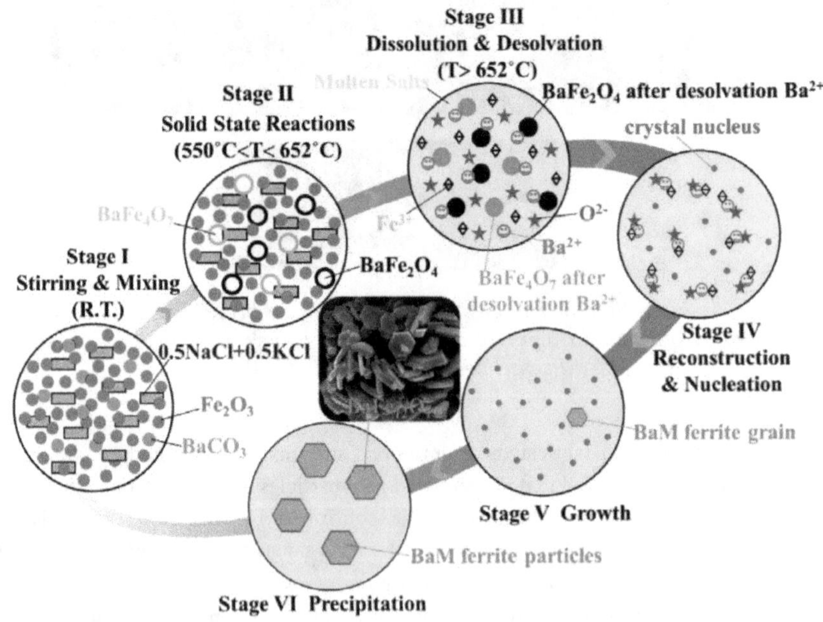

Figure 8: shows fformation mechanism of BaM ferrite via MSS with $BaCO_3$ and α-Fe_2O_3 powders as raw materials and (0.5NaCl+0.5KCl) as molten salt[59](Copyright with the permission of license number: 5087600526257).

BaM hexaferrites with near-theoretical saturation magnetization (Ms) and large remanent magnetization (Mr) were prepared using a low-temperature MSM technique, with Ms, Mr, and Hc of 71.9 emu/g, 39.5 emu/g, 3849.1 Oe and 71.8 emu/g, 38.9 emu/g, 2890.3 Oe, respectively. The Ms value of the prepared samples BaM (MSS-S) achieved 104.8 emu/g at a low temperature of 5 K. The (0.5NaCl0.5KCl) combination salts were crucial in the establishment of hexagonal plate shape and the acquisition of considerably increased magnetic characteristics[59]. Melt-flux precipitation is a modification on this approach in which less fluxing material is introduced, resulting in only microregions of partially melted salt in which crystallisation occurs instead of a true liquid phase system. The fluxing agents $7BaCl_2.3B_2O_3$ and γ-Fe_2O_3, $BaCO_3$, $CoCO_3$, TiO_2, have been used to generate BaCoTiM [60]. Pure ferrite was generated at a lower temperature and with a grain size one third that of under 100 nm when the melt-flux was heated radio thermally by an electron beam instead of conventional thermal heating [61].

3.7 Industrial manufacture

Commercial M ferrites are made from α-Fe_2O_3, which can be found naturally or manufactured from iron pyrite (FeS_2), iron chloride or organo-metallic pigments, and barium or strontium carbonates. Iron oxides rescued from waste steel pickling liquors, a cheap supply of raw material in sulphuric acid liquor with a high Fe^{2+} content, can be used to make ferrites. All three minerals, goethite, haematite, and magnetite, have been effectively recovered [62], and BaM with excellent magnetic characteristics has been produced from iron oxide recycled materials [63]. The vacuum hot steam process can be used to granulate ferrites, producing a material that is drier, denser, more homogeneous, and more abrasion resistant than conventional granulation methods. Shearing forces are used to homogenise and mix the ceramic in a high-speed mixer, and the material is subsequently granulated by drying under vacuum while superheated steam is pumped through it [64]. Mechanochemical processing, in which mixtures of iron and barium chlorides, in a non-stoichiometric ratio of Fe:Ba = 10, and a substantial excess of NaOH are milled together, is a straightforward way of producing small grain size and high coercivity ferrites. A chemical reaction is triggered, resulting in a homogenous combination of metal oxides/hydroxides and NaCl that may be washed off when the powder is annealed at 800 °C to produce BaM grains 100 nm wide and 20 nm thick [65]. Ms and Hc decrease with milling duration in dry milling of BaM due to two factors: increased lattice distortions from generated stress and development of the magnetically weak BaO and α-Fe_2O_3 phases. As the particle size is decreased, while Ms and Mr reduced and with the increase milling time, Hc also increases. However, if surfactants are added to the liquid, a significant Hc value is still formed, but the reduction in Mr is decreased, resulting in a material with a square loop and higher energy [66].

4. Magnetic properties of BaM and substituted BaM hexaferrites

Magnetic properties of hexaferrites are sensitive to intrinsic structure and symmetries of the structure. e.g. BaM synthesized through different techniques have emerged out with different magnetic properties. In a recent reported work by Guanghui Han et al. have synthesized Al doped Ba hexaferrite through solution combustion method and the prepared materials was quite porous in nature [67]. Bare Ba ferrite exhibit significant Mr/Ms ratio and ratio increases with the Al doping. The enhanced value of coercivity with Al doping has been linked with the enhancement on the magneto-crystalline anisotropy constant and shape morphology of particles. In an earlier work, Al doped Ba ferrites are fabricated using a chemical process followed by sodium citrate as chelating agent with further thermal treatment of precursor solution [68]. The saturation magnetization value was found to be 51.43 emu/g and consequently decreases up to 28.32 emu/g with the increase in Al content. However, anisotropy field increases 16.21 kOe to 25.01 kOe. Anisotropy plays a significant role in determining the magnetic properties of such M-type hexaferrites.

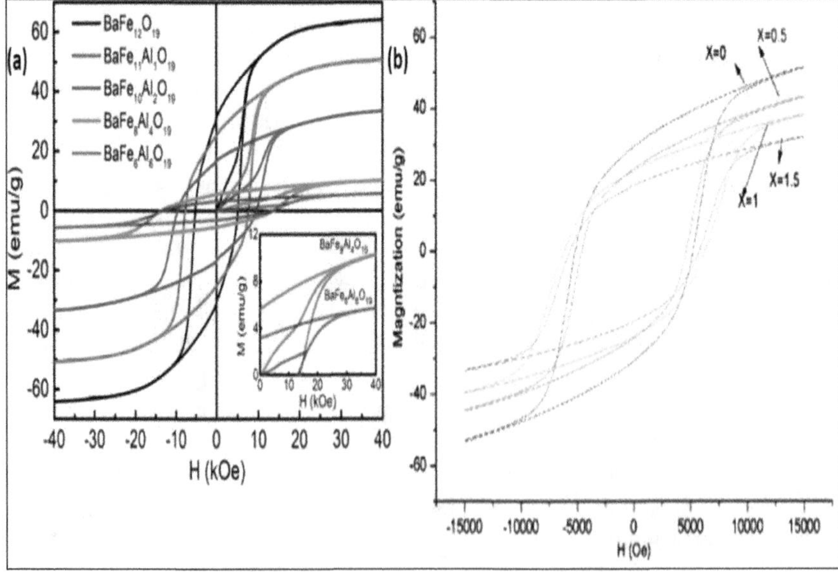

Figure 9: (a) M-H loops of the $BaFe_{12-x}Al_xO_{19}$ annealed at 1000 °C (Reprinted with the permission [67]), (b) The M-H curves of $BaAl_xFe_{12-x}O_{19}$ obtained at 900 °C[68] (Copyright permission with the licence number: 5087650685018).

However, the magnetic properties of BaM ferrite have been found to vary with the sintering temperature. In a recent report, BaM ferrite has been fabricated using citrate precursor method and effect of varied sintering temperature has been studied for 1000, 1100, 1200, 1300 ˚C for 24, 62, 124 and 148 hours of sintering [69]. At the sintering temperature of 1100 ˚C , formation of single crystalline BaM take place, and for 1300 ˚C sintering temperature, polycrystalline BaM ferrite forms. In addition, crystallite size also varied for different sintering temperature and sintering time. Figure 10 presents the obtained structure and the corresponding magnetic properties of samples. One can easily observe the drastic changes in magnetic properties of BaM with sintering temperature and time. These results shows the sensitivity of magnetic parameters towards the crystallite structure of samples which varies with synthesis process variations.

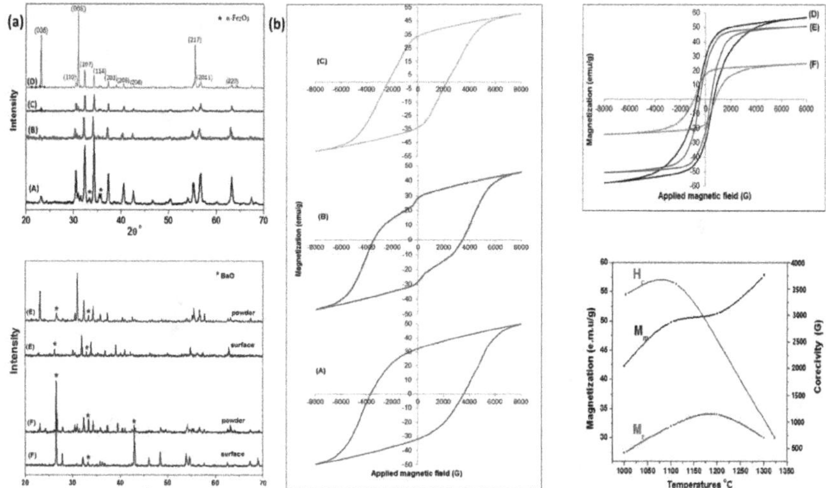

Figure 10: (a) XRD patterns for BaM (A, B, C and D) sintered at 1000, 1100, 1200 °C for 24 h, 1300 °C for 62 h respectively and surface XRD of the two samples (E and F) sintered at 1300 °C for 124 and 148 h respectively; (b) M-H loops for A,B,C,D,E,F samples and variation of magnetic parameters with sintering temperature and sintering time [69] (Copyright permission with the licence number: 5087650894770).

Table 5: Some reported BaM type ferrites and their magnetic parameters.

Composition	**Synthesis route**	**Annealing Temperature**	**Particle size**	**Magnetic parameters**	**Ref.**
$Ba_{0.5}Sr_{0.5}Fe_2O_4$	sol-gel auto-combustion	700 °C	81nm (h00)	M_S-2.20emu/g M_r- 0.66 emu/g H_c-424 Oe	[70]
		1050 °C	336 nm (h00)	M_S -3.83 emu/g M_r- 1.92 emu/g H_c-3438 Oe	
$BaFe_{11.6}Co_{0.2}Ni_{0.2}O_{19}$	co-precipitation method	750 °C	29.73 nm	M_S - 34.83 emu/g M_r- 18.03 emu/g H_c-2099 Oe	[71]
$BaFe_{12-x}Al_xO_{19}$(0.50)	solution combustion technique	900 °C	70 nm	M_S - 30.874 emu/g M_r- 18.861emu/g H_c- 5169 Oe	[72]
$Ba_{0.5}Sr_{0.5}Co_xLa_xFe_{12-2x}O_{19}$(x-0.4)	conventional ceramic method	1000 °C	47.75 nm	M_S – 59.82 emu/g H_c- 780.57 Oe	[73]
$Ba_{0.5}Sr_{0.5}Co_xW_xFe_{12-2x}O_{19}$(x-0.2)	ceramic method	1000 °C	21.66 ± 1.08 nm	M_S - 56.0 emu/g M_r- 17.5 emu/g H_c- 841.6Oe	[74]
$Ba_{0.5}Sr_{0.5}Co_xAl_xFe_{12-2x}O_{19}$(x-0.4)	ceramic method	1000 °C	24.37 ± 1.22 nm	M_S – 49.9 emu/g M_r- 7.28 emu/g H_c- 247.6 Oe	[75]
$BaCu_xFe_{12-x}O_{19}$(x-0.8)	co-precipitation technique	1200 °C	37.8 nm	M_S – 34.9 emu/g M_r- 26 emu/g H_c- 856.65 Oe	[76]

5. Applications of hexaferrites

Hexagonal ferrites are taking a lot scientific attention by the researchers and scientists because of its potential applications in different research areas of science and technology. Some of the applications of hexagonal ferrites are explained below.

5.1 Permanent Magnets

The finest permanent magnets are fabricated with the usage of rare earth metals-based alloys, mostly made up of neodymium. But, on the other hand, China sold 97% of the rare earth metals globally and nowadays, China raises the price of rare earth metals up to 500% which will result in the increase of shortage of production of permanent magnets. Recently, the hard ferrites are also utilized in the fabrication of permanent magnets but they are not so much superior in comparison to those permanent magnets which one gets from alloys of rare earth metals. The main advantage is that the production of permanent magnets from hexagonal ferrites is cheaper and very easy to fabricate [1]. For the production of excellent permanent magnets, magnetic materials with large retentivity and coercivity along with resistant to demagnetisation will be largely preferable. The permanent magnets of hexagonal ferrites are fabricated in two ways, first is sintered magnets which is actually a hard-pressed ceramic powder and second way is bonded magnets which is actually a moulded composite. Barium M-type hexaferrite (BaM) is used in fabricating bonded dust magnets and moreover, this type of bonded dust magnets is used to make plastoferrite. The bonded dust magnets can be easily censored into any type of form and can be used as fridge magnets. Out of all hexagonal ferrites, M-type hexagonal ferrite is considered as one of the most excellent magnetic material for making permanent magnet and it can be used as a magnet material for potential applications in loudspeakers, clocks, microwave dynamos, magnetic separators etc. The common use of hexagonal ferrites as permanent magnets in refrigerators, microphones, loud speakers and automobile applications which make the M-type hexaferrite for account over 90 percent of the fabricated material worldwide as a permanent magnet material [77]. Therefore, with the increase in the price of neodymium, hexagonal ferrites grow as an alternative valuable magnetic material for the fabrication of permanent magnets worldwide.

5.2 Electrical and microwave devices application

Today, the requirement for the signal processing devices has been increasing in radar instruments, GPS location, satellites, for wireless signals, defence system etc. and also, the frequencies at which these devices operate are increasing from microwave to millimetre wave. Therefore, for this, the demand of hexagonal ferrite will go on increasing. Moreover, the microwave dielectric losses must be lowered for their use at microwave frequencies. For this, the hexagonal ferrite should behave like an insulator for a complete penetration of the electric field but that was a big issue with the hexagonal ferrite based magnetic materials. The solution for this issue is that the hexaferrite should be synthesize on a superconducting substratum for the elimination of conductor losses [78]. Few of the microwave applications of hexagonal ferrites are antennas, circulators and isolators. The hexagonal ferrites have high value of resistivity with low dielectric losses which makes them more suitable for high frequency and microwave devices applications. The total dielectric losses in microwave devices can be determined with the utilization of magnetic damping from ferromagnetic resonance. Thus, our main objective is to synthesize the hexagonal ferrites with minimum ferromagnetic resonance line widths to reduce the dielectric losses in microwave devices. From the last few years, Barium M-type hexagonal ferrite (BaM) has been widely utilized for microwave devices applications. The integration of ferrite passive devices (i.e., circulators, isolators, phase shifters, filters, etc.) with semiconductor active devices (i.e., amplifiers, switches, and signal processing devices) onto a single planar system platform such as a Si wafer is a current trend in microwave technology. This satisfies the need for greater system integration while simultaneously decreasing device profile, volume, and weight. This necessitates the formation of high-quality ferrite films on semiconductor substrates. It has been demonstrated that reduced microwave losses in ferrite films are caused by good crystal quality. The development and processing of highly-oriented BaM films on semiconductor substrates remains a substantial problem due to chemical reactivity, interfacial diffusion, poor lattice matching, and problems linked to the high temperatures required for ferrite deposition [6].

5.3 Data Storage and recording media application

A bit in a digital memory is preserved as 0 or 1, therefore, a square loop with a high retentivity is required to ensure the data remains preserved and is not unintentionally deleted, and moreover, a large signal to noise ratio is essential to decrease mistake rates. In case of re-recordable memory, small saturation magnetizations along with coercivity are requirable. This means low fields can be beneficial for rewriting the data. In both the situations, chemical stability and permanency of the magnetization are critical. Audio tape, VCR tape, Professional Hi-8 mm data tape, floppy discs, and stiff or hard discs were all traditional commercial items. Because they are chemically stable, mechanically strong, and have large saturation magnetisations, BaM and SrM are employed for high density magnetic recording and magneto-optical recording. Oriented ferrites are employed in both longitudinal and perpendicular recording, in which the material's magnetic anisotropy is aligned parallel or perpendicular to the recording media. Moreover, without any substitution, BaM or SrM films have encountered several problems for high-density magnetic recording: they have a large positive temperature coefficient of coercivity, the grain size in such films is in the order of 102 nm, and as the grain size decreases, the magnetisation becomes thermally unstable, necessitating a larger magnetocrystalline anisotropy. The M ferrites have high coercivity for many recording applications since a significant magnetic field is required to record and re-record data, but it was discovered that in doped M ferrites, all of these characteristics may be customised by changing doping. Toshiba developed cobalt–titanium doped M-type hexagonal ferrites for recording in 1982, and they are particularly suited for use in recording media such as magnetic tape when coated as a thin film onto a substrate, as the grain size and coercivity can be reduced significantly without much loss of magnetisation in the ferrite [79]. Therefore, it was found that cobalt–titanium doped M-type hexagonal ferrites fulfilled the criteria of materials with switching fields, low magnetic losses, and good signal-to-noise ratios [80]. Recent breakthroughs in M-type hexagonal ferrites with ferroelectric-like behaviour at ambient temperature and polarisation switching by modest magnetic fields might lead to innovative magnetically-rewritable ferroelectric memories and electrically rewritable magnetic memories in the future.

5.4 RAM and microwave/EM wave absorption application

The wide usage of microwave-frequency equipment in our immediate surroundings, like radar and wireless and mobile communications, has led to a significant rise in electromagnetic interference, and the intensity of this ever-present non-ionising radiation. As a result, there is an increasing demand for microwave absorbing materials to decrease interference, as well as shielding devices, shield rooms, and chambers for electromagnetic compatibility testing and to prevent the damaging effects of electromagnetic waves on biological tissues [81]. Ferrites are employed as microwave absorbers because they can absorb microwave energy at the same frequency as ferromagnetic resonance. They're already being employed as electromagnetic absorbers in rooms and chambers where new items and equipment are being tested for electromagnetic compatibility at microwave frequencies. Absorbers are commonly made of plasto-ferrite tiles for electromagnetic applications, although they can also be made of ceramics as composites. By altering the volume fractions of the component phases, the electromagnetic characteristics of composites may be successfully modified. In some composites, a synergistic impact between the characteristics of the component phases has also been found. The X and K bands (8–40 GHz), which are typically pulsating over a range of frequencies, are used by current military radar because the short wavelength allow for better resolution [82]. Radar-absorbing paints or coatings composed of ferrites have been employed to coat military aircraft for stealth operations, as well as ships and other naval and ground-based assets. They absorb a few of the radar signal that is directed towards the target, decreasing or hiding the radar cross section signature that is received at the other end; however, the object is not completely radar "invisible". This technique is referred to be "stealth" or "low observable". Hexaferrites as radar absorbing materials used in electronic warfare are gaining a lot of attention, especially for their tuneable properties or wide absorption range to covering a wide spectrum of frequencies, as well as increased directional capabilities in oriented or fibrous hexaferrites. Because cost is less of a factor in these applications, most of the research into more unusual forms of hexaferrites is now being driven by aerospace and defence demands. The majority of this material, however, is confidential or classified, and most of it is not available in the public domain.

5.5 Magnetoelectric (ME) and multiferroic (MF) applications

The utilization of the connection between electric and magnetic characteristics is a major importance in magnetoelectric and multiferroic materials. This could contribute to plenty of new devices, including dual electrical and magnetic field tuneable devices at microwave and millimetre wave frequencies, extremely sensitive magnetic field sensors, switches and activators, high-frequency signal processing devices and gyrators [83,84]. The composite materials comprised of phases with good electrical and magnetic characteristics can have significantly stronger magnetoelectric interactions, and they have already shown tremendous potential. Although there has been less research on hexagonal ferrites in this sector, it is suitable for investigation; devices including yttrium iron garnet and a piezoelectric phase such as PZT, BT, or BST have showed strong magnetoelectric coupling constants in the hundreds of V cm^{-1} Oe^{-1} range. Enhancing magnetoelectric effects by resonance coupling offers a lot of promise, particularly at high GHz frequencies, and the low magnetic fields required substantially expands the variety of real-world applications. They might potentially be employed as current sensors or transformers, particularly during resonance, when the voltage gain effect is enhanced. Most significantly, all of these phenomena, especially in composites, can occur at room temperature.

5.6 Plastoferrites

Plastoferrites, often known as bonded magnets, are magnetic composite materials composed of a powdered ceramic ferrite integrated into a resin or thermoplastic matrix. Because of their composite nature, they have diluted magnetic properties, yet the products are not brittle and fragile, and are difficult to shatter or shatter. The plastoferrites can be cut in any form of shape in a very mass-productive and hence, it is a low-cost process. One of the examples of plastoferrites are flexible sheets either long or in thin form. Plastoferrites are well known as refrigerator door magnets, both decorative on the exterior and magnetic sealing strips within the door's rim [85].

5.7 Other important applications

Several designs for ferrites-based sensors have been suggested, including a humidity sensor based on resistance changes in a copper–zinc ferrite [86]. Hexaferrites can be used in biomedical purposes such as nuclear magnetic resonance imaging and magnetomotive biomedical implants. However, the toxicity of a few of the constituent elements notably barium, limits their usage in applications where particles or fluids must be injected into the human body. Like SrM grains were surface modified with oleic acid to increase their dispersibility in water with nearly negligible loss of magnetism, and they were then processed with oleic acid to form large polymeric microspheres with a magnetic core that should be biocompatible [87]. Similarly, Co_2Z coated magnetic force microscopy cantilevers were employed to provide better high-frequency measurements than normal CoCr coated tips, with imaging up to 2 GHz and 20 nm resolution. The imaging, characterization, and quality control of high-density magnetic disc drives and recording heads at high frequencies might be of special importance to industry [88].

Conclusions

In the current chapter, we have tried to summarize the basics of hexa-ferrites including the classification of all ferrites. We have categorically focussed on M-type of hexa-ferrites especially the magnetic properties and their linkage with the crystalline structure of hexaferrites. Various potential applications of hexaferrites constitute a part of this chapter. Owing to the higher crystalline field of hexagonal ferrites, their suitability of applications in GHz range is appreciated and proposed.

References

[1] P. Dhiman, T. Mehta, A. Kumar, G. Sharma, M. Naushad, T. Ahamad, G.T. Mola, $Mg_{0.5}Ni_xZn_{0.5-x}Fe_2O_4$ spinel as a sustainable magnetic nano-photocatalyst with dopant driven band shifting and reduced recombination for visible and solar degradation of Reactive Blue-19, Advanced Powder Technology, (2020). https://doi.org/10.1016/j.apt.2020.10.010

[2] S. Kour, R. Jasrotia, P. Puri, A. Verma, B. Sharma, V.P. Singh, R. Kumar, S. Kalia, Improving photocatalytic efficiency of $MnFe_2O_4$ ferrites via doping with Zn^{2+}/La^{3+} ions: photocatalytic dye degradation for water remediation, Environmental Science and Pollution Research, (2021). https://doi.org/10.1007/s11356-021-13147-7

[3] T. Kikuchi, T. Nakamura, T. Yamasaki, M. Nakanishi, T. Fujii, J. Takada, Y. Ikeda, Magnetic properties of La–Co substituted M-type strontium hexaferrites prepared by polymerizable complex method, Journal of Magnetism and Magnetic Materials, 322 (2010) 2381-2385. https://doi.org/10.1016/j.jmmm.2010.02.041

[4] J. Kasahara, T. Katayama, S. Mo, A. Chikamatsu, Y. Hamasaki, S. Yasui, M. Itoh, T. Hasegawa, Room-Temperature Antiferroelectricity in Multiferroic Hexagonal Rare-Earth Ferrites, ACS Applied Materials & Interfaces, 13 (2021) 4230-4235. https://doi.org/10.1021/acsami.0c20924

[5] R. Tang, H. Zhou, W. You, H. Yang, Room-temperature multiferroic and magnetocapacitance effects in M-type hexaferrite $BaFe_{10.2}Sc_{1.8}O_{19}$, Applied Physics Letters, 109 (2016) 082903. https://doi.org/10.1063/1.4961615

[6] R.C. Pullar, Hexagonal ferrites: A review of the synthesis, properties and applications of hexaferrite ceramics, Progress in Materials Science, 57 (2012) 1191-1334. https://doi.org/10.1016/j.pmatsci.2012.04.001

[7] H. Bayrakdar, Fabrication, magnetic and microwave absorbing properties of $Ba_2Co_2Cr_2Fe_{12}O_{22}$ hexagonal ferrites, Journal of Alloys and Compounds, 674 (2016) 185-188. https://doi.org/10.1016/j.jallcom.2016.03.055

[8] C.A. Stergiou, G. Litsardakis, Y-type hexagonal ferrites for microwave absorber and antenna applications, Journal of Magnetism and Magnetic Materials, 405 (2016) 54-61. https://doi.org/10.1016/j.jmmm.2015.12.027

[9] A. Geiler, A. Daigle, J. Wang, Y. Chen, C. Vittoria, V. Harris, Consequences of magnetic anisotropy in realizing practical microwave hexaferrite devices, Journal of magnetism and magnetic materials, 324 (2012) 3393-3397. https://doi.org/10.1016/j.jmmm.2012.02.050

[10] J. Smit, H. Wijn, Ferrites, Eindhoven, the Netherlands: Philips Tech, in, Library, 1959.

[11] M. Sugimoto, The past, present, and future of ferrites, Journal of the American Ceramic Society, 82 (1999) 269-280. https://doi.org/10.1111/j.1551-2916.1999.tb20058.x

[12] R. Jotania, Crystal structure, magnetic properties and advances in hexaferrites: A brief review, in: AIP Conference Proceedings, American Institute of Physics, 2014, pp. 596-599. https://doi.org/10.1063/1.4898528

[13] M. Chandel, V.P. Singh, R. Jasrotia, K. Singha, M. Singh, P. Thakur, S. Kalia, Fabrication of Ni^{2+} and Dy^{3+} substituted Y-Type nanohexaferrites: A study of

structural and magnetic properties, Physica B: Condensed Matter, 595 (2020) 412378. https://doi.org/10.1016/j.physb.2020.412378

[14] O. Aleshko-Ozhevskiĭ, R. Sizov, I. Yamzin, V. Lubimtsev, Helicoidal antiphase spin ordering in hexagonal ferrites of the $BaSc_x Fe_{12-x} O_{19}$ (M) system, Soviet Journal of Experimental and Theoretical Physics, 28 (1969) 425.

[15] R. Sizov, K. Zaitsev, Noncollinear spin ordering in a hexagonal ferrite of type W, Zh. Eksp. Teor. Fiz, 66 (1974) 368-373.

[16] U. Enz, Magnetization process of a helical spin configuration, Journal of Applied Physics, 32 (1961) S22-S26. https://doi.org/10.1063/1.2000413

[17] N. Momozawa, Y. Yamaguchi, Field-Induced Commensurate Intermediate Phases in Helimagnet $(Ba_{1-x} Sr_x)_2 Zn_2 Fe_{12} O_{22}$ (x= 0.748), Journal of the Physical Society of Japan, 62 (1993) 1292-1304. https://doi.org/10.1143/JPSJ.62.1292

[18] C. Fang, F. Kools, R. Metselaar, R. De Groot, Magnetic and electronic properties of strontium hexaferrite $SrFe_{12}O_{19}$ from first-principles calculations, Journal of Physics: Condensed Matter, 15 (2003) 6229. https://doi.org/10.1088/0953-8984/15/36/311

[19] V. Adelskold, Arkiv Kemi Miner, in, Geol, 1938.

[20] Z. Haijun, Y. Xi, Z. Liangying, The preparation and microwave properties of $Ba_2Zn_xCo_{2-x}Fe_{28}O_{46}$ hexaferrites, Journal of magnetism and magnetic materials, 241 (2002) 441-446. https://doi.org/10.1016/S0304-8853(01)00447-4

[21] F. Leccabue, R. Panizzieri, G. Bocelli, G. Calestani, C. Rizzoli, N.S. Almodovar, Crystal structure and magnetic characterization of $Sr_2Zn_2Fe_{28}O_{46}$ (SrZn− X) hexaferrite single crystal, Journal of magnetism and magnetic materials, 68 (1987) 365-373. https://doi.org/10.1016/0304-8853(87)90015-1

[22] H. Elkady, M. Abou-Sekkina, K. Nagorny, New information on Mössbauer and phase transition properties of Z-type hexaferrites, Hyperfine Interactions, 128 (2000) 423-432. https://doi.org/10.1023/A:1012612405813

[23] D. Lisjak, D. Makovec, M. Drofenik, Formation of U-type hexaferrites, Journal of materials research, 19 (2004) 2462-2470. https://doi.org/10.1557/JMR.2004.0317

[24] T. Kimura, Magnetoelectric hexaferrites, Annu. Rev. Condens. Matter Phys., 3 (2012) 93-110. https://doi.org/10.1146/annurev-conmatphys-020911-125101

[25] S. Munir, I. Ahmad, A. Laref, H.M.T. Farid, Synthesis, structural, dielectric and magnetic properties of hexagonal ferrites, Applied Physics A, 126 (2020) 722. https://doi.org/10.1007/s00339-020-03809-7

[26] R. Sagayaraj, T. Dhineshkumar, A. Prakash, S. Aravazhi, G. Chandrasekaran, D. Jayarajan, S. Sebastian, Fabrication, microstructure, morphological and magnetic properties of W-type ferrite by co-precipitation method: Antibacterial activity, Chemical Physics Letters, 759 (2020) 137944. https://doi.org/10.1016/j.cplett.2020.137944

[27] J. Mahapatro, S. Agrawal, Effect of Eu^{3+} ions on electrical and dielectric properties of barium hexaferrites prepared by solution combustion method, Ceramics International, (2021). https://doi.org/10.1016/j.ceramint.2021.04.062

[28] L. Wang, L. He, J. Li, Y. Yu, H. Li, Effects of Al and Ca ions co-doping on magnetic properties of M-type strontium ferrites, Journal of Materials Science: Materials in Electronics, 31 (2020) 22375-22384. https://doi.org/10.1007/s10854-020-04739-z

[29] Y. Yang, J. Shao, F. Wang, X. Liu, D. Huang, Impacts of MnZn doping on the structural and magnetic properties of M-type SrCaLa hexaferrites, Applied Physics A, 123 (2017) 309. https://doi.org/10.1007/s00339-017-0950-1

[30] Y. Yang, F. Wang, J. Shao, D. Huang, H. He, A. Trukhanov, S. Trukhanov, Influence of Nd-NbZn co-substitution on structural, spectral and magnetic properties of M-type calcium-strontium hexaferrites $Ca_{0.4}Sr_{0.6-x}Nd_xFe_{12.0-x}(Nb_{0.5}Zn_{0.5})_xO_{19}$, Journal of Alloys and Compounds, 765 (2018) 616-623. https://doi.org/10.1016/j.jallcom.2018.06.255

[31] S. Díaz-Castañón, J. Faloh-Gandarilla, F. Leccabue, G. Albanese, The optimum synthesis of high coercivity Pb–M hexaferrite powders using modifications to the traditional ceramic route, Journal of magnetism and magnetic materials, 272 (2004) 2221-2223. https://doi.org/10.1016/j.jmmm.2003.12.923

[32] K. Haneda, C. Miyakawa, H. Kojima, Preparation of High-Coercivity $BaFe_{12}O_{19}$, Journal of the American Ceramic Society, 57 (1974) 354-357. https://doi.org/10.1111/j.1151-2916.1974.tb10921.x

[33] M.M. Barakat, D.E.-S. Bakeer, A.-H. Sakr, Structural, Magnetic Properties and Electron Paramagnetic Resonance for $BaFe_{12-x}Hg_xO_{19}$ Hexaferrite Nanoparticles Prepared by Co-Precipitation Method, Journal of Taibah University for Science, 14 (2020) 640-652. https://doi.org/10.1080/16583655.2020.1761676

[34] P. Dhiman, S. Sharma, A. Kumar, M. Shekh, G. Sharma, M. Naushad, Rapid visible and solar photocatalytic Cr(VI) reduction and electrochemical sensing of dopamine using solution combustion synthesized ZnO–Fe_2O_3 nano heterojunctions: Mechanism

Elucidation, Ceramics International, 46 (2020) 12255-12268. https://doi.org/10.1016/j.ceramint.2020.01.275

[35] P. Dhiman, J. Chand, A. Kumar, R.K. Kotnala, K.M. Batoo, M. Singh, Synthesis and characterization of novel Fe@ZnO nanosystem, Journal of Alloys and Compounds, 578 (2013) 235-241. https://doi.org/10.1016/j.jallcom.2013.05.015

[36] R. Pullar, M. D TAYLOR, A. Bhattacharya, Novel aqueous sol–gel preparation and characterization of barium M ferrite, $BaFe_{12}O_{19}$ fibres, Journal of materials science, 32 (1997) 349-352. https://doi.org/10.1023/A:1018593014378

[37] R.C. Pullar, Combinatorial bulk ceramic magnetoelectric composite libraries of strontium hexaferrite and barium titanate, ACS combinatorial science, 14 (2012) 425-433. https://doi.org/10.1021/co300036m

[38] R. Pullar, M. D TAYLOR, A. Bhattacharya, Magnetic Co_2Y ferrite, $Ba_2Co_2Fe_{12}O_{22}$ fibres produced by a blow spun process, Journal of materials science, 32 (1997) 365-368. https://doi.org/10.1023/A:1018549232125

[39] R. Pullar, M. D TAYLOR, A. Bhattacharya, Aligned hexagonal ferrite fibres of Co_2W, $BaCo_2Fe_{16}O_{27}$ produced from an aqueous sol–gel process, Journal of materials science, 32 (1997) 873-877. https://doi.org/10.1023/A:1018541314320

[40] R. Pullar, S. Appleton, M. Stacey, M. Taylor, A. Bhattacharya, The manufacture and characterisation of aligned fibres of the ferroxplana ferrites Co_2Z, 0.67% CaO-doped Co_2Z, Co_2Y and Co_2W, Journal of Magnetism and Magnetic materials, 186 (1998) 313-325. https://doi.org/10.1016/S0304-8853(98)00098-5

[41] I. Ali, M. Islam, M. Awan, M. Ahmad, M.N. Ashiq, S. Naseem, Effect of Tb^{3+} substitution on the structural and magnetic properties of M-type hexaferrites synthesized by sol–gel auto-combustion technique, Journal of Alloys and Compounds, 550 (2013) 564-572. https://doi.org/10.1016/j.jallcom.2012.10.121

[42] S. Chawla, R. Mudsainiyan, S. Meena, S. Yusuf, Sol–gel synthesis, structural and magnetic properties of nanoscale M-type barium hexaferrites $BaCo_xZr_xFe_{(12-2x)}O_{19}$, Journal of Magnetism and Magnetic Materials, 350 (2014) 23-29. https://doi.org/10.1016/j.jmmm.2013.09.007

[43] A.V. Trukhanov, K.A. Darwish, M.M. Salem, O.M. Hemeda, M.I. Abdel Ati, M.A. Darwish, E.Y. Kaniukov, S.V. Podgornaya, V.A. Turchenko, D.I. Tishkevich, T.I. Zubar, K.A. Astapovich, V.G. Kostishyn, S.V. Trukhanov, Impact of the heat treatment conditions on crystal structure, morphology and magnetic properties

evolution in BaM nanohexaferrites, Journal of Alloys and Compounds, 866 (2021) 158961. https://doi.org/10.1016/j.jallcom.2021.158961

[44] P. Dhiman, K.M. Batoo, R.K. Kotnala, J. Chand, M. Singh, Room temperature ferromagnetism and structural characterization of Fe,Ni co-doped ZnO nanocrystals, Applied Surface Science, 287 (2013) 287-292. https://doi.org/10.1016/j.apsusc.2013.09.144

[45] P. Dhiman, G. Kumar, K. Batoo, A. Kumar, G. Sharma, M. Singh, Effective Degradation of Methylene Blue using ZnO: Fe: Ni Nanocomposites, Materials Research Foundations, 29.

[46] P. Dhiman, M. Naushad, K.M. Batoo, A. Kumar, G. Sharma, A.A. Ghfar, G. Kumar, M. Singh, Nano $Fe_xZn_{1-x}O$ as a tuneable and efficient photocatalyst for solar powered degradation of bisphenol A from aqueous environment, Journal of Cleaner Production, 165 (2017) 1542-1556. https://doi.org/10.1016/j.jclepro.2017.07.245

[47] V.N. Dhage, M. Mane, A. Keche, C. Birajdar, K. Jadhav, Structural and magnetic behaviour of aluminium doped barium hexaferrite nanoparticles synthesized by solution combustion technique, Physica B: Condensed Matter, 406 (2011) 789-793. https://doi.org/10.1016/j.physb.2010.11.094

[48] S.S. Afshar, M. Hasheminiasari, S. Masoudpanah, Structural, magnetic and microwave absorption properties of $SrFe_{12}O_{19}/Ni_{0.6}Zn_{0.4}Fe_2O_4$ composites prepared by one-pot solution combustion method, Journal of Magnetism and Magnetic Materials, 466 (2018) 1-6. https://doi.org/10.1016/j.jmmm.2018.06.061

[49] B. Abasht, S.M. Mirkazemi, A. Beitollahi, Solution combustion synthesis of Ca hexaferrite using glycine fuel, Journal of Alloys and Compounds, 708 (2017) 337-343. https://doi.org/10.1016/j.jallcom.2017.03.036

[50] K. Byrappa, M. Yoshimura, Handbook of hydrothermal technology, William Andrew, 2012. https://doi.org/10.1016/B978-0-12-375090-7.00002-5

[51] A. Kumar, G. Sharma, A. Kumari, C. Guo, M. Naushad, D.-V.N. Vo, J. Iqbal, F.J. Stadler, Construction of dual Z-scheme $g-C_3N_4/Bi_4Ti_3O_{12}/Bi_4O_5I_2$ heterojunction for visible and solar powered coupled photocatalytic antibiotic degradation and hydrogen production: Boosting via I^-/I^{3-} and Bi^{3+}/Bi^{5+} redox mediators, Applied Catalysis B: Environmental, 284 (2021) 119808. https://doi.org/10.1016/j.apcatb.2020.119808

[52] A. Kumar, S.K. Sharma, G. Sharma, M. Naushad, F.J. Stadler, $CeO_2/g-C_3N_4/V_2O_5$ ternary nano hetero-structures decorated with CQDs for enhanced photo-reduction capabilities under different light sources: Dual Z-scheme mechanism, Journal of

Alloys and Compounds, 838 (2020) 155692. https://doi.org/10.1016/j.jallcom.2020.155692

[53] M. Shandilya, R. Rai, J. Singh, hydrothermal technology for smart materials, Advances in Applied Ceramics, 115 (2016) 354-376. https://doi.org/10.1080/17436753.2016.1157131

[54] T.B. Ghzaiel, W. Dhaoui, A. Pasko, F. Mazaleyrat, Effect of non-magnetic and magnetic trivalent ion substitutions on BaM-ferrite properties synthesized by hydrothermal method, Journal of Alloys and Compounds, 671 (2016) 245-253. https://doi.org/10.1016/j.jallcom.2016.02.071

[55] A. Xia, C. Zuo, L. Chen, C. Jin, Y. Lv, Hexagonal $SrFe_{12}O_{19}$ ferrites: Hydrothermal synthesis and their sintering properties, Journal of magnetism and magnetic materials, 332 (2013) 186-191. https://doi.org/10.1016/j.jmmm.2012.12.035

[56] V. Banihashemi, M. Ghazi, M. Izadifard, Structural, optical, dielectric and magnetic properties of Ce-doped strontium hexaferrite synthesized by a hydrothermal process, Journal of Materials Science: Materials in Electronics, 30 (2019) 17374-17381. https://doi.org/10.1007/s10854-019-02086-2

[57] T.-S. Chin, S. Hsu, M. Deng, Barium ferrite particulates prepared by a salt-melt method, Journal of magnetism and magnetic materials, 120 (1993) 64-68. https://doi.org/10.1016/0304-8853(93)91288-I

[58] S. Dursun, R. Topkaya, N. Akdoğan, S. Alkoy, Comparison of the structural and magnetic properties of submicron barium hexaferrite powders prepared by molten salt and solid state calcination routes, Ceramics International, 38 (2012) 3801-3806. https://doi.org/10.1016/j.ceramint.2012.01.028

[59] H. Li, X. Yi, Y. Wu, X. Wei, D. Deng, L. Zheng, W. Luo, X. Luo, R. Gong, M. Zhang, Molten salt synthesis, formation mechanism and greatly enhanced magnetic properties of randomly oriented BaM ferrite, Journal of Alloys and Compounds, 827 (2020) 154083. https://doi.org/10.1016/j.jallcom.2020.154083

[60] N. Borisova, Z. Golubenko, T. Kuz'micheva, L. Ol'khovik, V. Shabatin, Optimization principles for preparation methods and properties of fine ferrite materials, Journal of magnetism and magnetic materials, 114 (1992) 317-328. https://doi.org/10.1016/0304-8853(92)90274-R

[61] L. Ol'Khovik, N. Borisova, A. Kamzin, O. Fisenko, Radiothermal synthesis of fine barium ferrite powders and their properties, Journal of magnetism and magnetic materials, 154 (1996) 365-368. https://doi.org/10.1016/0304-8853(95)00593-5

[62] J. Dufour, L. Lopez, A. Formoso, C. Negro, R. Latorre, F. Lopez-Mateos, Mathematical model of goethite synthesis by oxyprecipitation of steel pickling liquors, The Chemical Engineering Journal and The Biochemical Engineering Journal, 59 (1995) 287-291. https://doi.org/10.1016/0923-0467(94)02950-4

[63] J. Dufour, R. Latorre, C. Negro, E. Alcalá, A. Formoso, F. López-Mateos, Protocol for the synthesis of Ba-hexaferrites with prefixed coercivities, Journal of magnetism and magnetic materials, 172 (1997) 308-316. https://doi.org/10.1016/S0304-8853(97)00145-5

[64] S. Han, Part~ I. The synthesis and thermal rearrangements of 7, 7-dihalo-trans-bicyclo (4.1. 0) hept-3-enes. Part~ II. Synthesis of electron deficient cyclopropene derivatives and investigation of cyclopropenyl anions, University of Minnesota, 1997.

[65] J. Ding, T. Tsuzuki, P. McCormick, Ultrafine $BaFe_{12}O_{19}$ powder synthesised by mechanochemical processing, Journal of magnetism and magnetic materials, 177 (1998) 931-932. https://doi.org/10.1016/S0304-8853(97)00858-5

[66] W. Kaczmarek, B. Ninham, Application of mechanochemistry in ferrite materials technology, Le Journal de Physique IV, 7 (1997) C1-47-C41-48. https://doi.org/10.1051/jp4:1997106

[67] G. Han, R. Sui, Y. Yu, L. Wang, M. Li, J. Li, H. Liu, W. Yang, Structure and magnetic properties of the porous Al-substituted barium hexaferrites, Journal of Magnetism and Magnetic Materials, 528 (2021) 167824. https://doi.org/10.1016/j.jmmm.2021.167824

[68] D. Chen, Y. Liu, Y. Li, K. Yang, H. Zhang, Microstructure and magnetic properties of Al-doped barium ferrite with sodium citrate as chelate agent, Journal of Magnetism and Magnetic Materials, 337-338 (2013) 65-69. https://doi.org/10.1016/j.jmmm.2013.02.036

[69] R.E. El Shater, E.H. El-Ghazzawy, M.K. El-Nimr, Study of the sintering temperature and the sintering time period effects on the structural and magnetic properties of M-type hexaferrite BaFe12O19, Journal of Alloys and Compounds, 739 (2018) 327-334. https://doi.org/10.1016/j.jallcom.2017.12.228

[70] Z.K. Heiba, A.M. Wahba, M.B. Mohamed, Phase analysis and cation distribution correlated with magnetic properties of spinel $Ba_{1-x}Sr_xFe_2O_4$ ferrites prepared at different annealing temperatures, Journal of Materials Science: Materials in Electronics, 31 (2020) 12482-12492. https://doi.org/10.1007/s10854-020-03795-9

[71] M. Ginting, P. Sebayang, M. Rianna, M. Situmorang, H. Fujiati, A.P. Tetuko, E.A. Setiadi, C. Kurniawan, A.M.S. Sebayang, Effect of Co and Ni additions as doping materials on the micro-structures and the magnetic properties of barium hexa-ferrites, Case Studies in Thermal Engineering, 18 (2020) 100589. https://doi.org/10.1016/j.csite.2020.100589

[72] V.N. Dhage, M.L. Mane, A.P. Keche, C.T. Birajdar, K.M. Jadhav, Structural and magnetic behaviour of aluminium doped barium hexaferrite nanoparticles synthesized by solution combustion technique, Physica B: Condensed Matter, 406 (2011) 789-793. https://doi.org/10.1016/j.physb.2010.11.094

[73] H. Kaur, A. Marwaha, C. Singh, S.B. Narang, R. Jotania, S. Jacobo, A.S.B. Sombra, S.V. Trukhanov, A.V. Trukhanov, P. Dhruv, Investigation of structural, hysteresis and electromagnetic parameters for microwave absorption application in doped Ba–Sr hexagonal ferrites at X-band, Journal of Alloys and Compounds, 806 (2019) 1220-1229. https://doi.org/10.1016/j.jallcom.2019.07.032

[74] R. Joshi, C. Singh, D. Kaur, H. Zaki, S. Bindra Narang, R. Jotania, S.R. Mishra, J. Singh, P. Dhruv, M. Ghimire, Structural and magnetic properties of Co^{2+}-W^{4+} ions doped M-type Ba-Sr hexaferrites synthesized by a ceramic method, Journal of Alloys and Compounds, 695 (2017) 909-914. https://doi.org/10.1016/j.jallcom.2016.10.192

[75] J. Singh, C. Singh, D. Kaur, H. Zaki, I.A. Abdel-Latif, S.B. Narang, R. Jotania, S.R. Mishra, R. Joshi, P. Dhruv, M. Ghimire, S.E. Shirsath, S.S. Meena, Elucidation of phase evolution, microstructural, Mössbauer and magnetic properties of $Co^{2+}Al^{3+}$ doped M-type BaSr hexaferrites synthesized by a ceramic method, Journal of Alloys and Compounds, 695 (2017) 1112-1121. https://doi.org/10.1016/j.jallcom.2016.10.237

[76] S. Vadivelan, N. Victor Jaya, Investigation of magnetic and structural properties of copper substituted barium ferrite powder particles via co-precipitation method, Results in Physics, 6 (2016) 843-850. https://doi.org/10.1016/j.rinp.2016.07.013

[77] Q. Mohsen, Factors affecting the synthesis and formation of single-phase barium hexaferrite by a technique of oxalate precursor, American Journal of applied sciences, 7 (2010) 914. https://doi.org/10.3844/ajassp.2010.914.921

[78] G.F. Dionne, D.E. Oates, D.H. Temme, J.A. Weiss, Ferrite-superconductor devices for advanced microwave applications, IEEE Transactions on Microwave theory and Techniques, 44 (1996) 1361-1368. https://doi.org/10.1109/22.508241

[79] R. Gerber, R. Atkinson, Z. Šimša, Magnetism and magneto-optics of hexaferrite layers, Journal of magnetism and magnetic materials, 175 (1997) 79-89. https://doi.org/10.1016/S0304-8853(97)00151-0

[80] O. Kubo, T. Ido, H. Yokoyama, Properties of Ba ferrite particles for perpendicular magnetic recording media, IEEE transactions on magnetics, 18 (1982) 1122-1124. https://doi.org/10.1109/TMAG.1982.1062007

[81] R. Jasrotia, V.P. Singh, B. Sharma, A. Verma, P. Puri, R. Sharma, M. Singh, Sol-gel synthesized Ba-Nd-Cd-In nanohexaferrites for high frequency and microwave devices applications, Journal of Alloys and Compounds, 830 (2020) 154687. https://doi.org/10.1016/j.jallcom.2020.154687

[82] R. Meena, S. Bhattachrya, R. Chatterjee, Complex permittivity, permeability and microwave absorbing studies of ($Co_{2-x}Mn_x$) U-type hexaferrite for X-band (8.2–12.4 GHz) frequencies, Materials Science and Engineering: B, 171 (2010) 133-138. https://doi.org/10.1016/j.mseb.2010.03.086

[83] C.-W. Nan, M. Bichurin, S. Dong, D. Viehland, G. Srinivasan, Multiferroic magnetoelectric composites: Historical perspective, status, and future directions, Journal of applied physics, 103 (2008) 1. https://doi.org/10.1063/1.2836410

[84] G. Srinivasan, Magnetoelectric composites, Annual Review of Materials Research, 40 (2010) 153-178. https://doi.org/10.1146/annurev-matsci-070909-104459

[85] P. Muth, E. P. Wohlfarth (ed.). Ferromagnetic Materials, vol. 2. North-Holland Publ. Co. Amsterdam 1980 592 Seiten. Preis US £ 102,50, Dfl. 210,00, Kristall und Technik, 16 (1981) 127-127. https://doi.org/10.1002/crat.19810160127

[86] A. Vaingankar, S. Kulkarni, M. Sagare, Humidity sensing using soft ferrites, Le Journal de Physique IV, 7 (1997) C1-155-C151-156. https://doi.org/10.1051/jp4:1997155

[87] S. Kong, P. Zhang, X. Wen, P. Pi, J. Cheng, Z. Yang, J. Hai, Influence of surface modification of $SrFe_{12}O_{19}$ particles with oleic acid on magnetic microsphere preparation, Particuology, 6 (2008) 185-190. https://doi.org/10.1016/j.partic.2008.03.004

[88] M. Koblischka, M. Kirsch, J. Wei, T. Sulzbach, U. Hartmann, Preparation of ferrite-coated MFM cantilevers, Journal of Magnetism and Magnetic Materials, 316 (2007) e666-e669. https://doi.org/10.1016/j.jmmm.2007.03.075

Keyword Index

About the Editors

Dr. Gaurav Sharma

International Research Centre of Nanotechnology for Himalayan Sustainability (IRCNHS), Shoolini University, Solan, 173229, Himachal Pradesh, India
Email: Gaurav.541@shooliniuniversity.com

Dr. Gaurav Sharma research activity started in 2009 at Shoolini University (India) as master of philosophy student, and then, He continued his research work as PhD student with the preparation and characterization of diverse multifunctional nanomaterials, and their composites, specially focused on potential applications in environmental remediation (as photocatalysts and adsorbents). For four years He worked as assistant professor in the School of chemistry at Shoolini University (India), where He carried out diverse research lines, interrelated to each other based on synthesis and characterization of nanocomposites, hydrogels, bi and trimetallic nanoparticles, ion exchangers, adsorbents and photocatalysts etc. Moreover, He performed and taught different courses as nanochemistry, polymer chemistry, spectroscopy and natural products, among others. On the other hand, He supervised 3 PhD, 5 Master of Philosophy, and more than 25 Master and Bachelors students. He established collaborative research with various professors in countries as Finland, Saudi Arabia, China, Spain and South Africa. In this context, He was invited as visiting research professor from University of KwaZuklu-Natal (South Africa) in 2017 and 2019. In 2017, He joined as postdoctoral fellow at college of materials science and engineering, Shenzhen University. He got project from china postdoctoral science foundation in 2018. The outcome of his research work was depicted in more than 140 publications, in various journals such as Renewable and Sustainable Energy Reviews, Chemical Engineering Journal, Journal of Cleaner Production, Carbohydrate Polymers, ACS Applied Materials and Interfaces, Journal of Hazardous Materials, Applied Catalysis B, and International Journal of Biological Macromolecules etc, 9 book chapters and 6 edited books. He is also serving as Director, International Research Centre of Nanotechnology for Himalayan Sustainability (IRCNHS), Shoolini University, India.

He is Highly Cited Researcher -2020 Crossfield (Web of Science); and also Ranked among World top 2% Scientists (Current year 2019 category) as per Stanford.

His h-index is 50, citations: 6214 (web of science); Google Scholar: h-index is 53, citations: 6628. He is Associate Editor of International Journal of Environmental Science and Technology (Springer) IF: 2,031. Editorial Board member of Current Organic Chemistry (IF:1.9), Current analytical Chemistry (IF:1.3), Materials-MDPI (IF:3.057) Innovations in Corrosion and Materials Science, Journal of Nanostructure in Chemistry (IF:4.0), Nanotechnology for Environmental Engineering(Springer), Letters in Applied NanoBioScience etc, and Academic Editor of Journal of Nanomaterials (IF: 1.980) Advances in Polymer Technology (IF: 1.539).

Dr Amit Kumar

International Research Centre of Nanotechnolog for Himalayan Sustainability (IRCNHS), Shoolii University, Solan, 173229, Himachal Pradesh, Indi
Email: amitkumar@shooliniuniversity.com

Dr Amit Kumar started his research career wi Ph.D in Chemistry at Himachal Pradesh Universit Shimla, India in 2008. He was awarded with tl prestigious predoctoral scholarship of the Council of Scientific and Industrial Research (Government of India) in 2007. After his PhD, he worked as assistant professor at Shoolini University (India) for 3.5 years, where he was involved in research and teaching at undergraduate, postgraduate and doctoral level (2014-2017). He has experience of supervising three Ph.D students (all graduated). Furthermore, he has supervised 10 Masters students. He has worked as post doctoral fellow in the College of Materials Science and Engineering of Shenzhen University (2017-2019). He is involved in field of optical designing of semiconductor heterojunctions as catalysts for environmental remediation and energy production. Currently He is a senior researcher at College of Materials Science and Engineering of Shenzhen University, PR China (01.01.2020 onwards). In addition he holds position of Co-director, International Research Centre of Nanotechnology for Himalayan Sustainability (IRCNHS), Shoolini University, India. He is also a Visiting Professor at School of technology, Glocal University, India. Moreover, he is a visiting faculty at Department of Chemistry and Physics from University of KwaZulu Natal, South Africa.

He has diverse research collaborations in Instituto de Catálisis y Petroleoquímica (Spain), King Saud University (Saudi Arabia), Chinese Research Academy of Environmental Sciences (China) and Universidade Federal do Rio Grande do Sul (Brazil). He is also ranked among World top 2% Scientists (Current year 2019 category) as per Stanford University rankings, 2020.

His scientific production encompasses 110 papers (WoS) including Applied Catalysis B: Environmental, Chemical Engineering Journal, ACS Applied Materials and Interfaces (I.F 8.75), Journal of Hazardous Materials, Journal of Cleaner Production etc. He has received h-index 47 and more than 5000 citations.

Dr. Pooja Dhiman

School of Physics and Materials Science, Shoolini University of Biotechnology and Management Sciences, Solan, Himachal Pradesh,173229, India

Email id:poojadhiman@shooliniuniversity.com

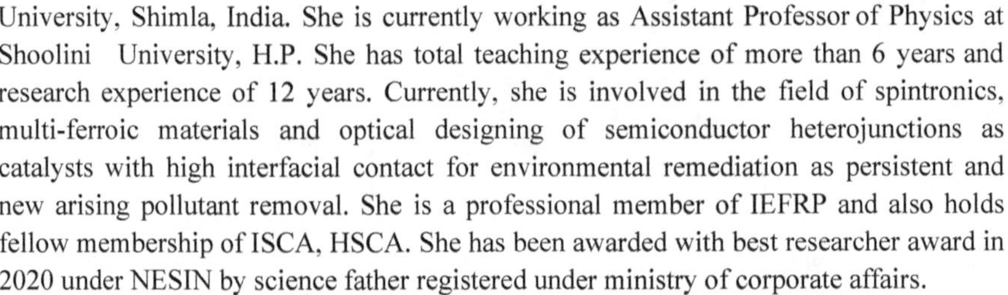

Dr. Pooja Dhiman started her research career started with Ph.D. and M.Phil. in Physics from Himachal Pradesh University, Shimla, India. She is currently working as Assistant Professor of Physics at Shoolini University, H.P. She has total teaching experience of more than 6 years and research experience of 12 years. Currently, she is involved in the field of spintronics, multi-ferroic materials and optical designing of semiconductor heterojunctions as catalysts with high interfacial contact for environmental remediation as persistent and new arising pollutant removal. She is a professional member of IEFRP and also holds fellow membership of ISCA, HSCA. She has been awarded with best researcher award in 2020 under NESIN by science father registered under ministry of corporate affairs.

She has published 40 publications in journals of repute and has published 4 book chapters. Her highest impact factor publication (7.246) is in journal of cleaner production. She has a h-index of 15 (SCOPUS) and 16 (Google scholar). The total number of citations achieved is 957 (Google Scholar). She has experience of supervising many research candidates.

CPSIA information can be obtained
at www.ICGtesting.com
Printed in the USA
BVHW012046241021
619770BV00006B/126

9 781644 901588